Lecture Notes in Artificial Intelligence   9287

Subseries of Lecture Notes in Computer Science

LNAI Series Editors

Randy Goebel
University of Alberta, Edmonton, Canada
Yuzuru Tanaka
Hokkaido University, Sapporo, Japan
Wolfgang Wahlster
DFKI and Saarland University, Saarbrücken, Germany

LNAI Founding Series Editor

Joerg Siekmann
DFKI and Saarland University, Saarbrücken, Germany

More information about this series at http://www.springer.com/series/1244

Clare Dixon · Karl Tuyls (Eds.)

Towards Autonomous Robotic Systems

16th Annual Conference, TAROS 2015
Liverpool, UK, September 8–10, 2015
Proceedings

 Springer

Editors
Clare Dixon
University of Liverpool
Liverpool
UK

Karl Tuyls
University of Liverpool
Liverpool
UK

ISSN 0302-9743   ISSN 1611-3349   (electronic)
Lecture Notes in Artificial Intelligence
ISBN 978-3-319-22415-2   ISBN 978-3-319-22416-9   (eBook)
DOI 10.1007/978-3-319-22416-9

Library of Congress Control Number: 2015945155

LNCS Sublibrary: SL7 – Artificial Intelligence

Springer Cham Heidelberg New York Dordrecht London

Printed on acid-free paper

Springer International Publishing AG Switzerland is part of Springer Science+Business Media
(www.springer.com)

Preface

The 16th edition of the Towards Autonomous Robotic Systems (TAROS) conference was held September 8–10, 2015, at the University of Liverpool, UK. The event included an academic conference, industry exhibitions and talks, robot demonstrations, and other satellite events.

The TAROS series was initiated by Ulrich Nehmzow in Manchester in 1997 under the name TIMR (Towards Intelligent Mobile Robots). In 1999, Chris Melhuish and Ulrich formed the conference Steering Committee, which was joined by Mark Witkowski in 2003 when the conference adopted its current name. The Steering Committee has provided a continuity of vision and purpose to the conference over the years as it has been hosted by robotics centers throughout the UK. Under their stewardship, TAROS has become the UK's premier annual conference on autonomous robotics, while also attracting an increasing international audience. Sadly, Ulrich died in 2010, but his contribution is commemorated in the form of the Ulrich Nehmzow Best Student Paper Award sponsored by his family.

TAROS 2015 attracted 59 submissions (46 full papers and 13 short papers) from institutions in 17 countries. In the final volume 16 full papers (12 pages in length) were accepted, corresponding to an acceptance rate of 27 % and 18 short papers (6 pages in length) for poster presentation.

We were delighted to be able to attract three world-renown researchers in robotics to present keynote speeches. Professor Tone Belpaeme (Plymouth University, UK) gave the public lecture sponsored by the IET (the Institution of Engineering and Technology) entitled "The Friendly Face of Robots". Professor Marco Dorigo (Université Libre de Bruxelles, Belgium) spoke about "Swarm Robotics Research at IRIDIA" and Professor Wolfram Burgard (University of Freiburg, Germany) gave a talk entitled "Probabilistic Techniques for Mobile Robot Navigation". We thank them for their inspirational talks and their contribution to the success of TAROS 2015.

We would like to thank the Program Committee for their careful and prompt reviewing of papers, the authors for choosing to submit their work to TAROS, and all authors and participants at TAROS 2015 for their contributions. We would also like to thank the following people from the University of Liverpool who worked hard to make the event a success: Michael Fisher (General Chair); Katie Atkinson and Lynn Dwyer (Industry Exhibition and Demonstrations); Dave Shield and Ken Chan (Technology and Media); Rebekah Martin, Elaine Smith, and Lisa Smith (Finance and Administration); and Robert Merrifield, Mike Jump, and Michael Fisher (UK-RAS Network). The organizers also thank Mark Witkowski and Chris Melhuish for providing guidance and support.

Additionally we would like to thank the Institution of Engineering Technology (IET) for sponsoring an invited lecture and providing a paper prize, Springer for sponsoring the best paper award, and the family of Ulrich Nehmzow for sponsoring the best student paper award. Finally we thank the UK Robotics and Autonomous Systems

Network for sponsorship of the event and both the Department of Computer Science and the Virtual Engineering Centre at the University of Liverpool for their support.

June 2015   Clare Dixon
                                                                              Karl Tuyls

Organization

General Chair

Michael Fisher   University of Liverpool, UK

Conference Chairs

Clare Dixon   University of Liverpool, UK
Karl Tuyls   University of Liverpool, UK

Industry Exhibition and Demonstrations

Katie Atkinson   University of Liverpool, UK
Lynn Dwyer   University of Liverpool, UK

Administration, Technology, and Media

Ken Chan   University of Liverpool, UK
Rebekah Martin   University of Liverpool, UK
Dave Shield   University of Liverpool, UK
Elaine Smith   University of Liverpool, UK
Lisa Smith   University of Liverpool, UK

UK-RAS Network Workshop

Robert Merrifield   Imperial College London, UK
Mike Jump   University of Liverpool, UK
Michael Fisher   University of Liverpool, UK

TAROS Steering Committee

Mark Witkowski   Imperial College London, UK
Chris Melhuish   Bristol Robotics Laboratory, UK

Sponsoring Organizations

Institution of Engineering and Technology (IET)
Springer UK
UK Network for Robotics and Autonomous Systems (UK-RAS)
The University of Liverpool

Program Committee

Rob Alexander	University of York, UK
Joseph Ayers	Northeastern University, USA
Robert Babuska	Delft University of Technology, The Netherlands
Tony Belpaeme	University of Plymouth, UK
Daan Bloembergen	University of Liverpool, UK
Stephen Cameron	University of Oxford, UK
Angelo Cangelosi	Plymouth University, UK
Anders Christensen	Lisbon University Institute, Portugal
Kerstin Dautenhahn	University of Hertfordshire, UK
Clare Dixon	University of Liverpool, UK
Tony Dodd	University of Sheffield, UK
Sanja Dogramadzi	University of the West of England, UK
Stéphane Doncieux	Université Pierre et Marie Curie, France
Marco Dorigo	Université Libre de Bruxelles, Belgium
Kerstin Eder	University of Bristol, UK
Michael Fisher	University of Liverpool, UK
Roderich Gross	The University of Sheffield, UK
Dongbing Gu	University of Essex, UK
Heiko Hamann	University of Paderborn, Germany
Marc Hanheide	University of Lincoln, UK
Nick Hawes	University of Birmingham, UK
Daniel Hennes	European Space Agency, France
Ioannis Ieropoulos	University of the West of England, UK
Raul Marin	Jaume I University, Spain
Ben Mitchinson	The University of Sheffield, UK
Paolo Paoletti	University of Liverpool, UK
Simon Parsons	University of Liverpool, UK
Martin Pearson	University of the West of England, UK
Jacques Penders	Sheffield Hallam University, UK
Anthony Pipe	University of the West of England, UK
Robert Richardson	University of Leeds, UK
Ferdinando Rodriguez y Baena	Imperial College London, UK
Thomas Schmickl	University of Graz, Austria
Elizabeth Sklar	University of Liverpool, UK
Matthew Studley	University of the West of England, UK
Jon Timmis	University of York, UK
Kagan Tumer	Oregon State University, USA
Karl Tuyls	University of Liverpool, UK
Bram Vanderborght	Vrije Universiteit Brussel, Belgium
Sandor Veres	University of Sheffield, UK
Gerhard Weiss	University of Maastricht, The Netherlands
Myra Wilson	Aberystwyth University, UK

Alan Winfield   University of the West of England, UK
Mark Witkowski   Imperial College London, UK

Additional Reviewers

Jonathan M. Aitken   Melvin Gauci   Lenka Mudrova
Dejanira Araiza-Illan   Jianjun Gui   Andrea Paoli
Bastian Broecker   Giacomo Innocenti   Jaime Pulido Fentanes
Jen Jen Chung   Wesam Jasim   Hongyang Qu
Mitchell Colby   Ferdian Jovan   Carrie Rebhuhn
Elisa Cucco   Yuri Kaszubowski Lopes   Emmanuel Senft
Louise Dennis   James Law   Gabriele Valentini
Christian Dondrup   Owen McAree   Minlue Wang
Joscha Fossel   Rico Mockel   Logan Yliniemi

Contents

Endurance Optimisation of Battery-Powered Rotorcraft

Analiza Abdilla[1,2(✉)], Arthur Richards[1,2], and Stephen Burrow[1]

[1] Department of Aerospace Engineering, University of Bristol, Bristol BS8 1TR, UK
{analiza.abdilla,arthur.richards,stephen.burrow}@bristol.ac.uk
http://www.bristol.ac.uk
[2] Bristol Robotics Laboratory, Bristol BS16 1QY, UK

Abstract. This paper proposes a technique to extend the endurance of battery-powered rotorcraft by sub-dividing the monolithic battery into multiple smaller capacity batteries which are sequentially discharged and released. The discarding of consumed battery mass reduces the propulsive power required, thereby contributing towards increased endurance. However, the corresponding implementation introduces additional parasitic mass due to the required battery switching and attachment and release mechanism, which, together with the decrease in battery efficiency with decreasing size, results in endurance improvements only being achieved beyond a threshold payload and which scale with rotorcraft size. An endurance model for battery-powered rotorcraft is presented, together with a technique to determine the maximum endurance and corresponding battery combination, by solving the Knapsack Problem by Dynamic Programming. The theoretical upper bound on rotorcraft endurance, which may be obtained from an ideal "infinitely-divisible" battery, is derived from the Breguet-Range Equation. Theoretical derivations and model predictions are validated through experimental flight tests using a popular commercial quadrotor.

Keywords: Endurance · Optimisation · Power · Battery · LiPo · UAV · Rotorcraft · Quadrotor · Knapsack problem · Dynamic programming

1 Introduction

Electric Unmanned Aerial Vehicles (UAVs) offer advantages in terms of low acoustic signature, high system efficiency and reliability [1], as well as more precise and faster control, which is essential for differential thrust control in rotorcraft [2]. However, battery-powered vehicles suffer from major limitations with respect to endurance capabilities, due to the low specific energy of batteries as compared to other energy sources, as well as the fact that, unlike chemical propulsion, where the mass is reduced as fuel is burnt, the battery mass is fixed irrespective of its State of Charge (SoC) [1]. Moreover, rotary-wing vehicles have higher power consumption as compared to other configurations. Thus, whilst attractive to use in exploration missions due to their hovering and low-speed

© Springer International Publishing Switzerland 2015
C. Dixon and K. Tuyls (Eds.): TAROS 2015, LNAI 9287, pp. 1–12, 2015.
DOI: 10.1007/978-3-319-22416-9_1

flight capabilities and high manoeuvrability, the endurance of such expeditions is significantly constrained [3].

This paper proposes a technique to mitigate the limitation due to fixed battery mass by replacing the monolithic on-board battery with multiple smaller batteries which are discharged and released as required. Discarding consumed battery mass results in a reduction in the propulsive power required and thus contributes towards increased endurance. The corresponding implementation, however, introduces additional parasitic mass due to the required battery switching circuitry and attachment and release mechanism. Furthermore, battery efficiency reduces with decreasing size due to the battery packaging and connectors [4]. This results in a trade-off between the endurance improvements that may be achieved by discarding consumed battery mass as compared with the penalties due to the additional parasitic mass, which depends on the available payload, and thus rotorcraft size. The conditions for which battery sub-division results in enhanced endurance are therefore investigated and discussed.

This technique has been proposed by [5] to extend the endurance of fixed-wing aircraft, and the effect of the parasitic mass of the Battery Dumping System (BDS) on the potential endurance improvements is investigated. Analytical results indicate that endurance benefits scale with BDS weight ratio, and only occur at ratios exceeding 15% [5]. A similar technique is employed in the Flight of the Century project [6], which aims to complete a transatlantic flight in an all-electric airplane by having multiple battery packs which are discharged individually and jettisoned on GPS-guided parachutes to land on recovery and recharge stations upon depletion. Further improvements may be obtained by launching fully-charged battery packs to replace the jettisoned ones, rather than taking off with all the battery packs on board [6].

In order to extend endurance of a UAV, [7] developed an automated battery changing/charging station to "hot swap" its depleted battery. Whilst enhancing operation in terms of endurance, such a solution requires the UAV to return to the station, and thus, no gain is obtained in range. In a similar manner, [8] proposes to increase the endurance of a multi-agent persistence surveillance mission by utilising a team of autonomous UAVs and multiple charging stations. The latter can be mounted on-board mobile Unmanned Ground Vehicles (UGV) which are used to transport the UAVs, thereby enabling their charging during transit. The down-time of each UAV may be further reduced by replacing the battery recharging system with a battery swapping mechanism [8].

The endurance of aerial exploration may also be increased by utilising "perch and stare" techniques, as proposed by [9], in which the AV perches onto a location whilst maintaining a "bird's-eye-view", thereby reducing the power from its propulsive value to its idle value. Another widely-employed technique makes use of energy harvesting, by complementing the battery with solar and/or photovoltaic cells [10]. Wireless remote recharging of the UAV's battery through power beaming of Radio Frequency (RF) [10] or laser [11] enables in-flight recharging of electric UAVs, offering unlimited flight endurance. Mission endurance may also be enhanced by adopting a mother-ship daughter-ship configuration, in which small, light-weight and highly manoeuvrable daughter-ship UAVs are deployed from a long endurance mother-ship UAV to collect data from a designated area and return to re-dock, transfer the acquired data and recharge [12].

[13] proposes to extend the endurance of quadrotors by altering their traditional design to one with a large main rotor located at the centre of a Y-shaped frame having a small rotor at the end of each boom, termed the "Y4" or "triangular quadrotor" ("triquad"). Combining the simplicity of a quadrotor with the increased efficiency of a helicopter, this configuration results in 15% decrease in hover power consumption and an equal increase in endurance [13].

This paper also presents a technique to determine the maximum endurance of a rotorcraft, together with a possible combination of batteries which can provide this flight time. This is achieved by solving the Knapsack Problem [14], which is a key problem in combinatorial optimisation seeking to determine the optimal selection of items with a given value and volume to be placed in a knapsack with limited capacity such that its total value is maximised. A Dynamic Programming implementation is adopted, based on the principles of overlapping sub-problems and optimality, which provides a pseudo-polynomial time solution for all payloads up to the maximum [15]. A sensitivity study analysing the effect of the parasitic mass on the maximum endurance and corresponding optimal battery selection, as well as the threshold mass for endurance improvements, is also performed. In contrast, [5] use Genetic Algorithms to determine the maximum endurance of the UAV.

Furthermore, the theoretical absolute upper bound on the endurance, which may be achieved by an ideal "infinitely-divisible" battery, is derived based on an adaptation of the Breguet-Range Equation [16] to battery-powered rotorcraft.

Theoretical analysis and model predictions are validated by experimental flight tests on the Parrot ARDrone2.0 [17], a hobby-grade quadrotor which is widely used for hobby flight and as a research platform due to its ease of use.

2 Rotorcraft Endurance Model

This section reviews the rotorcraft endurance model derived in [18], a crucial part of the endurance optimisation. Particular values are quoted for the ARDrone.

As presented in [18], the endurance of battery-powered rotorcraft may be expressed as:

$$t = \frac{k_V k_u (m, \eta_{ps}) k_B E_{nom}}{m^{3/2}} \tag{1}$$

where:

– k_V is a rotorcraft constant defined as:

$$k_V = \frac{\eta_{ps} r_p \sqrt{2 n_R \rho_a \pi}}{g^{3/2}} \tag{2}$$

where $k_V = 4.9 \times 10^{-3}$ for the ARDrone;

- η_{ps} is the rotorcraft propulsion system efficiency [dimensionless], a Figure of Merit characterising the rotor (i.e. motor, gearbox (if applicable) and propeller) efficiencies as well as their interaction. For the ARDrone, η_{ps} was experimentally determined based on the average results obtained from five such vehicles, as discussed in [18], and is estimated by $\eta_{ps}{=}0.27$.
- r_p is the propeller blade radius in metres, where $r_p{=}0.1016$m for the ARDrone;
- n_R is the number of rotors, where $n_R = 4$ for the ARDrone;
- ρ_a is the density of air [$1.2041 kg/m^3$];
- g is the acceleration due to gravity [$9.81 m/s^2$];
- $k_u(m, \eta_{ps})$ is the Battery "Usable Capacity Factor", characterising the "usable" capacity of the battery as a function of the rotorcraft mass and propulsion system efficiency. As discussed in [18], this accounts for the increase in battery cut-off voltage at higher rotorcraft mass to ensure stable operation, due to the saturation of the pulse width modulation in the motor controller. $k_u(m, \eta_{ps})$ is sensitive to the rotorcraft model and firmware, as well as the vehicle itself and its State of Health (SoH). As presented in [18], this term was experimentally determined through flight tests on the ARDrone, and based on the average results obtained from five such vehicles is estimated by:

$$k_u\left(m, \eta_{ps}\right) = \begin{cases} 1, & m \leq 0.525kg \\ -17.2m^2 + 16.7m - 3, & \text{otherwise} \end{cases}$$

- k_B is the Battery "Available Capacity Factor", characterising the inherent variability in Commercial Off-The-Shelf (COTS) LiPo batteries, and is in the range 0.8 to 1.0.
- E_{nom} is the Battery Nominal Energy [Wh], defined as: $E_{nom} = C_{nom}V_{nom}$, where:
- C_{nom} is the Battery Nominal Capacity [Ah]; and
- V_{nom} is the Battery Nominal Voltage [3.7V/cell for LiPo's], where $V_{nom} = 11.1V$ for the ARDrone;
- m is the mass in kg.

Using the LiPo battery model presented in [18], namely:

$$C_{nom} = 14.4m_b - 0.14 \tag{3}$$

$$\Rightarrow E_{nom} = 160m_b - 1.6 \tag{4}$$

where m_b is the battery mass in kg, the endurance model may be generalised as:

$$t = \frac{k_V k_u\left(m, \eta_{ps}\right) k_B \left(160\left[m - m_d\right] - 1.6\right)}{m^{3/2}} \tag{5}$$

where m_d is the rotorcraft "dry" mass, that is, without any batteries, such that $m_d = m - m_b$, where $m_d = 0.36kg$ for the ARDrone.

Thus, the Parrot ARDrone endurance model is:

$$t = 7.8 \times 10^{-3} \frac{k_u\left(m, \eta_{ps}\right) k_B \left(100m - 37\right)}{m^{3/2}} \tag{6}$$

3   Benchmark: Rotorcraft Endurance Upper Bound

In order to establish the range of achievable endurance, the theoretical absolute upper-bound $t_{UB}[h]$ was determined by deriving an adaptation of the Breguet-Range Equation [16] for battery-powered rotorcraft with take-off mass m_{TO}.

This is obtained by assuming an ideal "infinitely-divisible" battery, with specific energy E_D [Wh/kg] corresponding to the specific fuel consumption, in which each unit gram of battery mass is able to provide the power/current required by the rotorcraft at the respective mass and is discarded upon being consumed.

$$m_{b\_tot} = m_{TO} - m_d \tag{7}$$

$$\frac{dm}{dt} = -\frac{P}{E_D} \Rightarrow dt = -\frac{E_D}{P} dm \tag{8}$$

$$\Rightarrow t_{UB} = \int_{t_i}^{t_f} dt = \int_{m_{TO}}^{m_d} -\frac{E_D}{P} dm \tag{9}$$

$$= \int_{m_{TO}}^{m_d} -\frac{k_V E_D}{m^{\frac{3}{2}}} dm = -k_V E_D \int_{m_{TO}}^{m_d} m^{-\frac{3}{2}} dm \tag{10}$$

$$= -k_V E_D \lfloor \frac{m^{-\frac{1}{2}}}{-\frac{1}{2}} |_{m_{TO}}^{m_d} = 2k_V E_D \lfloor \frac{1}{\sqrt{m}} |_{m_{TO}}^{m_d} \tag{11}$$

$$\Rightarrow t_{UB} = 2k_V E_D \left(\frac{1}{\sqrt{m_d}} - \frac{1}{\sqrt{m_{TO}}} \right) \tag{12}$$

This endurance is unattainable in practice due to the maximum power ratings of batteries, resulting in a maximum continuous discharge current rating which limits the corresponding vehicle All Up Weight (AUW). Similarly, for a specific rotorcraft mass, there is a minimum battery size, corresponding to the power/current rating, below which a battery is not able to provide the power required by the vehicle.

This practical limitation of batteries is taken into account in the determination of the maximum achievable rotorcraft endurance using COTS LiPo batteries which are sequentially discharged and released, as presented in Section 4.

From (12), taking the LiPo Energy Density $E_D = 160Wh/kg$, which is the average obtained from the experimental results presented in [18], the endurance upper bound for the ARDrone is given by:

$$t_{UB\,ARDrone}[h] = 1.57 \left(\frac{1}{\sqrt{0.36}} - \frac{1}{\sqrt{m_{TO}}} \right) \tag{13}$$

This expression provides a benchmark for the endurance optimisation by establishing the absolute upper bound and enables gauging of the difference in the maximum achievable endurance that may be obtained in practice by the sequential discharge and release of COTS batteries as compared with ideal ones, as displayed in Fig. 1.

4   Endurance Optimisation with Practical Batteries

The maximum practically-achievable endurance may be obtained by sub-dividing the monolithic on-board battery into multiple smaller batteries which are discharged and released as required. This endurance lies between a minimum value obtained from a monolithic battery which is on-board throughout using (1), and the upper bound obtained from an ideal "infinitely-divisible" battery using (12).

For this purpose, the practical limitations of batteries are taken into account, that is, the battery's maximum power rating limits the maximum continuous discharge current and thus rotorcraft mass. In order to avoid basing results on specific battery data, the optimisation was performed using the presented LiPo battery model (3), assuming a maximum continuous discharge current rating of 20C, i.e. $20 \times C_{nom}[A]$. The additional parasitic mass of the corresponding implementation, namely, the battery switching circuitry, connectors, and attachment and release mechanism, is also considered. A Sensitivity Analysis investigating the effect of varying this parasitic mass on the maximum achievable endurance as well as the threshold mass for endurance improvements is carried out.

4.1   Optimisation Technique

This maximum achievable endurance, together with a possible combination of batteries which can provide this flight time, is determined by solving the Knapsack Problem [14] using a Dynamic Programming technique [15]. Based on the principles of Overlapping Sub-problems and Optimal Sub-Structure, a Dynamic Programming approach recursively defines the solution in terms of sub-problems, that is, by considering fewer batteries and/or smaller payloads, with the optimal solution consisting of the optimal solutions to each of these sub-problems. Since this maximum endurance may be achieved by several battery combinations, one such combination may be determined by back-tracking [15].

Define $T(n, m)$ as the maximum endurance using mass m made up of n batteries. Then the boundary conditions are:

$$T(0, m) = 0 \qquad \forall\, m \geq 0 \tag{14}$$

$$T(n, 0) = 0 \qquad \forall\, n = 0, 1, 2... \tag{15}$$

Also define a set of available batteries B. Let the mass of a battery $b \in B$ be m_b and the endurance flying at mass m using battery b be $t_b(m)$.

Now $T(n, m)$ is calculated by backwards recursion:

$$T(n, m) = \max_{b \in B} \Big(T(n-1, m), T(n-1, m - m_b) + t_b(m) \Big) \tag{16}$$

and the maximum endurance at any mass is given by:

$$T^*(m) = \max_n T(n, m) \tag{17}$$

4.2 ARDrone Optimisation Predictions

Fig. 1 displays the maximum achievable endurance for the ARDrone, for parasitic mass varying between 0 and 5 grams, compared with the endurance obtained from an ideal "infinitely-divisible" battery and a monolithic battery which is on board throughout. It can be observed that endurance improvements from battery sub-division can only be achieved beyond a threshold mass, below which a monolithic battery results in higher endurance due to the reduced battery efficiency with decreasing battery size. This threshold mass for endurance benefits depends on the parasitic mass, and increases by 10 grams for each gram increase in parasitic mass. Moreover, model predictions indicate an endurance gain of 0.25 minutes per gram decrease in parasitic mass. Endurance improvements also increase with available payload, thus AUW, and are hence expected to scale with rotorcraft size, as discussed in Section 5.

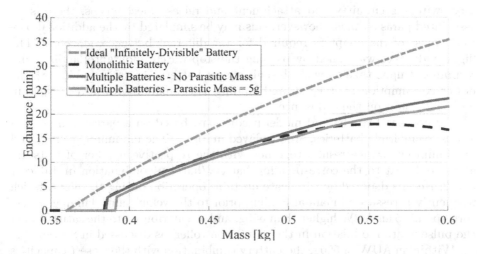

Fig. 1. ARDrone Endurance Model Predictions: maximum achievable endurance obtained from multiple batteries considering a parasitic mass of 0 and 5g, compared with the endurance obtained from a monolithic battery and the upper bound derived from an ideal "infinitely-divisible" battery

Flight tests carried out with five Parrot ARDrones indicate that the maximum AUW for stable operation is sensitive to the model and firmware version as well as the vehicle itself, including its SoH. In practice, an AUW of 550g is a fairly robust limit. Whilst taking off at a higher mass is possible, this generally results in instability during the flight, which is attributed to the saturation of the pulse width modulation in the motor controller, and thus depends on the voltage level, as discussed in detail in [18].

As can be seen from Fig. 1, at an AUW of 550g the model predicts an endurance improvement of 2 minutes from having multiple batteries as compared to a monolithic one. This can be obtained with two batteries of mass 99 and 89g respectively, which, based on the LiPo battery model (3), correspond to nominal capacities of 1286 and 1142 mAh. Whilst other battery combinations may still provide increased endurance as compared with a monolithic battery, this may be lower than the maximum practically-achievable endurance.

4.3 Experimental Validation and Verification

Model predictions were verified by performing flight tests on five ARDrones with both monolithic and multiple batteries, as specified in Tables 1 and 2 respectively. Multiple battery experiments were carried out by connecting the first battery to be discharged to the quadrotor and attaching the other batteries on top of the indoor hull. Upon complete discharge of the first battery, the ARDrone auto-lands, and the depleted battery is manually removed and the subsequent one connected. Due to the fact that this process is not automated using battery switching circuitry and attachment and release mechanisms, there is no associated parasitic mass; however, this may be simulated by the additional parasitic mass of the adaptors required for the different battery connectors. The flight endurance was timed by pausing the stop-watch upon auto-landing, and resuming it upon take-off with the subsequent battery. As presented in [18], the power consumption for take-off and landing as compared with hover results in negligible effect on the endurance.

Experimental results and model predictions, based on a parasitic mass of 3 grams for multiple batteries, are displayed in Fig. 2. The minimum, average and maximum endurance results together with the respective number of tests are also listed next to the corresponding battery/battery combination in Tables 1 and 2. Results denoted by ‡ indicate unstable operation during the flight, which frequently necessitated manual landing prior to the vehicle auto-landing. These correspond to an AUW higher than 550g, and are attributed to the saturation of the pulse width modulation in the motor controller, as discussed in Section 4.2.

Within an AUW of 550g, the battery combination with the closest capacities to the optimal combination for maximum endurance resulting from the optimisation model prediction presented in Section 4.2, namely (1286,1142mAh), is (T1300,T1000), since the (T1300,T1100) combination exceeds this AUW. The monolithic battery providing an equivalent AUW is the T2200. As can be seen from Fig. 2, several repeated flight tests with different batteries on multiple ARDrones result in a fairly consistent advantage of about 2 minutes obtained from this combination of multiple batteries as compared with the monolithic one. Thus, experimental results confirm the model and prove that endurance improvements may be achieved by sub-dividing the monolithic battery into multiple smaller ones. However, the low maximum stable flight weight severely limits the endurance gains that may be obtained for this size of vehicle. In this respect, the variation of such endurance improvements with rotorcraft scale is investigated in Section 5.

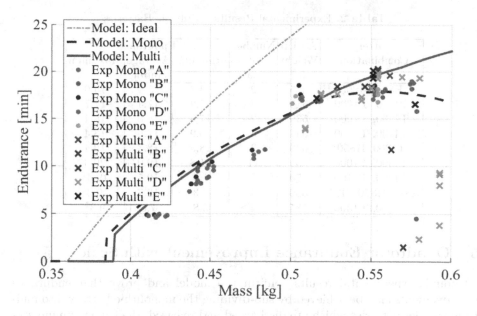

Fig. 2. ARDrone Endurance: model predictions (nominal case) for an ideal "infinitely-divisible" battery ("Model: Ideal"), monolithic battery ("Model: Mono") and multiple batteries ("Model: Multi"), assuming a parasitic mass of 3 grams, with superimposed experimental results for both monolithic ("Exp Mono") and multiple ("Exp Multi") batteries for five ARDrones ("A","B","C","D","E")

Table 1. Experimental Results - Monolithic Batteries

Nominal Capacity [mAh]	Brand	Reference	Battery Mass [g]	All Up Weight [g]	Number of Trials	Endurance Minimum [min]	Average	Maximum
450	Polypower	P450	47	407	5	4.8	4.8	5.0
500	Polypower	P500	51	411	5	4.6	4.7	4.9
850	Hyperion	H850	72	432	5	7.3	7.8	8.2
850	Zippy	Z850	75	435	5	8.4	8.7	8.9
1000	Turnigy	T1000	78	438	5	9.5	9.8	10.1
1100	Turnigy	T1100	93	453	3	9.6	10.0	10.4
1300	Turnigy	T1300	112	472	5	10.8	11.4	11.8
2000	Turnigy	T2000	153	513	13	16.6	17.2	18.5
2200	Turnigy	T2200	191	551	15	16.8	17.8	18.5
2600	Hyperion	H2600	223	583	2	18.7	18.8	18.8‡
2650	Polypower	P2650	225	585	2	18.2	18.6	19.0‡
3350	Overlander	O3350	228	588	2	4.5	10.2	15.8‡

Table 2. Experimental Results - Multiple Batteries

Battery Combinations	All Up Weight [g]	Number of Trials	Endurance [min]		
			Minimum	Average	Maximum
H850, H850	505	2	13.8	13.9	14.0
T1000, T1000	515	3	17.2	17.4	17.7
T1100, T1000	530	2	17.6	18.0	18.4
T1100, T1100	545	3	16.9	17.6	18.4
T1300, H850	548	3	18.3	18.3	18.4
T1300, T1000	550	11	11.8	19.0	20.3
T1300, T1100	570	2	1.5	10.5	19.4‡
H850, H850, H850	580	3	2.3	12.7	19.3‡
T1300, T1300	585	4	3.8	7.6	9.3‡

5   Quadrotor Endurance Improvement with Scale

Although experimental results confirm the model and prove that endurance improvements may be achieved by sub-dividing the monolithic battery into multiple smaller batteries which are discharged and released, the low maximum stable flight weight severely limits the endurance gains that may be obtained with this size of vehicle. In this respect, this section investigates how these endurance benefits vary with rotorcraft scale.

For this purpose, a general quadrotor endurance model was derived based on the general rotorcraft endurance model, (5), with $n_R = 4$. It is assumed that the battery comprises 33% of the quadrotor's mass, as determined in the average mass distribution study for quadrotors presented in [10], such that $m_b = 0.33m$. It is further assumed that at the respective mass, the batteries can be completely discharged down to their cut-off voltage, whilst maintaining stable rotorcraft operation. This results in an ideal unity battery "usable" capacity factor, $k_u(m, \eta_{ps}) = 1$.

Calculation of the rotorcraft constant k_V requires estimations of the variation of both the propeller blade radius r_p and propulsion system efficiency η_{ps} with rotorcraft mass.

The former was derived from dimensional analysis, as follows [10]:

$$T \propto r_p{}^4 \Rightarrow r_p \propto m^{\frac{1}{4}} \tag{18}$$

Based on a least-squares fit to the parameters of some popular commercial quadrotors listed in [10], the model adopted is:

$$r_p = 0.16m^{\frac{1}{4}} - 0.04 \tag{19}$$

The corresponding electrical power consumption during hover enables estimation of the propulsion system efficiency. The least-squares fit to the respective data presented in [10] results in:

$$\eta_{ps} = 0.3m^{\frac{1}{4}} \tag{20}$$

Thus, from (5), the general quadrotor endurance model may be expressed as:

$$t = \frac{k_B \left(2.1 \times 10^{-3} m^{\frac{1}{4}}\right) \left(3.97 m^{\frac{1}{4}} - 1\right) (52.8m - 1.6)}{m^{3/2}} \tag{21}$$

The corresponding maximum achievable endurance, together with the endurance from a monolithic battery as well as the upper bound benchmark that may be obtained from an ideal "infinitely-divisible" battery, for quadrotors with mass varying between 0.1 and 2kg, are displayed in Fig. 3. It can be observed that endurance improvements that may be achieved from battery sub-division as compared with a monolithic battery follow a logarithmic growth trend.

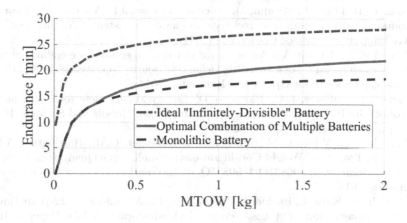

Fig. 3. Rotorcraft Endurance against Maximum Take-Off Weight (MTOW) with a Monolithic Battery, Optimal Combination of Multiple Batteries and Ideal "Infinitely-Divisible" Battery (Upper Bound)

6 Conclusion

Endurance improvements may be achieved by sub-dividing the monolithic on-board battery into multiple smaller batteries which are sequentially discharged and released. These improvements are limited by the additional parasitic mass of the required battery switching circuitry, connectors, attachment and release mechanism, as well as the reduced battery efficiency with decreasing size. For a given rotorcraft, endurance gains are only obtained beyond a threshold payload, which depends on this parasitic mass, and increase with increasing available payload and thus All Up Weight. Furthermore, endurance extensions are expected to scale with rotorcraft size, following a logarithmic growth trend.

Acknowledgments. This work is supported by the Defence Science and Technology Laboratory (DSTL).

References

1. Schoemann, J.: Hybrid-electric propulsion systems for small unmanned aircraft. München, Technische Universität München, PhD diss. (2014)
2. Saadé Latorre, E.: Design Optimization of a Small Quadrotor's Electrical Propulsion System. Masters diss., Universitat Politècnica de Catalunya (2011)
3. Pereira, J.L.: Hover and wind-tunnel testing of shrouded rotors for improved micro air vehicle design. PhD diss., University of Maryland (2008)
4. Dell, R.M., Rand, D.A.J.: Understanding batteries. Royal Society of Chemistry (2001)
5. Chang, T., Hu, Y.: Improving electric powered UAVs' endurance by incorporating battery dumping concept. In: 2014 Asia-Pacific International Symposium on Aerospace Technology, APISAT 2014 (2014)
6. Flight Of The Century. http://www.flightofthecentury.com/
7. Toksoz, T., Redding, J., Michini, M., Michini, B., How, J.P., Vavrina, M., Vian, J.: Automated battery swap and recharge to enable persistent UAV missions. In: AIAA Infotech@Aerospace Conference (2011)
8. Mulgaonkar, Y., Kumar, V.: Autonomous charging to enable long-endurance missions for small aerial robots. In: SPIE Defense+ Security, pp. 90831S-1–90831S-15. International Society for Optics and Photonics (2014)
9. Roberts, J.F., Zufferey, J.-C., Floreano, D.: Energy Management for indoor hovering robots. In: International Conference on Intelligent Robots and Systems, IROS 2008, pp. 1242–1247. IEEE/RSJ (2008)
10. Mulgaonkar, Y., Whitzer, M., Morgan, B., Kroninger, C.M., Harrington, A.M., Kumar, V.: Power and Weight Considerations in Small, Agile Quadrotors. In: SPIE Defense+ Security, pp. 90831Q-1–90831Q-16. International Society for Optics and Photonics (2014)
11. Nugent, Jr, T., Kare, J., Bashford, D., Erickson, C., Alexander, J.: 12-Hour Hover: Flight Demonstration of a Laser-Powered Quadrocopter. White Paper (2011), www.lasermotive.com/wp-content/uploads/2010/04/AUVSI-white-paper-8-11.pdf
12. NASA, Appendix B: Capabilities and Technologies (condensed version), December 2005. http://www.nasa.gov/centers/dryden/doc/139873main_Appendix%20B.doc
13. Driessens, S., Pounds, P.E.I.: Towards a more efficient quadrotor configuration. In: 2013 IEEE/RSJ International Conference on Intelligent Robots and Systems (IROS), pp. 1386–1392 (201)
14. Pisinger, D.: Algorithms for knapsack problems, PhD diss. University of Copenhagen, Denmark (1995)
15. Cormen, T.H., Leiserson, C.E., Rivest, R.L., Stein, C: Introduction to algorithms. MIT Press (2009)
16. Performance - Range and Endurance. Virginia Tech. http://www.dept.aoe.vt.edu/lutze/AOE3104/range&endurance.pdf
17. PARROT ARDrone2.0. http://ardrone2.parrot.com/
18. Analiza, A., Arthur, R., Stephen, B: Power and Endurance Modelling of Battery-Powered Rotorcraft. Submitted to IROS 2015 (2015)

An Overview of Anthropomorphic Robot Hand and Mechanical Design of the Anthropomorphic Red Hand – A Preliminary Work

Muhamad Faizal Abdul Jamil, Jamaludin Jalani$^{(\boxtimes)}$, Afandi Ahmad, and Amran Mohd Zaid

Universiti Tun Hussein Onn Malaysia, 86400, Parit Raja, Batu Pahat, Johor, Malaysia
faizal-jamil@engineer.com, {jamalj,afandia,amranz}@uthm.edu.my

Abstract. This paper provides a brief overview of the design of the past and current of anthropomorphic robot hands. The paper also introduces a new mechanical design of the anthropomorphic robotic hand known as the Red Hand. The Red Hand must be able to emulate the capability of the human hand by providing similar degrees of freedom (DOFs), ranges of motion, link and sizes. The provision of the Red Hand is important as a platform to study active compliance control. Generally, the Red Hand possesses four fingers and one thumb, with a total number of 15 degrees of freedom. The fingers are actuated by the Brushless DC Motor, with integrated speed controller. Two different force sensors namely the Tactile Pressure Sensing (TPS) and the Force-Sensitive Resistor (FSR) are considered for the Red Hand to produce active compliance control.

Keywords: Robotic hand · Multi-fingered hand · Anthropomorphic hand · Force sensor and tactile sensor

1   Introduction

The objective of the development of robot hand is to emulate the capability of the human hand by providing similar degrees of freedom (DOFs), ranges of motion, links and sizes as well as performing similar grasping. The main structure of human hand comprises of palm and fingers. There are five fingers on each hand, where each finger has three different phalanxes: Proximal, Middle and Distal Phalanxes (see Figure 1). These three phalanxes are connected by two joints, which are known as the Interphalangeal joints (IP joints). The IP joints function like hinges for bending and straightening the fingers and the thumb. The IP joint closest to the palm is called the Metacarpals joint (MCP). Next to the MCP joint is the Proximal IP joint (PIP) which is in between the Proximal and Middle Phalanx of a finger. The joint at the end of the finger is called the Distal IP joint (DIP). Both PIP and DIP joints have one Degree of Freedom (DOF) due to rotational movement [1].

© Springer International Publishing Switzerland 2015
C. Dixon and K. Tuyls (Eds.): TAROS 2015, LNAI 9287, pp. 13–18, 2015.
DOI: 10.1007/978-3-319-22416-9_2

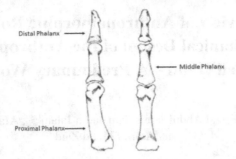

Fig. 1. Skeletal Structure of Human Finger

Over the years, a significant amount of research has been conducted to produce an anthropomorphic robotic hand. Shadow Hand [2] and ACT hand [3], are an example of an anthropomorphic robot hand that accurately mimic human hand in terms of structure and movement of the fingers. More significant effort to emulate as much as possible the functions and the size of a human hand can be found in [4] and [5] (Gifu Hand), [6] and [7] (DLR Hand), [8] (Presilia Hand) and [9] (Elumotion Hand). Refer [10] and [11] for the list of advanced robotic hands and general descriptions robot hand construction respectively.

There are two types of approach in the development of robot hand in particular to achieve active compliance robot. First is the underactuated approach that provides simple manipulations using the minimum of a single actuator, which serves the purpose of cost-effective grasping of a variety of objects. In contrast, the second approach pursued highly dexterous anthropomorphic designs, using more actuators. More actuators in the system will increase the grasping strength of the robot. Despite the strength of gripping or grasping of the robot hand is stronger; the cost of production of the robot hand can be significantly higher.

The choice of gear is also vital in determining the effectiveness the grasping robotic hand. In this study, worm wheel gears has been known to able to withstand high loads is considered to rotate the first joint of robot finger. They also operate smoothly with low noise and they have self-locking properties, and also have good meshing effectiveness, with low backlash. The use of worm gears can be considered new for the development of robotic hand since none of the existing robotic hand uses them. Other gear transmission by using bevel and spur gears are also employed in this study.

Taking into consideration other aspects such as size, weight, degrees of freedom, exerted force, load capacity, tactile sensor types/resolution/coverage, tendon/direct-driven, grasp types, precision, size, actuators, gears, functionalities, a new mechanical design of the Red Hand is developed. The details of the preliminary design are explained in this paper based on a simulation.

2   Design of the Red Hand

A Red Hand is developed to mimic a human hand. In early development, an illustration of the Red Hand is depicted in simulation as shown in Figure 2.

2.1   Mechanical Description

The Red Hand is relatively light-weight, anthropomorphic in the structure and capable of grasping and manipulation like human hands. Ideally, the size of an anthropomorphic should be within the range of the size of a human hand. In the case of the Red Hand, the size of the overall structure is largely dependent on the choice of motor or the actuator, which are located inside the fingers. The smallest size motor available is 26mm in diameter, which is about the same size as the width of human fingers. The motors are located in the finger; hence, the width of the finger has to be bigger than 26mm. In the case of the Red Hand, the width of the finger is 32mm.

The total DOF for the Red Hand is 16 includes 3 DOF for index, middle ring and small fingers and 4 DOF for thumb finger. The PIP (first joint) and the MCP (second joint) for all fingers move independently. Meanwhile, the DIP joint (third joint) depends on the movement of the second and the first joints. However, for the thumb all joints move independently.

2.2   Motor Selection

The Brushless DC Motor (BLDC) from Faulhaber is used for actuation. Its weight is considerably low; ranges between 29 g. Output speed and input speed are at 5000 rpm 4 rpm respectively. The output torque is 100 mNm in continuous operation and 180 mm in intermittent operation (detail information can be obtained from [12]). The motor is suitable for the Red Hand due to its high output torque and low rotational speed. In addition, it possesses integrated speed controller, which is important for the robot hand control.

2.3   Mechanism Description (Index, Middle, Ring and Small Fingers)

Apart from using the BLDC motor, the transmission mechanism is required for the Red Hand in particular for the first and second joint of the index, middle, ring and small fingers. It is to note that the human hand is coupling between the distal and middle phalanxes. Hence, in order to mimic similar structure and function, bevel and spur gears are used. Specifically, the BLDC motor is placed in joint 1 and joint 2 and both joints are connected via worm gears. Meanwhile, the shaft of the BLDC motor of joint 2 is connected to bevel gears. Then, to link between bevel and spur gears another shaft is utilized. Subsequently, a pulley belt system is used to connect spurs gear to joint 3 which allows a complete finger rotation.

2.4   Mechanism Description (Thumb Finger)

The mechanical design of the thumb finger has a slight modification as compared to other fingers. This is mainly due to the flexibility of the thumb during grasping (i.e. adduction and abduction) which causes a little complexity of the Red hand design. The thumb capable to move around the object and can refine its grip to hold an object. The BLDC motors are placed on all joints which considered the thumb as a fully actuated system. This allows a sufficient torque applied to the thumb during grasping.

Similar mechanism transmission (worm, bevel and spur gears) as used in four other fingers are applied on the thumb finger. The range of movement of the Red Hand is shown in Table 1.

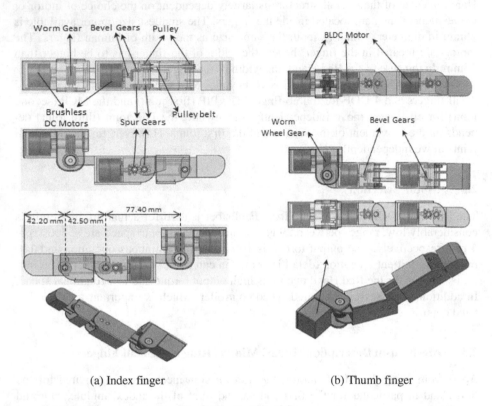

(a) Index finger   (b) Thumb finger

Fig. 2. The Red Hand finger

Table 1. Range of Red Hand Movement

Joints	Range of Movement for Index (Degrees)	Range of Movement for Thumb (Degrees)
1st (PIP)	± 36	± 30
2nd (MP)	-30 ~ +90	-35 ~ +100
3rd (DP)	0 ~ +110	0 ~ +110

2.5   The Palm

Figure 3 shows a complete structure of the Red Hand and the position of the palm. The palm is designed in such a way so that the fingers are positioned in a human finger like manner. Obviously, the middle finger appears to be the longest, followed by the point and ring finger and the shortest is the small and the thumb. It is to note that the appearance of the Red Hand is without the cosmetics cover. The cosmetic design will be developed in the future work.

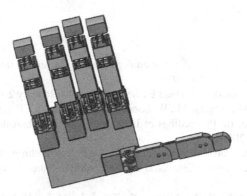

Fig. 3. The Red Hand Palm Assembly

2.6   Fingertip Sensor for Active Compliance Control

The Red Hand is designed as a platform to study of grasping and active compliance control. Active compliance control is a technique where sensors and proper control action are employed [13]. Usually, to achieve an active compliance control a force sensor will be mounted on the top of each fingertip. This allows a sensor to measure the grasped object during grasping.   There are two different types of a force sensor to be considered for the Red Hand. First is Tactile Pressure Sensing (TPS) and second is a Force-Sensitive Resistor (FSR).

In general, the tactile permits the detection and measurement of the spatial distribution of forces perpendicular to a predetermined sensory area, and the subsequent interpretation of the spatial information. In addition, a tactile-sensing array can be considered to be a coordinated group of touch sensors.   On the other hand, the FSR consists of a conductive polymer, which changes resistance in a predictable manner following application of force to its surface. Detail explanation of both sensors can found in [14] and [15].

3   Conclusion

The paper described an early development of the Red Hand in simulation. The robot hand will be fabricated, assembled and the grasping analysis will be carried out to show active compliance control in future. It is to note that the Red Hand design is still in the early phase. Obviously, the fabrication of the Red Hand is subject to change and improvement for a better grasping. In addition, the Tactile Pressure Sensing (TPS) and the Force-Sensitive Resistor (FSR) are to be considered to permit compliant control for future work.

Acknowledgement. The research leading to these results has received funding from the Ministry of Education Malaysia under the Exploratory Research Grant Scheme (ERGS), Vot E038. The authors also wish to thank the Faculty of Engineering Technology, Universiti Tun Hussein Onn Malaysia for providing a platform to carry out this research.

References

1. Davidoff, N.A., Freivalds, A.: A graphic model of the human hand using CATIA. Int. J. Ind. Ergon. **12**(4), 55–264 (1993)
2. Shadow, Shadow Dexterous Hand E1 Series, pp. 1–14, January 2013
3. Vande Weghe, M., Rogers, M., Weissert, M., Matsuoka, Y.: The ACT hand: design of the skeletal structure. In: Proceedings of IEEE Int. Conf. Robot. Autom., ICRA 2004, vol. 4, pp. 3375–3379 (2004)
4. Kawasaki, H., Komatsu, T., Uchiyama, K.: Dexterous anthropomorphic robot hand with distributed tactile sensor: Gifu hand II. IEEE/ASME Trans. Mechatronics **7**(3), 296–303 (2002)
5. Mouri, T., Kawasaki, H., Yoshikawa, K., Takai, J., Ito, S.: Anthropomorphic Robot Hand: Gifu Hand III. In: ICCAS 2002, pp. 1288–1293 (2002)
6. Liu, H., Butterfass, J., Knoch, S., Meusel, P., Hirzinger, G.: New control strategy for DLR's multisensory articulated hand. IEEE Control Syst. Mag. **19**(2), 47–54 (1999)
7. Butterfass, J., Grebenstein, M., Liu, H., Hirzinger, G.: DLR-Hand II: next generation of a dextrous robot hand. In: Proceedings 2001 ICRA IEEE International Conference on Robotics and Automation (Cat. No.01CH37164), vol. 1, pp. 109–114 (2001)
8. Controzzi, M.: The BioRobotics Institute, November 2014
9. Melhuish, C., Jalani, J., Herrmann, G.: Robust trajectory following for underactuated robot fingers. In: UKACC Int. Conf. Control 2010, pp. 495–500 (2010)
10. Uknown, "No Title". http://mindtrans.narod.ru/hands/hands.htm
11. Kaneko, K., Harada, K., Kanehiro, F.: Development of multi-fingered hand for life-size humanoid robots. In: Proc. - IEEE Int. Conf. Robot. Autom., pp. 913–920, April 2007
12. FAULHABER. http://www.faulhaber.com/en/global/
13. Jalani, J., Herrmann, G., Melhuish, C.: Robust active compliance control for practical grasping of a cylindrical object via a multifingered robot hand. In: Proc. IEEE Conf. Robot. Autom. Mechatronics (RAM), pp. 316–321 (2011)
14. Tegin, J., Wikander, J.: Tactile sensing in intelligent robotic manipulation – a review. Ind. Robot An Int. J. **32**(1), 64–70 (2005)
15. Tactile Sensing. http://www.southampton.ac.uk/~rmc1/robotics/artactile.htm

An Interactive Approach to Monocular SLAM

Erick Noe Amezquita Lucio[(✉)], Jun Liu, and Tony J. Dodd

Department of Automatic Control and Systems Engineering,
University of Sheffield, Mappin Street, Sheffield S1 3JD, UK
{erick.noe,j.liu,t.j.dodd}@sheffield.ac.uk
https://www.sheffield.ac.uk/acse

Abstract. A novel paradigm in SLAM involving user interaction without precise selection using snakes is presented allowing robot and landmark estimation. SLAM and particularly Vision-SLAM rely on automated and algorithm dependent features for robot and landmark estimation. Interactive user input is often not accounted for which could not only provide semantics through object differentiation, giving better map description as well as on-the-fly decision making for a person or robot. General Purpose computing on Graphics Processing Units produce Gradient Vector Flow forces, needed for real-time snake evolution.

Keywords: SLAM · Interactive · GVF · GPGPU

1 Introduction and Related Works

Simultaneous Localisation and Mapping (SLAM) offers many possibilities in robotics, from simple domestic chores to rescue tasks [1]. Current SLAM approaches acquire rigid and algorithm dependent features from the environment using sensors for localisation. However, no effort has been made to incorporate user input. This would allow for on-the-fly semantics, which might differentiate an object with marked importance or state, useful for a person or robot.

Cameras in SLAM have gained momentum due to price and vast image information. Real-time vision-SLAM first used a stereo camera building a sparse 3D map [3]. Simplification resulted in monocular SLAM [4]. Improvements in feature initialisation and non-linearities were obtained using the parallax effect [5]. However, all of these approaches remain automated and dependent on the feature detector used, leading to dense representations with limited map description.

New feature acquisition approaches use curve fitting, mapping places where points are hard to obtain [8], object recognition detects predefined models [6] or offline semantics offering an alternative to dense acquisition [7]. Whilst these show the trend of using simpler approaches in lieu of dense representations, useful semantics and on-the-fly user input is not accounted for, giving information of interest to both user and robot *in the moment* of map creation.

E.N. Amezquita Lucio—This work was supported by CONACYT grant 217509.

© Springer International Publishing Switzerland 2015
C. Dixon and K. Tuyls (Eds.): TAROS 2015, LNAI 9287, pp. 19–25, 2015.
DOI: 10.1007/978-3-319-22416-9_3

Here a novel paradigm in SLAM is introduced: interactive user selection that provides semantics through object differentiation. The key aspects are seen in Figure 1: An user creates an active contour (snake) around a shape of interest (rectangles are used for ease of demonstration). Gradient Vector Flow (GVF) generates forces driving the snake from the image, latching it onto the shape whilst feeding its tracking information to SLAM.

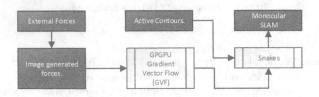

Fig. 1. A novel SLAM approach. A user surrounds a shape creating a snake which later is driven by the object's GVF forces, delivering data to SLAM.

To the best of our knowledge, this is a first interactive approach in SLAM:

- Feature selection is based on the person in lieu of an automated algorithm.
- User semantics in an object might indicate where it comes from.
- On-the-fly semantics differentiating a feature, offering decision making in the moment for a person or robot.
- A first implementation of GVF and snakes into the SLAM methodology.

The interactive methodology is detailed in Section 2 and Section 3 shows results and concludes the paper.

2   Methodology

The overall idea is to make use of an interface, in which an user sets an snake, which tracks a chosen object using attractive forces providing data to SLAM. In parallel General Purpose computing on Graphics Processing Units (GPGPU) produce GVF forces driving the snake. This approach is seen in Algorithm 1.

2.1   Active Contours, Feature Selection

A snake is a contour deformed by external forces. The initial position is set by curve fitting user coordinates, yielding a B-Spline $P(t)$ of given order k [10]:

$$P(t) = \sum_{i=1}^{n+1} B_i N_{i,k}(t), \qquad t_{\min} \leq t < t_{\max}, \qquad 2 \leq k < n+1, \quad (1)$$

where B_i are n number of control polygon points (see Figure 2); $N_{i,k}$ are recursively obtained basis functions and t is the parameter range approximated through chord lengths [9]. The B-Spline is evolved using [11]:

$$\partial C(p,t)/\partial t = \alpha(p,t)\boldsymbol{T}(p,t) + \beta(p,t)\boldsymbol{N}(p,t), \qquad (2)$$

```
Acquire new image;
Interactive user feature selection through user clicks;
Initialise active contour /* Section 2.1 */ ;
for i ← 0 to N do
 |  Gradient vector flow iteration /* from current image. Section 2.2 */
end
while User input do
 |  /* user decides when the snake is fully attached to the feature */
 |  Snake evolution /* Section 2.1 */
end
while SLAM set to start do
 |  Acquire new image;
 |  for i ← 0 to N do
 |   |  Gradient vector flow iteration /* from current image. Section 2.2*/
 |  end
 |  /* active contour follows the feature. Section 2.1 */
 |  Snake evolution;
 |  if High-level feature not initialised in SLAM then
 |   |  Add feature to state and covariance /* Section 2.3 */
 |  else
 |   |  Obtain observations and update EKF /* Section 2.3 */
 |  end
end
```

Algorithm 1. Interactive SLAM methodology pseudo code

where $C(p,t)$ is the snake evolution, B_i displacements $\partial C(p,t)/\partial t$ are given by forces affecting $\alpha(p,t)$ and $\beta(p,t)$, $\boldsymbol{T}(p,t)$ is the tangential component of the curve and $\boldsymbol{N}(p,t)$ its inward normal, obtained from the derivatives of (1).

Fig. 2. B-Spline. A cubic ($k = 4$) B-Spline (blue) and its points (purple crosses) lie in the convex hull (segmented line) created by the 9 control polygon points (red crosses). For $k = 2$ the convex hull disappears and the snake can only follow the black line.

2.2 Image Forces and Gradient Vector Flow

GVF forces are created independently in every pixel allowing the snake to become attached onto an object as in Figure 3. This is costly for CPUs but GPUs excel in parallel processing. A GPGPU implementation allows enough recursions at video speed, considering GVF as a function of time [12,13]:

$$u_t(x,y,t_k) = \mu \nabla^2 u_t(x,y,t_{k-1}) - [u_t(x,y,t_{k-1}) - f_x(x,y)] \times \left[f_x(x,y)^2 + f_y(x,y)^2 \right],$$
$$v_t(x,y,t_k) = \mu \nabla^2 v_t(x,y,t_{k-1}) - [v_t(x,y,t_{k-1}) - f_y(x,y)] \times \left[f_x(x,y)^2 + f_y(x,y)^2 \right],$$

where $u_t(x,y,t_k)$ and $v_t(x,y,t_k)$ are horizontal and vertical forces, with x and y pixel coordinates, these directly affect $\alpha(p,t)$ and $\beta(p,t)$ in (2); μ is a positive value set according image noise, $f_x(x,y)$ $f_y(x,y)$ are obtained from

Fig. 3. Active contour evolution: the yellow crosses are points of the snake (sometimes refered as snaxels). Note how it mimics the silhouette of the shape attracts it.

the input binary image's finite differences [13]. Finally $\nabla^2 u_t(x,y,t_{k-1})$ and $\nabla^2 v_t(x,y,t_{k-1})$ are Laplacians applied to previous GVF iterations $u_t(x,y,t_{k-1})$ and $v_t(x,y,t_{k-1})$.

2.3   Monocular SLAM

Once a snake follows a shape its information is used in lieu of algorithm dependent points, often seen in vision-SLAM. Inverse depth Monocular SLAM is used here which uses a camera motion model using constant velocity [5]:

$$\mathbf{x}_v = \left[\mathbf{r}_{k+1}^{wc}, \mathbf{q}_{k+1}^{wc}, \mathbf{v}_{k+1}^{w}, \omega_{k+1}^{c}\right]$$
$$= \left[\mathbf{r}_k^{wc} + (\mathbf{v}_k^w + \mathbf{V}^w)\Delta t, \mathbf{q}_k^{wc} \times \mathbf{q}((\omega_k^c + \Omega^c)\Delta t), \mathbf{v}_k^w + \mathbf{V}^w, \omega_k^c + \Omega^c\right],$$

where $\mathbf{r}^{wc}$, $\mathbf{q}^{wc}$ and $\mathbf{v}^w$ are respectively the camera position, the quaternion and the linear velocity w.r.t the world frame and ω^c is angular velocity w.r.t the camera frame; $\mathbf{q}((\omega_k^c + \Omega^c)\Delta t)$ is a quaternion defined by $(\omega_k^c + \Omega^c)\Delta t$. Finally $\mathbf{V}^w$ and Ω^c are zero mean and Gaussian linear and angular velocity impulses $\mathbf{a}^w \Delta t$ and $\alpha^c \Delta t$, produced by linear and angular accelerations $\mathbf{a}^w$ and α^c w.r.t the world and camera frame.

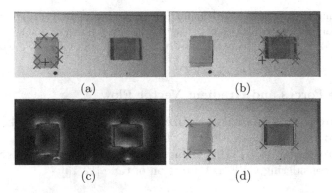

(a)   (b)

(c)   (d)

Fig. 4. Feature selection and snake evolution. Figure 4(a) shows a chosen object with red crosses indicating user clicks. Figure 4(b) shows a 2nd feature selected. GVF Forces (washed colours) are seen in Figure 4(c). Figure 4(d) depicts the snakes (green lines) with their control polygon points (red crosses) attracted to GVF's local minima.

Rectangular shapes are used for a simple demonstration of the idea. A snake is set of order $k = 2$ and 4 control polygon points, which fall in the objects' corners (recall Figure 2). Their coordinates are used first as new features $\mathbf{y}_{f_i}$ and then as observations $\mathbf{z}_{f_i}$. The full state vector contains camera and feature states, i.e. $\mathbf{x} = [\mathbf{x}_v \quad \mathbf{y}_{f_1} \quad \mathbf{y}_{f_2} \quad \cdots \quad \mathbf{y}_{f_i}]^\mathsf{T}$.

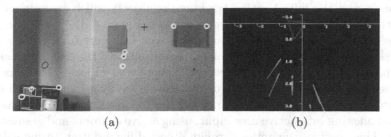

(a)   (b)

Fig. 5. Inverse depth monocular-SLAM. Figure 5(a) shows sparse features (yellow circles) from the environment. Map feature representation (green) and camera position (red triangle) is seen in Figure 5(b). Elongated ellipses show big depth uncertainty.

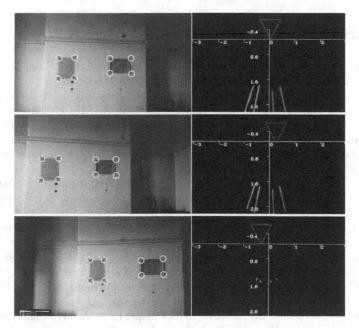

Fig. 6. Camera motion. The left hand images show control polygon points (red crosses) of the snake (green contour), with position uncertainty (yellow). Right hand pictures depict camera (red triangle) and selected objects (green and pink) estimated positions in X and Z planes. Colours allow for semantics through colour differentiation without dense representations or complex recognition.

3   Results

The algorithm was tested on a PC with an AMD FX8350 CPU, AMD R7950 GPU and a Logitech C920 camera. A user is able to judge and surround without precision 2 objects in a wall, about 2 metres from the camera as seen in Figures 4(a) and 4(b). The interface also shows GPGPU GVF forces generated at video rate, with $N = 300$ over an image with 800 by 448 pixels, Figure 4(c). Commands start the initial snakes tracking the objects which are seen as green lines with red crosses for control polygon points in Figure 4(d). This is compared to a baseline inverse depth monocular SLAM seen in Figure 5.

The snakes keep track with camera movements on the left hand side of Figure 6, decreasing uncertainty between camera and objects. The generated map is seen on the right hand side, displaying on-the-fly semantics with different feature colours (green and pink). Thus a novel SLAM paradigm has been presented allowing interactive user input, using active contours and gradient vector flow. This provides semantics through object differentiation (using colours), which can help both robot an user in real-time decision making, whilst avoiding ambiguity of dense mapping. Performance is comparable to a baseline inverse depth monocular-SLAM implementation. Future research will focus on further user on-the-fly interaction in SLAM.

References

1. Durrant-Whyte, H., Bailey, T.: Simultaneous Localization and Mapping: Part I. IEEE Robotics & Automation Magazine **13**(2), 99–110 (2006)
2. Munguia, R., Grau, A.: Camera localization and mapping using delayed feature initialization and inverse depth parametrization. In: IEEE Conf. Emerging Technologies and Factory Automation, Patras, Greece, pp. 981–988, September 2007
3. Davison, A.J., Murray, D.W.: 3D Simultaneous Localisation and Map-Building Using Active Vision. IEEE Trans. on Pattern Analysis and Machine Intelligence **24**(7), 865–880 (2002)
4. Davison, A.J., Reid, I.D., Molton, N.D.: MonoSLAM: Real-Time Single Camera SLAM. IEEE Trans. on Pattern Analysis and Machine Intelligence **29**(6), 1052–1067 (2007)
5. Montiel, J.M.M., Civera, J., Davison, A.J.: Inverse depth parametrization for monocular SLAM. IEEE Trans. on Robotics **24**(5), 932–945 (2008)
6. Salas-Moreno, R.F., Newcombe, R.A., Strasdat, H., Kelly, P.H.J., Davison, A.J.: SLAM++: simultaneous localisation and mapping at the level of objects. In: Proc. IEEE Computer Vision and Pattern Recognition, Portland, OR, pp. 1352–1359, June 2013
7. Siddiqui, J.R., Khatibi, S.: Semantic indoor maps. In: 28th Int. Conf. of Image and Vision Computing, Wellington, New Zealand, pp. 465–470, November 2013
8. Rao, D., Chung, S.-J., Hutchinson, S.: CurveSLAM: An approach for vision-based navigation without point features. In: IEEE/RSJ Int. Conf. on Intelligent Robots and Systems, Vilamoura, Portugal, pp 4198–4204, October 2012
9. Rogers, D.F.: An introduction to NURBS with historical perspective. Morgan Kaufmann Publishers, San Francisco (2001)

10. Cox, M.: The numerical evaluation of B-splines. IMA Journal of Applied Mathematics **10**(2), 134–149 (1972)
11. Srikrishnan, V., Chaudhuri, S.: Stabilization of parametric active contours using a tangential redistribution term. IEEE Trans. on Image Processing **18**(8), 1859–1872 (2009)
12. Smistad, E., Elster, A.C., Lindseth, F.: Real-time gradient vector flow on GPUs using OpenCL. Journal of Real-Time Image Processing, 1–8 (2012). ISSN 1861–8200, Online ISSN 1861–8219
13. Xu, C., Prince, J.L.: Snakes, shapes, and gradient vector flow. IEEE Trans. on Image Processing **7**(3), 359–369 (1998)

Symmetry Reduction Enables Model Checking of More Complex Emergent Behaviours of Swarm Navigation Algorithms

Laura Antuña[1], Dejanira Araiza-Illan[2(✉)], Sérgio Campos[1], and Kerstin Eder[2]

[1] Computer Science Department, Universidade Federal de Minas Gerais, Belo Horizonte, Brazil
{laura.antuna,scampos}@dcc.ufmg.br
[2] Computer Science Department, University of Bristol, Bristol, UK
{dejanira.araizaillan,kerstin.eder}@bristol.ac.uk

Abstract. The emergent global behaviours of robotic swarms are important for them to achieve their navigation task goals. These emergent behaviours can be verified to assess their correctness, through techniques like model checking. Model checking exhaustively explores all possible behaviours, based on a discrete model of the system, such as a swarm in a grid. A common problem in model checking is the state-space explosion that arises when the states of the model are numerous. We propose a novel implementation of symmetry reduction, in the form of encoding navigation algorithms relatively with respect to a reference, exploiting the symmetrical properties of swarms in grids. We applied the relative encoding to a swarm navigation algorithm, *Alpha*, modelled for the NuSMV model checker. A comparison of the state-space and verification results with an absolute (or global) and a relative encoding of the *Alpha* algorithm highlights the advantages of our approach, allowing model checking both larger grid sizes and higher numbers of robots, and consequently verifying more complex emergent behaviours. For example, a property was verified for a grid with 3 robots and a maximum allowed size of 8×8 cells in a global encoding, whereas this size was increased to 16×16 using a relative encoding. Also, the time to verify a property for a swarm of 3 robots in a 6×6 grid was reduced from almost 10 hours to only 7 minutes. Our approach is transferable to other swarm navigation algorithms.

1 Introduction

Robotic swarms consist of a set of robots with simple individual behaviour rules, working together in cooperation to achieve a more complex or emergent final behaviour. Appealing characteristics of swarms are the low cost incurred in producing the robots, which have a simple hardware design, scalability, and fault tolerance [7]. Examples of their application to real-life tasks include nanorobotics, disaster rescue missions, and mining or agricultural foraging tasks.

The emergent behaviours of a swarm of robots need to be *verified*, with respect to safety and liveness requirements [19,20], and *validated* to determine

© Springer International Publishing Switzerland 2015
C. Dixon and K. Tuyls (Eds.): TAROS 2015, LNAI 9287, pp. 26–37, 2015.
DOI: 10.1007/978-3-319-22416-9_4

whether it is fit for purpose in the target environment. Safety requirements are the allowed behaviours of the system, and liveness requirements specify the dynamic behaviours expected to happen during the execution of the system [20]. Verification methods include testing over the real robotic platforms or in simulation, and formal, such as model checking and theorem proving. In model checking all the possible behaviours of a system are exhaustively explored to determine whether the requirements are satisfied or violated. Model checking has previously been employed to verify robotic swarms [6–8,13,15].

For model checking the system is modelled in a finite-state manner. The continuous space in which the robots in the swarm move represents a challenge, since it can cause an infinite state model, which translates into a state-space explosion problem when this model is used for verification; i.e., the number of states to explore is beyond computational capabilities. The discretization of the continuous space into cells of fixed size —i.e., a grid— is a solution that has been applied in swarms, to enable model checking [8,14]. Even with the discretization of the environment into a "small" grid (e.g., 4×4 cells), the state-space explosion problem can occur due to the presence of other variables, which results in too many possible configurations of the robots in the grid.

Symmetry reduction techniques have been used to reduce the size of the models in model checking [1,2,5,9–12,18]. These techniques compute a subset of representatives of all the states, after the user provides the classification criteria for the grouping. Alternatively, the classification is computed by analysing similarities amongst the states. Our proposed solution to the state-space explosion problem for swarms in a grid is to exploit the symmetry of the configurations of the robots in the grid, implementing symmetry reduction in a novel manner.

In this paper, we explore the vertical and horizontal symmetry in the grid to reduce the size of the finite-state model. We implemented a *relative encoding* of a swarm environment model that eliminates symmetrically equivalent states from the state space. The swarm is assumed to be homogeneous; i.e., all the robots are considered identical in capabilities and rank. In the relative encoding, one robot is set as the "reference", with a fixed location and direction of motion. The other robots' locations and directions are defined based on this reference. In a global or absolute encoding, if all the robots in the grid are simultaneously rotated in the same direction and shifted horizontally or vertically by the same distance, the robots' new configurations change in location and direction. In a relative encoding, the locations and directions would remain the same since they are encoded relative to the reference robot, resulting in a reduction of the state space of direction and position configurations.

We applied our approach to the *Alpha* swarm algorithm [17], modelled in the NuSMV model checker [3]. An abstraction of the *Alpha* algorithm is proposed in [7,8] and verified through model checking against Linear Temporal Logic (LTL) properties. However, further analysis for grids of more than 8×8 cells and 3 robots [7,8] could not be performed due to state-space explosion. Verification results for larger grids and number of robots had to be extrapolated. In [16], a different abstraction for the *Alpha* algorithm is proposed, using parametrized interleaved interpreted systems, a semantics developed for the verification of multi-agent systems. Expressive temporal-epistemic specifications are verified by parametrizing the model on the number of robots. A "cut-off" that represents the behaviour of swarms of any size can be identified. Although our solution does not scale as well as [16] in terms of the number of robots, it scales very well with respect to the grid size, and thus complements [16] in that aspect.

Firstly, we modified the models in [7,8] to be relative. The state reduction allowed us to check the same swarm property for a larger number of robots and grid size. Secondly, we abstracted the *Alpha* algorithm in our own terms, employing an explicit collision avoidance mechanism, and modelled it using the relative encoding approach. This new abstraction, despite having more variables, was of a reduced order of states compared to the global encoding in [7,8]. The reduction allowed us to check the swarm for the same swarm configurations used in [7,8], but obtaining different verification results. The encoding of this new abstraction shows the potential of applying the relative concept to other swarm navigation algorithms in the same manner, which we will be exploring in the future.

The structure of this paper is as follows. Section 2 introduces grid-based discrete models for swarm robotics. Section 3 presents an overview of model checking, the state-space explosion problem and related symmetry reduction techniques. Section 4 presents the relative encoding model. Section 5 introduces the *Alpha* algorithm. In Section 6 we present, compare and discuss the results of the three encodings of the *Alpha* algorithm, using the abstraction in [7,8] and ours. The concluding remarks are presented in Section 7.

2   Swarms in Grids

When modelling navigating algorithms for swarms, a critical aspect is the continuous space in which the robots in the swarm act. A common approach is to discretize this environment into squared cells of the same size, forming a grid. This grid can "wrap around", i.e., it works as if it was projected over a sphere.

Another aspect in the modelling of a swarm is the concurrency of its elements. Four main types have been proposed [8]: *synchrony*, where all robots move at the same time in each step; *strict turn taking*, where only one robot moves at a time, following a strict order; *non-strict turn taking*, where only one robot moves at a time in a random order, but all the robots get the chance to move after a number of steps; and *fair asynchrony*, where robots move at different time in a random order, and the only guarantee is that a robot will always eventually move.

3   Model Checking and the State-Space Explosion Problem

Model checking is a formal verification method. An exhaustive traversal of the reachable states of a discrete model of the system is performed to check the validity of some desired properties, such as liveness and safety. The reachability depends on the allowed transitions from state to state. The properties to be verified are defined using a temporal logic, for example Linear Temporal Logic (LTL). Model checking is fully automatic. Counterexamples can be produced in the case of a property being false, which helps discovering the reason of the failure [4].

Model checkers can be either explicit or symbolic, the latter having an internal structure such as a Binary Decision Diagram (BDD) or Boolean functions that implicitly represent the transitions within the states of the system. The BDDs are constructed before the traversal of the model. Explicit-state model checkers traverse the model whilst verifying it, which may lead to running out of memory before finishing the traversal. Symbolic model checkers allow an initial compression of the model, at the cost of an overhead and memory usage before the checking. NuSMV is an open-source symbolic model checker, with its own input language [3].

The number of states and transitions to traverse in the model can cause problems for model checking (called state-space explosion). Different techniques have been incorporated to alleviate this issue, for instance the use of BDDs that led to the branch of symbolic model checking, symmetry reduction, and abstractions of the model to reduce the number of states [4].

The symmetrical properties of a finite-state model [4] can be identified and the state space of the model reduced before model checking. For example, a "static channel diagram" of a Promela model is computed in [9]. In [12], symmetrical components are reduced by hand, by annotating the model with the directive TRANS to eliminate equivalent transitions from state to state. This manual approach is not trivially transferable to reducing the model of a swarm in a grid.

Symmetry reduction can be applied to BDDs [4]. "Quotient models" are proposed in [5,11]. Automorphisms that preserve the same transition relations in a BDD (or "orbit relations") are computed from permutations of the states, and a chosen representative state substitutes all the states in each orbit relation, forming a reduced quotient model instead of the original BDD.

Symmetry reduction techniques have also been applied to explicit-state model checkers. In [18], a new data type, "scalarsets", is added to the input language of a model checker, to create automorphisms and a "quotient graph" (representatives of groups of states) whilst traversing the model. Scalarsets are similar to the static channel diagram model in [9]. The automorphisms can be computed on-the-fly along with the traversal of the model, as in [2] based on heuristics. The main disadvantage of all these explicit-state and BDD based symmetry reduction methods is that they are applied into the algorithms of the model checker software or BDD computation, which becomes a non-trivial software re-implementation task.

Our approach to avoid the state-space explosion problem in model checking is based on exploiting symmetrical properties of a swarm in a grid. The encoding of the model in a relative manner with respect to a reference point, as opposed to a global or absolute encoding, can be interpreted as the combination of symmetry reduction and abstraction. The proposed approach is analogous to finding orbit relations or representatives in the global model. The relative encoding employs representatives of the global encoding as possible states, i.e., configurations of the robots in the grid.

4 Relative Encoding of a Swarm in a Grid

In swarms, the focal point is the interaction of the elements through time. With only the swarm's overall behaviour in mind, the swarm moving north or south, with the same distance and orientation between all the robots and environment elements, such as bounds and obstacles, represents essentially the same situation. Furthermore, these two behaviours correspond to the same swarm configuration if the swarm is encoded in a relative manner. A simple relative encoding is to set a robot as the reference, with fixed location and direction, and to base the location of other robots and environment elements on this reference. A grid populated with robots, rotated and shifted horizontally and vertically, results in different values for the direction and position of each robot in a global or absolute encoding. However, in a relative encoding some rotations and shifting motions correspond to the same grid configurations.

If a model with r robots in a $m \times m$ size grid (locations), with d possible directions, p other robots' variables of domain sizes $v_i, i = 1, ..., p$, and q global variables of domain sizes $s_j, j = 1, ..., q$, is globally encoded, the size of the state space to be explored is $(d \times m^2 \times v_1 \times v_2 \times ... \times v_p)^r \times (s_1 \times s_2 \times ... \times s_q)$. In a relative encoding, the reference robot will have fixed location and direction, and the resulting state space will be of size $(v_1 \times v_2 \times ... \times v_p) \times (d \times m^2 \times v_1 \times v_2 \times ... \times v_p)^{r-1} \times (s_1 \times s_2 \times ... \times s_q)$. This corresponds to a reduction of the state space of $d \times m^2$. In practice this bound changes according to the variables used in the relative encoding of an algorithm.

This decrease in the state space would improve the performance of model checking in terms of time and memory usage, when verifying emergent behaviours of the swarm. Moreover, a counterexample in the relative model is equivalent to a class of counterexamples in the global model.

The update of the direction and location of the robots based on the reference robot's location motion must be performed in two situations: when the reference robot makes a move to an adjacent cell, and when it combines the motion with a change in its direction (rotation). In the first case, the equivalent of the reference robot moving in a direction is the other robots moving in the opposite direction. For example, if the reference moves north, the other robots move south instead. By following this update rule, the distance and direction relation between the swarm of robots remains the same. When the reference robot moves to another cell and also rotates, the other robots make a turn to an opposing direction and their locations need to be updated, as summarized in Table 1.

The relative encoding in NuSMV input language has been designed to have the following structure, to facilitate modularity, employing MODULE constructions:

Table 1. Location and orientation update after the reference robot changed direction

Reference's change	Direction change	Location change
$n \to e$	$n \to w,\ s \to e,\ e \to n,\ w \to s$	$x' = m - y,\ y' = x$
$n \to s$	$n \to s,\ s \to n,\ e \to w,\ w \to e$	$x' = m - x,\ y' = m - y$
$n \to w$	$n \to e,\ s \to w,\ e \to s,\ w \to n$	$x' = y,\ y' = m - x$

(a) we used distances between the reference robot and other robots, instead of specific locations; *(b)* we defined and encoded the update of the distance variables in the `main` module, along with the concurrency mode logic (e.g., a `turn` variable); and *(c)* we defined and encoded the motion algorithm (next motion, step size, rotations) of each robot in `robot` modules, which update the values of the variables in the `main` module. This procedure can be partially automated through a script to create the `main` module, from selecting the number of robots and concurrency mode, which we implemented to generate the NuSMV code for the experiments in Section 6. The `robot` modules are encoded manually, as they depend on the navigation algorithm.

5   Alpha Algorithm

The *Alpha* algorithm has been used as a case study to demonstrate how to verify emergent behaviours in swarms through model checking tools, as in [6–8,16]. In the *Alpha* algorithm, the robots in the swarm navigate the environment trying to maintain connectivity, defined as a wireless range. This is achieved by the following rules: *(a)* the default movement of a robot is forward, maintaining its current direction. *(b)* When a robot loses connection with another robot, if the remaining number of connected robots is smaller than a value α, the robot makes a 180° turn. *(c)* Every time a robot regains connectivity with another, it performs a random turn.

A critical requirement for a swarm is: *All robots shall eventually be connected.* Expressed formally in LTL, this property was proved to be false in [7,8].

6   Results

We compared a global encoding of the *Alpha* algorithm based on the abstraction proposed in [7,8], against a relative encoding of it, both implemented in the input language of NuSMV. Also, we implemented our own abstraction of the *Alpha* algorithm from the ground up, which employs more variables than in [7,8] for the concurrency modes (strict turn taking, non-strict turn taking, and fair asynchrony), encoded in a relative manner. We measured the state-space reduction of the relative models, compared to the global model.

Table 2 shows the reduction in the reachable states for the *Alpha* algorithm, with grid size 8 × 8, 3 robots, and $\alpha = 1$. The state space was computed automatically when compiling the models in NuSMV. Our new abstraction of the

Alpha algorithm also has fewer states than the global one from [7,8]. We decided not to consider the fully synchronous concurrency mode in the results, since, in the context of the *Alpha* algorithm, it allows for behaviours that are incorrect in the real world: the robots would be allowed to "swap" their cell locations.

For the strict turn taking concurrency mode, for example, all three encodings contain a variable `turn` declared in the `main` module (a global variable), of domain size 3, and variables `x`, `y` and `direction` declared in the `robot` module (robot variables), of domain sizes 8, 8 and 4, respectively. Both the global abstraction from [7,8] and its relative encoding contain a `motion` variable with domain size 2. The relative encoding also contains a global variable `random`. Our new abstraction includes the global variables `random_turn` and `random_move`, with domain sizes of 3 and 2, and robot variables `last_num_con`, with domain size 3. By representing D_v as the domain size of variable v we derive the total number of states for each encoding as being $(D_x \times D_y \times D_{direction} \times D_{motion})^3 \times D_{turn}$ for the global abstraction in [7,8], $(D_x \times D_y \times D_{direction} \times D_{motion})^2 \times (D_{motion}) \times D_{turn} \times D_{random}$ for our relative encoding of the same abstraction, and $(D_x \times D_y \times D_{direction})^2 \times D_{turn} \times D_{rand\_turn} \times D_{rand\_motion} \times (D_{last\_num\_con})^3$ for our new abstraction. The result of those calculations are mirrored by the values shown in Table 2.

Table 2. State space size (approximate) for different encodings of the *Alpha* algorithm

Concurrency	Statistic	Abstraction from [7,8]		Our new abstraction
		Global	Relative	Relative
Fair	Total States	134.2×10^6	1.1×10^6	31.9×10^6
asynchrony	Reachable States	68.1×10^6	0.5×10^6	1.6×10^6
Strict	Total States	402.7×10^6	3.1×10^6	31.9×10^6
turn taking	Reachable States	1.1×10^6	0.2×10^6	0.4×10^6
Non-strict	Total States	2818.5×10^6	22.0×10^6	223.0×10^6
turn taking	Reachable States	48.4×10^6	0.7×10^6	1.3×10^6

Figure 1 and Figure 2 show a comparison of the state-space for global and relative encodings from the abstraction implemented in [7,8], and our new abstraction when varying the grid size or the number of robots, when compiling the models in NuSMV. These experiments consider a strict turn taking concurrency mode. We repeated these experiments for other concurrency modes, non-strict turn taking and fair asynchrony, with similar results in the state-space reduction. In the experiments of Figure 1, the number of robots is set to 3, and the size of the grid is varied. In the experiments of Figure 2, the size of the grid is set to 6 × 6, and the number of robots is varied. The final points in the graph correspond to the limit in the memory for the verification to be possible.

We verified the LTL property mentioned previously, *all the robots will eventually be connected*, in all these models, to validate the relative encoding of the

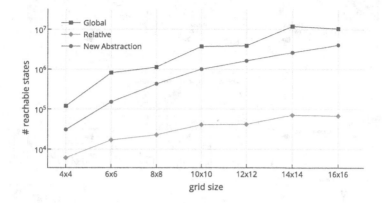

Fig. 1. Reachable states as the grid size increases, for strict turn taking concurrency mode and 3 robots

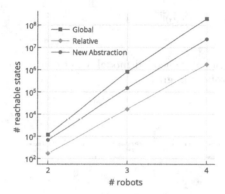

Fig. 2. Reachable states as number of robots increases, for strict turn taking concurrency mode and a 6 × 6 grid

abstraction in [7,8], with respect to their original global model. The verification results were identical, indicating our relative encoding preserved the properties of the global encoding. For the verification, we used NuSMV version 2.5.4, running on a PC with Ubuntu 14.10 and 4 GB RAM.

A counterexample (failing trace) provided by the model checker, with strict turn taking concurrency mode, 3 robots, and a grid of size 5 × 5, is shown in Figure 3, for the global encoding. From state 5 onwards, all robots move south in a loop and robot C (circle) never reconnects to the swarm. In the global model, it takes 15 states until the loops starts. However, the same pattern is observed within each 3 steps from the moment when robot C (circle) changes its direction. This repetitiveness was eliminated in the relative model, as illustrated by Figure 4, achieving a reduction of 12 states. In the relative encoding, the location of the reference (circle) is fixed to cell (0, 0) and its direction to North. Subsequently, other robots update their position and orientation according to

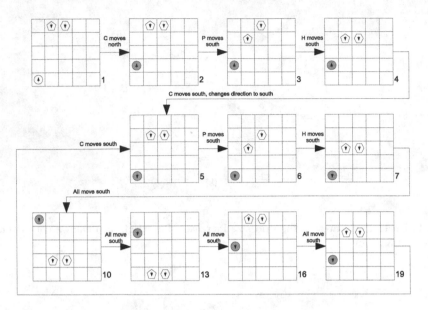

Fig. 3. Failing trace of globally encoded model. C: circle (reference), P: pentagon, H: hexagon. Disconnected robots in gray.

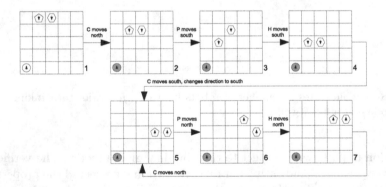

Fig. 4. Failing trace of relatively encoded model. C: circle (reference), P: pentagon, H: hexagon. Reference: circle. Disconnected robots in gray.

the decisions of the reference robot, or according to their individual decisions, as explained in Section 4.

Posteriorly, we compared the verification results using different abstractions of the *Alpha* algorithm, the relative version of the one proposed in [7,8] and our new abstraction, also relative. These experiments were conducted for $\alpha = 1$, and a strict turn taking concurrency mode. We found that the LTL property is true for some settings, such as 3 robots in grids of 2×2, 3×3 and 4×4, and two robots in a 5×5 grid. For other settings, such as 3 robots in a 5×5 grid, the property is false and the swarm will not regain connection. This is caused by the randomness of the individual robots decisions, that can be repeated infinitely, i.e., a robot can infinitely often perform the same patterns of motion but they do not lead to regain connectivity.

Figure 5 and Figure 6 show the verification time for the LTL property in Section 5 for global and relative encodings from the abstraction [7,8], and the new abstraction when varying the grid size or the number of robots. These experiments, as before, consider a strict turn taking concurrency mode. In the experiments of Figure 5, the number of robots is set to 3, and the size of the grid is varied. In the experiments of Figure 6, the size of the grid is set to 6 × 6, and the number of robots is varied. The final points in the graph correspond to a stipulated time limit of 5 days for the verification.

We observed the same time reduction patterns between the global and relative encodings of the abstraction in [7,8], as a consequence of the state-space reduction. The number of robots is a more significant constraint to the verification time than the grid size. Although it was not possible to verify the relative encoding with 4 robots due to time restrictions, applying some constraints to the initial configuration of the swarm allowed the generation of a counterexample, a proof that the property is false for that number of robots and grid size. In contrast, the global encoding could not be verified, even when applying the same constraints. The relative encoding of the new abstraction of the *Alpha* algorithm takes longer to be verified than the previous abstraction, as it is more complex. Nevertheless, it allowed us to obtain different verification results compared to [7,8], which we believe are closer to the intention of the *Alpha* algorithm. The difference between the two abstractions of the *Alpha* algorithm need to be further investigated to determine if any are incorrect, given the verification results.

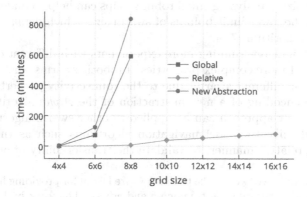

Fig. 5. Verification time for the LTL property in Section 5 as the grid size increases, for strict turn taking concurrency mode and 3 robots

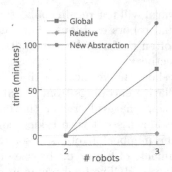

Fig. 6. Verification time for the LTL property in Section 5 as number of robots increases, for strict turn taking concurrency mode and a 6 × 6 grid

7 Conclusions

We presented an approach that achieves significant state-space reduction and thus allows model checking more complex emergent behaviours of swarms. We propose the use of symmetry reduction to eliminate symmetrical states (i.e., configurations of robots in the grid), implemented through a relative encoding of the swarm, where one robot is the reference, and the others' navigation is encoded relative to it. This encoding, compared to a global one, helps to reduce the state space of the model, as demonstrated by our results in Section 6. Thus, verification through model checking can be performed over larger grid sizes and higher numbers of robots, and also for more detailed abstractions that model navigation algorithms using more variables. Although the state space reduction is more significant in terms of the grid size, and not as expressive if considering realistic swarm sizes, analysing small robot groups can help to understand larger swarms within the lower limit bounds of swarm size, which demands more from the navigation algorithm [7,8].

Future work includes running more experiments with different α values, and the verification of more complex properties of robotic swarms, such as the emergence of teams or different swarms due to the connectivity properties. The successful relative encoding of a new abstraction of the *Alpha* algorithm gives us confidence that our approach can be applied to other swarm algorithms. In the future we would like to model navigation algorithms such as the *Beta* algorithm [17] in a relative manner, to validate the transferability of our approach.

Acknowledgments. We would like to thank Clare Dixon for providing her *Alpha* algorithm models, and her invaluable comments and advice. The work by D. Araiza-Illan and K. Eder was partially supported by the EPSRC, grants EP/J01205X/1 RIVERAS: Robust Integrated Verification of Autonomous Systems and EP/K006320/1 Trustworthy Robotic Assistants.

References

1. Appold, C.: Efficient symmetry reduction and the use of state symmetries for symbolic model checking. In: Proc. GandALF, pp. 173–187 (2010)
2. Bosnacki, D., Dams, D., Holenderski, L.: Symmetric SPIN. In: SPIN Model Checking and Software Verification, Stanford, CA, USA, pp. 1–19 (2000)
3. Cimatti, A., Clarke, E., Giunchiglia, E., Giunchiglia, F., Pistore, M., Roveri, M., Sebastiani, R., Tacchella, A.: NuSMV version 2: an opensource tool for symbolic model checking. In: Computer-Aided Verification, Copenhagen, Denmark. pp. 359–364 (2002)
4. Clarke Jr, E.M., Grumberg, O., Peled, D.A.: Model Checking. MIT Press, Cambridge (1999)
5. Clarke, E., Enders, R., Filkorn, T., Jha, S.: Exploiting symmetry in temporal logic model checking. Formal Methods in System Design 9(1–2), 77–104 (1996)
6. Dixon, C., Winfield, A.F., Fisher, M.: Verification of swarm robots: the Alpha algorithm. In: Proc. ARW, Westminster, UK, pp. 10–11 (2010)
7. Dixon, C., Winfield, A., Fisher, M.: Towards temporal verification of emergent behaviours in swarm robotic systems. In: Groß, R., Alboul, L., Melhuish, C., Witkowski, M., Prescott, T.J., Penders, J. (eds.) TAROS 2011. LNCS, vol. 6856, pp. 336–347. Springer, Heidelberg (2011)
8. Dixon, C., Winfield, A.F., Fisher, M., Zeng, C.: Towards temporal verification of swarm robotic systems. In: Robotics and Autonomous Systems, Bristol, UK, pp. 1429–1441 (2012)
9. Donaldson, A.F., Miller, A.: Automatic symmetry detection for model checking using computational group theory. In: Fitzgerald, J.S., Hayes, I.J., Tarlecki, A. (eds.) FM 2005. LNCS, vol. 3582, pp. 481–496. Springer, Heidelberg (2005)
10. Emerson, E.A., Wahl, T.: On combining symmetry reduction and symbolic representation for efficient model checking. In: Geist, D., Tronci, E. (eds.) CHARME 2003. LNCS, vol. 2860, pp. 216–230. Springer, Heidelberg (2003)
11. Emerson, E.A., Wahl, T.: Dynamic symmetry reduction. In: Halbwachs, N., Zuck, L.D. (eds.) TACAS 2005. LNCS, vol. 3440, pp. 382–396. Springer, Heidelberg (2005)
12. Gomes, P.d.C.: Verification of symmetric models using semiautomatic abstractions. Master's thesis, Computer Science, Belo Horizonte, Brazil (2010)
13. Juurik, S., Vain, J.: Model checking of emergent behaviour properties of robot swarms. Proceedings of the Estonian Academy of Sciences 60(1), 48–54 (2011)
14. Kloetzer, M., Belta, C.: Temporal logic planning and control of robotic swarms by hierachical abstractions. IEEE Transactions on Robotics 23(2), 320–330 (2007)
15. Konur, S., Dixon, C., Fisher, M.: Analysing robot swarm behaviour via probabilistic model checking. Robotics and Autonomous Systems 60, 199–213 (2012)
16. Kouvaros, P., Lomuscio, A.: A counter abstraction technique for the verification of robot swarms. In: Proceedings of the Twenty-Ninth AAAI Conference on Artificial Intelligence, January 25–30, 2015, Austin, Texas, USA, pp. 2081–2088 (2015)
17. Nembrini, J.: Minimalist Coherent Swarming of Wireless Networked Autonomous Mobile Robots. Ph.D. thesis, Bristol, UK (2005)
18. Norris Ip, C., Dill, D.L.: Better verification through symmetry. Formal Methods in System Design 9(1–2), 41–75 (1996)
19. Rouff, C., Hinchey, M., Truszkowski, W., Rash, J.: Formal methods for autonomic and swarm-based systems. In: Proc. ISoLA, Paphos, Cyprus, pp. 100–102 (2004)
20. Winfield, A.F., Sa, J., Fernandez-Gago, M.C., Dixon, C., Fisher, M.: On formal specification of emergent behaviours of swarm robotic systems. International Journal of Advanced Robotic Systems 2(4), 363–370 (2005)

Inertia Properties of a Prosthetic Knee Mechanism

Mohammed I. Awad[1(✉)], Abbas A. Dehghani-Sanij[1], David Moser[2],
and Saeed Zahedi[2]

[1] Institute of Design, Robotics and Optimization (iDRO),
University of Leeds, Leeds LS2 9JT, UK
{m.i.awad,a.a.dehghani-sanij}@leeds.ac.uk
[2] Institute of Design, Robotics and Optimization (iDRO),
Chas a Blatchford and Sons Ltd, Lister Road, Basingstoke, Hants RG22 4AH, UK
{David.Moser,Saeed.Zahedi}@blatchford.co.uk

Abstract. The prosthetic knee mechanism should be able to assist amputees during activities of daily living and improve their quality of life. The inertia asymmetry between intact and the prosthetic sides is one of the reasons for amputee gait asymmetry. This paper shows how to calculate the overall inertia properties during the design process.

1   Introduction

Every year, thousands of lower limb amputations are carried out around the world due to complications of diabetes, circulatory and vascular disease, trauma, or cancer in limb segments [1]. The loss in mobility following amputation results in a degradation of the quality of life of the amputees as it affects many aspects of their personal and professional lives. Lower limb prostheses are used to replace the lost limbs and assist amputees in restoring their missing mobility functions. The current commercial state-of-the-art prostheses can be divided into three main categories: *purely mechanical, microprocessor damping control* and *powered prostheses.*

Purely mechanical prostheses depend only on mechanical components and require significant mental and voluntary control effort during walking. Microprocessor-damping-controlled prostheses are developed with the objective of approximating the human gait functions. These prostheses are equipped with integrated sensors to supply information to a microprocessor which is used to control the prosthesis in real-time. The main task of this type of prosthetic is to support the body weight in the stance phase and provide dynamic control during the swing phase. This group of prostheses was introduced during the 1990s with the release of the Intelligent Knee (Nabtesco), the Intelligent Prosthesis (IP) (Chas. A. Blatchford & Sons), and the C-Leg (Otto Bock). These prostheses are still passive and cannot contribute any net positive power to the gait which limits the amputee ability. *Actively powered prostheses*, such as the Victhom knee [2-4], commercially known as the Power Knee (Ossur) are fully actuated. This group of prostheses can be powered using either DC motors [5-7], or pneumatic actuators [8] to provide positive power to the prosthetic limb.

© Springer International Publishing Switzerland 2015
C. Dixon and K. Tuyls (Eds.): TAROS 2015, LNAI 9287, pp. 38–43, 2015.
DOI: 10.1007/978-3-319-22416-9_5

Despite the current technological advances in prosthetics, lower limb amputees still suffer from noticeable gait asymmetry [9-11] and high metabolic energy costs [12] in comparison to healthy subjects. This gait asymmetry pattern occurs due to the inertia asymmetry between the intact and the prosthetic leg [13] and other factors. The prosthesis inertia plays an important role in the gait dynamics especially during the swing phase as in passive dynamic walkers. In order to improve the prostheses performance and design efficient lower limb prostheses, the inertia properties of lower limb prosthesis should be altered without increasing the prosthetic weight to allow more energetic amputee locomotion. This paper focuses on developing a prosthetic knee actuation mechanism and studying the inertia parameters in the mechanism.

2   Kinematics Analysis of the Proposed Prosthetic Knee Mechanism

In this section, the analysis of the proposed prosthetic knee based on closed kinematics chain mechanism, shown in Fig. 1a, is presented. According to the geometry of the closed loop configuration in Fig. 1a, the screw length is calculated by applying Cosine rule in $\triangle ABC$ as shown in the following equation:

$$L_{screw} = \sqrt{L_1^2 + x^2 - 2L_1 x \cos\beta} \tag{1}$$

It is clear that the knee's torque (T_k) and speed $(\dot{\theta}_k)$ relative to the motor's torque (T_m) and speed $(\dot{\theta}_m)$ are functions of the mechanism transmission ratio (r), the ball screw pitch (p) and the overall efficiency of the mechanism (η_o) as shown in Equations (2) and (3). These relations show that the knee's torque (T_k) will increase when the transmission arm (r) increases, and the knee's speed $(\dot{\theta}_k)$ will increase when the transmission arm (r) decreases.

$$T_k = F_{screw} \cdot r = \frac{2\pi\eta_o r}{p} \cdot T_m \tag{2}$$

$$\dot{\theta}_k = \frac{p}{2\pi r} \cdot \dot{\theta}_m \tag{3}$$

The transmission arm (r) of the torque arm can be calculated based on the mechanism geometry. By applying the sine rule in $\triangle ABC$, the transmission arm (r) is calculated as:

$$r = \frac{xL_1}{\sqrt{L_1^2 + x^2 - 2xL_1\cos\beta}}\sin\beta \tag{4}$$

Where:
r: the transmission ratio (torque arm length) of the mechanism, x: the length between joints A and C in Fig. 1a, L_1: the length between joints A and B in Fig. 1a, β: the angle of $\angle CAB$ in Fig. 1a.

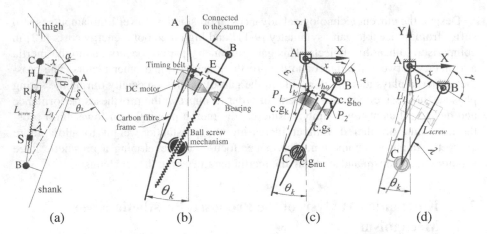

Fig. 1. Schematic diagram for closed loop kinematics chain of the knee prototype

Based on these selected parameters and the physical constrains of the mechanical components, the prosthetic knee design shown in Fig. 1b was proposed to produce the maximum torque and speed profiles. This prosthetic knee mechanism consists of seven main links (*n*) as shown in Fig. 1b. These links are, respectively, the fixed link (**AB**) that is attached to the amputee's socket as shown in Fig. 1c, the carbon fibre frame for the knee, the ball screw (**CD**), the ball screw nut (**C**), the ball screw housing (**BD**), the motor shaft and the timing belt (**E**). Furthermore, there are three types of joints used in the mechanism for kinematics and physical constraints: five lower pairs revolute joints (**A, B, C, D, E**), one screw joint, and two wrapping pair joints (higher pairs) between the belt and the pulleys. These joints create constrains that restrict the mechanism mobility to one degree of freedom (DoF).

3   Sources of Inertia and Impedance Torques in the Mechanism

The mechanism components have passive elements to mechanically dissipate the energy such as friction and damping while the inertias and the potential energy of the knee weight are used as energy storage elements. Equation (5) specifies the equivalent dynamics characteristics of the prosthetic knee prototype relative to joint A, which are shown in Fig. 1a and b.

$$J_{eq \to A}\ddot{\theta}_k + D_{eq \to A}\dot{\theta}_k + T_{f_{eq \to A}} = T_k + T_w + T_R \tag{5}$$

Where:

$T_w = m_{k_{eff}} g l_k \sin(\theta_k - \varepsilon)$, $J_{eq \to A}$: equivalent reflected inertia referred to knee revolute joint A, $\ddot{\theta}_k$: angular acceleration of the knee at joint A (rad/sec^2), $D_{eq \to A}$: equivalent reflected damping coefficient referred to revolute joint A (Nm/(rad/sec)), $\dot{\theta}_k$: angular velocity of the knee at joint A (rad/sec), $T_{f_{eq \to A}}$: equivalent reflected friction torque referred to joint A (Nm), T_k : active knee torque generated by the motor at joint A (Nm), T_w: equivalent torque produce by the prosthetic weight (Nm), T_R: external resistance torque applied on the prosthetic knee (Nm), $m_{k_{eff}}$: effective prosthetic weight (kg), l_k: the distance between the joint A and the prosthetic centre of gravity (m), θ_k: prosthetic knee angle (rad), ε: static equilibrium angle of the prosthesis.

3.1   Inertia Properties of the Developed Prosthetic Knee

The equivalent reflected inertia is the sum of all inertias in the knee mechanism reflected to joint A as follows:

$$J_{eq \to A} = J_{m \to A} + J_{p_1 \to A} + J_{p_2 \to A} + J_{s \to A} + J_{nut \to A} + J_k + J_{ho \to A} \tag{6}$$

Where:
$J_{m \to A}$: motor inertia reflected to the revolute joint A (kg.m^2), $J_{p_1 \to A}$: inertia of pulley 1 in timing belt arrangement reflected to the revolute joint A (kg.m^2), $J_{p_2 \to A}$: inertia of pulley 1 in timing belt arrangement reflected to the revolute joint A (kg.m^2), $J_{s \to A}$: ball screw inertia reflected to the revolute joint A (kg.m^2), $J_{nut \to A}$: the inertia of the ball screw nut reflected to the revolute joint A (kg.m^2), J_k: carbon fibre frame reflected to the revolute joint A (kg.m^2), $J_{ho \to A}$: ball screw and bearing holder reflected to the revolute joint A (kg.m^2).

The effective reflected inertia is determined using the kinetic energy, as the kinetic energy referred to any point must provide the same kinetic energy plus the losses. Hence, equation (7) is used to derive the reflected inertia based on the constant kinetic energy concept.

$$K.E_{m \to A} = \frac{K.E_m}{\eta_{m \to A}} \quad \to \quad \frac{1}{2} J_{m_r \to A} \dot{\theta}_k^2 = \frac{1}{2} J_{m_r} \dot{\theta}_m^2 \left(\frac{1}{\eta_{m \to A}} \right) \tag{7}$$

Where:
$K.E_{m \to A}$: kinetic energy of the motor's rotor shaft referred to joint A, $K.E_m$: kinetic energy of the motor's rotor shaft, $J_{m_r \to A}$: the mass moment of inertia of the motor shaft referred to A, J_{m_r}: the mass moment of inertia of the motor shaft around the rotation axis of the motor shaft, $\eta_{m \to A}$: transmission efficiency from the motor to joint A. The conversion ratio of the angular velocity from the knee ($\dot{\theta}_k$) to the motor ($\dot{\theta}_m$) is calculated based on equation (3) as the timing belt arrangement between the motor and the ball screw has 1:1 reduction ratio. Hence, the reflected inertia of the motor shaft referred to joint A is calculated as follows:

$$J_{m_r \to A} = \frac{J_{m_r}}{\eta_{m \to A}} \left(\frac{2\pi}{p} \right)^2 r^2 \tag{8}$$

The motor shaft not only rotates around the motor shaft axis, but also the motor with the ball screw holder arrangement rotates around joint B. Hence, the motor, pulley 1 (P_1), pulley 2 (P_2) and the ball screw have two inertial values which should be reflected to joint A as shown in the following equation:

$$J_{m \to A} + J_{p_1 \to A} + J_{p_2 \to A} + J_{s \to A} + J_{ho \to A} = \left[\frac{J_{m_r}}{\eta_{m \to A}} + \frac{J_{p1_r}}{\eta_{p1 \to A}} + \frac{J_{p2_r}}{\eta_{p2 \to A}} + \frac{J_{s_r}}{\eta_{s \to A}} \right] \left(\frac{2\pi r}{p} \right)^2 +$$

$$[J_{m_B} + J_{p1_B} + J_{p2_B} + J_{s_B} + J_{ho_B}] \left(\frac{1}{\eta_{B \to A}} \right) \left(\frac{\dot{\gamma}}{\dot{\theta}_k} \right)^2 \tag{9}$$

Where:

J_{p1_r}: the mass moment of inertia of the pulley 1 around the rotation axis, $\eta_{p1 \to A}$: transmission efficiency from the pulley 1 to joint **A**, J_{p2_r}: the mass moment of inertia of the pulley 2 around the rotation axis, $\eta_{p2 \to A}$: transmission efficiency from the pulley 2 to joint **A**, J_{s_r}: the mass moment of inertia of the ball screw around the rotation axis, $\eta_{s \to A}$: transmission efficiency from the ball screw to joint **A**, J_{m_B}: the mass moment of inertia of the motor around the joint **B**, J_{p1_B}: the mass moment of inertia of the pulley 1 around the joint **B**, J_{p2_B} : the mass moment of inertia of the pulley 2 around the joint **B**, J_{s_B}: the mass moment of inertia of the ball screw around the joint **B**, J_{ho_B}: the mass moment of inertia of the ball screw holder around the joint **B**, $\eta_{B \to A}$: transmission efficiency from joint **B** to joint **A**, $\dot{\gamma}$: angular velocity of joint **B** (rad/sec).

Similarly, the reflected mass moment of the ball screw nut is calculated as follows:

$$J_{nut \to A} = \frac{m_{nut}}{\eta_{nut \to A}} r^2 + \frac{J_{nut}}{\eta_{C \to A}} \left(\frac{\lambda}{\dot{\theta}_k} \right)^2 \tag{10}$$

Where:

m_{nut}: the ball screw nut mass in kg, $\eta_{nut \to A}$: transmission efficiency from the pulley the ball screw nut to joint **A**, J_{nut}: the mass moment of inertia of the ball screw nut around the joint **C**, $\eta_{C \to A}$: transmission efficiency from the joint **C** to joint **A**, λ: angular velocity of joint **C**.

Further work is required to optimize the distribution of the masses and inertia parameters to provide the prosthetic knee ($J_{eq \to A}$) with inertia properties similar to the natural limb without increase its weight. Also, these inertia parameters are important during the selection process of the actuator.

4 Conclusions

This paper presented a study of the kinematics and the inertia properties of the developed knee mechanism. The parameters considered here are important to optimise and improve the performance of the knee mechanism and as a result to reduce the amputee gait asymmetry. The significant sources of inertia in the proposed knee mechanism are highlighted and studied in this paper.

Acknowledgement. This work is linked to a current research sponsored by EPSRC (EP/K020462/1).

References

1. Cristian, A.: Lower Limb Amputation: A Guide to Living a Quality Life. Demos Health (2005)
2. Bedard, S.: Control system and method for controlling an actuated prosthesis, United States (2004)
3. Bedard, S.: Control device and system for controlling an actuated prosthesis, United States (2006)
4. Bédard, S., Roy, P.-o.: Actuated leg prosthesis for above-knee amputees. Victhom Human Bionics Inc., Saint-Augustin-de-Desmaures (2008)
5. Goldfarb, M.: Consideration of Powered Prosthetic Components as They Relate to Micro-processor Knee Systems. JPO: Journal of Prosthetics and Orthotics **25**, P65–P75 (2013). doi:10.1097/JPO.1090b1013e3182a8953e
6. Goldfarb, M., Lawson, B.E., Shultz, A.H.: Realizing the Promise of Robotic Leg Prostheses. Science Translational Medicine **5**, 210ps–215ps (2013)
7. Shultz, A., Lawson, B., Goldfarb, M.: Running with a Powered Knee and Ankle Prosthesis. IEEE Transactions on Neural Systems and Rehabilitation Engineering **PP**, 1–1 (2014)
8. Sup, F.C., Goldfarb, M.: Design of a pneumatically actuated transfemoral prosthesis. In: ASME Conference Proceedings 2006, pp. 1419–1428 (2006)
9. Jaegers, S., Arendzen, J.H., Dejongh, H.J.: Prosthetic Gait Of Unilateral Transfemoral Amputees - A Kinematic Study. Archives of Physical Medicine and Rehabilitation **76**, 736–743 (1995)
10. Seroussi, R.E., Gitter, A., Czerniecki, J.M., Weaver, K.: Mechanical work adaptations of above-knee amputee ambulation. Archives of Physical Medicine and Rehabilitation **77**, 1209–1214 (1996)
11. Vrieling, A.H., van Keeken, H.G., Schoppen, T., Otten, E., Halbertsma, J.P.K., Hof, L., Postema, K.: Gait initiation in lower limb amputees. Gait & Posture **27**, 423–430 (2008)
12. Sagawa Jr., Y., Turcot, K., Armand, S., Thevenon, A., Vuillerme, N., Watelain, E.: Bio-mechanics and physiological parameters during gait in lower-limb amputees: A systematic review. Gait & Posture **33**, 511–526 (2011)
13. Smith, J.D., Martin, P.E.: Short and longer term changes in amputee walking patterns due to increased prosthesis inertia. JPO: Journal of Prosthetics and Orthotics **23**, 114–123 (2011)

Development and Characterisation
of a Multi-material 3D Printed Torsion Spring

Andrew Barber$^{(\boxtimes)}$, Peter Culmer, and Jordan H. Boyle

School of Mechanical Engineering, University of Leeds, Leeds LS2 9JT, UK
{el08arb,p.r.culmer,j.h.boyle}@leeds.ac.uk

Abstract. Compliant actuation methods are popular in robotics applications where interaction with complex and unpredictable environments and objects is required. There are a number of ways of achieving this, but one common method is Series Elastic Actuation (SEA). In a recent version of their Unified Snake robot, Choset et al. incorporated a Series Elastic Element (SEE) in the form of a rubber torsional spring. This paper explores the possibility of using multi-material 3D printing to produce similar SEEs. This approach would facilitate the fabrication and testing of different springs and minimize the assembly required. This approach is evaluated by characterizing the behavior of two configurations of SEE, 3d printed with different dimensions. The springs exhibit predictable viscoelastic behavior that is well described by a five element Wiechert model. We find that individual springs behave predictably and that multiple copies of the same spring design exhibit good consistency.

Keywords: Series Elastic Actuation · 3D printing · Compliance

1 Introduction

Traditional robotic systems generally use stiff actuators, which have advantages such as increased precision and stability [1]. However there are situations where other factors have greater importance. Compliant actuators can increase shock tolerance [1], improve force control accuracy [2], simplify control systems [3] and improve safety [4]. There are a number of different approaches for achieving compliance, but one of the most common strategies is Series Elastic Actuation (SEA), in which a compliant elastic element (SEE) is connected in series with the actuator. This approach is used in the latest version of the Unified Snake robot developed by Choset et al [2]. Specifically, they developed a compact torsion spring consisting of a molded natural rubber element bonded between two machined metal plates. The rubber element has a tapered cross section designed to maintain uniform shear stress[5], as can be seen in Figure 1. Their manufacturing process is undoubtedly effective, but it is also quite complex. If a simpler process could achieve similar performance this would clearly be advantageous, particularly when testing multiple SEE variants. Multi-material 3D printers can incorporate a range of mechanical properties (from rigid to flexible) into a single

© Springer International Publishing Switzerland 2015
C. Dixon and K. Tuyls (Eds.): TAROS 2015, LNAI 9287, pp. 44–49, 2015.
DOI: 10.1007/978-3-319-22416-9_6

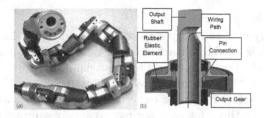

Fig. 1. (a): Unified Snake Robot [5]. (b) Cross-section of SEE [2]. Note the tapered shape of the rubber element. Bottom and top of rubber element fixed to gear output and output shaft respectively.

printed part [6], making it possible to print the entire SEE as an integrated unit. Of course, the viability of this approach depends on the mechanical properties of the resulting SEE. A 3D printed linear spring was successfully employed in Ref. [7] but it was not extensively characterized. Here we present the results from preliminary testing and characterization of the first 3D printed torsional SEEs.

2   Methods

2.1   Manufacture

We produced a total of four SEEs (two copies of two different variants) based on Rollinson's design [5]. Both variants have a similar tapered profile (see Fig. 2a) and have the same inner and outer diameter. Arms extend from both plates for attachment to a test rig. The first variant has an outer thickness of 10mm and inner thickness of 8mm, while the second has an outer thickness of 12mm and inner thickness of 10mm. The SEEs were printed using a Stratasys Objet1000 [6], using the rigid "VeroWhitePlus" material for the end plates and arms, and flexible "TangoPlus" (Shore 26 - 28) for the spring itself (see Fig. 2a,b).

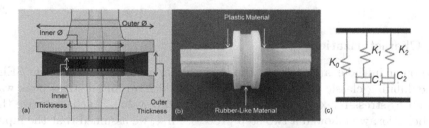

Fig. 2. Cross section of SEE CAD model. Inner Diameter=25mm (constant), Outer Diameter=45mm(constant). (b) 3D Printed SEE. (c) Diagram of 5 element Weichert model [9].

2.2 Testing

The SEEs were subjected to rotary step strain tests using an Instron E10000 [8]. One end of the SEE is fixed into the Instron's stationary base plate and the other to the rotating chuck. Each test starts from an equilibrium position (zero input angle) followed by a "step" rotation (over 100ms) to either 5° or 10° which is held for 20 seconds before returning to zero and holding for another 20 seconds. This is repeated three times for each spring, giving six samples of step test data (three up and three down). The time, angle and torque data are logged at a frequency of 50Hz.It was observed, however, that the instron's control system produces an imperfect step in which the angle overshoots the set point as shown in Figure 3a. This is taken into account during characterization.

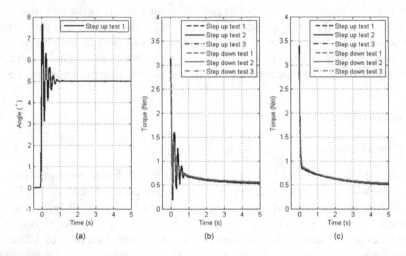

Fig. 3. (a) One sample of angle data from step test. The step input includes an initial oscillatory component. (b) Six samples of raw data from three iterations of 5° step test for 10mm thick SEE. (c) Double exponential fit for each sample. Note the repeatability between all six tests, whether they are step up or step down results.

2.3 Characterization

Based on an initial examination of the torque data, we concluded that our SEEs were exhibiting classic viscoelastic behavior in which an initial stress peak was followed by stress-relaxation down to a steady state value. To characterize the SEE behavior, we followed a two-step process. First, we assumed that the input angle was an ideal step and used Matlab to fit the stress-relaxation curve. We found that a double exponential produced a good fit, so we concluded that a five-element Weichert Model (Fig. 2c) was appropriate to model the viscoelastic behavior. For each spring, we performed this fitting process on each of the stress-relaxation curves (see Fig. 3), took the median across the six samples and used this to calculate the spring and damper constants for the Weichert Model. The second step in our characterization was to fine-tune the model parameters based on the real angle and torque data. We implemented a function that simulates the model's response to the actual recorded angle data for a given set of Weichert parameters before comparing this to the recorded torque data to calculate an error score. We then used Matlab's "fminsearch" function to optimize the model parameters from the first step. This step improved the model fit, as shown in Figure 4. This was done independently for each of the three experiments on each SEE and the results were used to obtain the median values and standard deviations given in Table 1.

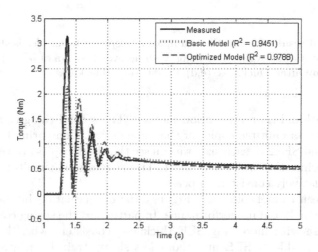

Fig. 4. Typical example of a sample of real SEE step test data compared to basic and optimized simulation

3   Results

As shown in Figure 3b, the step response of a single SEE is repeatable. While there is some minor difference between the up and down steps, the three repeats in each direction are virtually indistinguishable. Furthermore, as shown in Figure 4, the 5-element Weichert Model reproduces the SEE behavior well, particularly once the parameters have been fine-tuned. Table 1 summarizes our results for all four SEEs and demonstrates their repeatability. There are some anomalies relating to K_2 values, but these account for the fast adaptation which occurs in the first 100ms and are therefore of least significance. Here we will focus on the first three parameters. The first thing to note is the repeatability of behavior for a specific SEE, as indicated by the very small standard deviation values. Furthermore, the parameters obtained from 5° and 10° steps on the same SEE are virtually identical (as expected), which confirms the validity

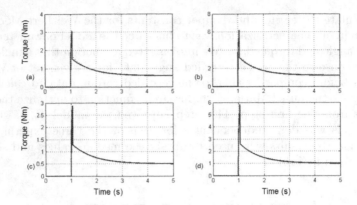

Fig. 5. Simulated results for two SEE samples in response to an idealised step input that ramps up to its final value over 50ms with no overshoot. (a,c) 5° step; (b,d) 10° step; (a,b) 10mm Outer Thickness. (c,d) 12mm Outer Thickness.

of our computational characterization method. The second thing to note is the repeatability across multiple copies of the same design. Comparing the parameters for SEE1 and SEE2 of each size, we can see that both copies are very similar, despite being printed several weeks apart. This suggests that multi-material 3D printing is a relatively consistent process.

As can be seen from Table 1 and Figure 5 there is a relationship between the thickness of the SEE and its performance. In particular, the spring constants K_0 and K_1 decrease with increasing SEE thickness, consistent with the fact that the material in a thicker SEE undergoes less shear strain for a given angular deflection.

Table 1. Spring and Damping Constants across all SEEs. Values given as median ± standard deviation.

| | | \multicolumn{5}{c}{10mm Outer Thickness} | | | | |
		K_0	K_1	C_1	K_2	C_2
SEE 1	5°	6.826±0.024	11.18±0.023	8.926±0.099	417.1±71.86	0.863±0.003
	10°	6.918±0.041	11.81±0.087	8.212±0.116	1345±341.5	0.855±0.001
SEE 2	5°	7.062±0.037	12.85±0.182	8.249±0.088	2339±3031	0.810±0.001
	10°	7.080±0.040	12.32±5.835	8.628±4.068	1116±540.0	0.809±2.825
		\multicolumn{5}{c}{12mm Outer Thickness}				
		K_0	K_1	C_1	K_2	C_2
SEE 1	5°	5.672±0.016	9.342±0.072	7.752±0.057	8924±2413	0.907±0.001
	10°	5.670±0.033	9.646±0.054	7.354±0.074	9046±32.76	0.912±0.001
SEE 2	5°	5.703±0.042	9.045±0.399	7.264±0.105	8945±24.70	0.898±0.002
	10°	5.800±0.009	9.511±0.077	7.458±0.049	8945±50.94	0.903±0.001

4   Conclusion

It has been shown in this paper that torsional SEEs can be created using multi-material 3D printing. These also perform consistently as shown in Figure 3, as well as repeatedly over separate prints (Fig. 5). It has proved possible to accurately simulate their behavior, allowing responses of SEEs under different circumstances to be computationally investigated. Another point of note is that the Rollinson SEE's single spring constant of 5.78 Nm/rad (0.101Nm/°) [5] is lower than the similar scale 10mm printed SEE steady state spring constant (6.826-7.080 Nm/rad) but the same as the 12mm SEE (5.67-5.8Nm/rad). This shows that while the material in these 3D printed SEEs is slightly stiffer than the original molded rubber, designs can be modified and manufactured easily and quickly to achieve desired behaviors. It would be worth altering other dimensions within the SEE design in future work to analyze how the performance can be altered.

References

1. Pratt, G.A., Williamson, M.M.: Series elastic actuators. In: Proceedings of the 1995 IEEE/RSJ International Conference on Intelligent Robots and Systems 1995. 'Human Robot Interaction and Cooperative Robots', vol. 1, pp. 399–406. IEEE (1995)
2. Rollinson, D., Bilgen, Y., Brown, B., Enner, F., Ford, S., Layton, C., Rembisz, J., et al.: Design and architecture of a series elastic snake robot. In: 2014 IEEE/RSJ International Conference on Intelligent Robots and Systems (IROS 2014), pp. 4630–4636. IEEE (2014)
3. Boyle, J.H., Johnson, S., Dehghani-Sanij, A.A.: Adaptive undulatory locomotion of a C. elegans inspired robot. IEEE/ASME Transactions on Mechatronics 18(2), 439–448 (2013)
4. Schiavi, R., Grioli, G., Sen, S., Bicchi, A.: VSA-II: a novel prototype of variable stiffness actuator for safe and performing robots interacting with humans. In: IEEE International Conference on Robotics and Automation, ICRA 2008, pp. 2171–2176. IEEE (2008)
5. Rollinson, D., Ford, S., Brown, B., Choset, H.: Design and modeling of a series elastic element for snake robots. In: ASME 2013 Dynamic Systems and Control Conference, pp. V001T08A002-V001T08A002. American Society of Mechanical Engineers (2013)
6. Objet1000 Information, Stratasys Website. http://www.stratasys.com/3d-printers/production-series/objet1000-plus
7. Kappassov, Z., et al.: Semi-anthropomorphic 3D printed multigrasp hand for industrial and service robots. In: 2013 IEEE International Conference on Mechatronics and Automation (ICMA). IEEE (2013)
8. E10000 Information, Instron Website. http://www.instron.com/products/testing-systems/dynamic-and-fatigue-systems/electropuls/e10000---linear-torsion
9. Wang, X., Di Natali, C., Beccani, M., Kern, M., Valdastri, P., Rentschler, M.: Novel medical wired palpation device: a validation study of material properties. In: 2013 Transducers & Eurosensors XXVII: The 17th International Conference on Solid-State Sensors, Actuators and Microsystems (TRANSDUCERS & EUROSENSORS XXVII), pp. 1653–1658. IEEE (2013)

Robotic Pseudo-Dynamic Testing (RPsDT) of Contact-Impact Scenarios

Mario Bolien[(✉)], Pejman Iravani, and Jonathan Luke du Bois

Robotics and Autonomous Systems Group, Department of Mechanical Engineering,
University of Bath, Bath BA2 7AY, UK
{M.Bolien,P.Iravani,J.L.du.Bois}@bath.ac.uk

Abstract. This paper presents a hybrid test method that enables the investigation of contact-impact scenarios in complex systems using kinematically versatile, off-the-shelf industrial robots. Based on the pseudo-dynamic test method, the technique conducts tests on an enlarged time scale, thereby circumventing control rate and response time limitations of the transfer system. An initial exploratory study of a drop test demonstrates that non-rate dependant effects including non-linear stiffness and structural hysteresis can be captured accurately while limitations result from the neglect of rate- and time-dependant effects such as viscous damping and creep. Future work will apply the new method to contact scenarios in air-to-air refuelling.

Keywords: Industrial robot · Robotic pseudo dynamic testing · RPsDT · Hardware in the loop · Hybrid testing · Contact dynamics · Impact testing · Pseudo-dynamic testing · Drop test

1 Introduction

The work in this paper builds upon the use of industrial robots for hybrid tests of pre-contact docking manoevres for satellites and air-to-air refuelling [1,2], and serves as a feasibility study into the extension of these tests into the contact phase of the manoeuvre. Challenges in the realisation of hybrid tests of contact dynamics predominately arise due to the non-linear and discontinuous nature of contact events. Crucial factors for successful hybrid testing are (i) precise manipulation of the position and orientation of the colliding structure(s) in 3D space, (ii) sufficiently fast response times for the contact event and (iii) compensation of the induced transfer dynamics. While current industrial robots satisfy the first factor, they fall short of the latter two, especially for high velocity impacts and particularly stiff collisions. The main limitations for satisfactory response-speeds and real-time (RT) performance result from the large link inertia as well as proprietary control architectures. The latter typically preclude low level access to the axis controller such that favourable RT control schemes like impedance control [3], passivity based control [4] or model inversion schemes [5] cannot be easily realised.

© Springer International Publishing Switzerland 2015
C. Dixon and K. Tuyls (Eds.): TAROS 2015, LNAI 9287, pp. 50–55, 2015.
DOI: 10.1007/978-3-319-22416-9_7

This paper contributes to preceding efforts of realising hybrid tests of contact scenarios with robots by grounding the hybrid test on the pseudo-dynamic (PsD) testing method. The PsD testing technique enables dynamic hybrid testing on an expanded time scale with actuators of suitable load ratings but inadequate response speeds and power ratings [6]. The application of this technique to contact testing circumvents the response-time and transfer-dynamics issues of industrial robots from the outset at the expense of neglecting time-dependent test characteristics. To the best of the authors' knowledge, neither the applicability of the PsD test method to contact-impact problems has been extensively discussed nor is the realisation of robot assisted pseudo-dynamic testing reported in literature and from this point onwards the test method will be referred to as Robotic Pseudo-Dynamic Testing (RPsDT).

2   Application of the Pseudo Dynamic Test Method to Robots and Contact Scenarios

System hybridisation for RPsDT is performed according to the same principle as for PsD testing: The system under investigation is broken up into an experimental and a numerically simulated substructure. For RPsDT of contact scenarios, the experimental substructure would typically consist of exactly those components that make physical contact in the real system or a representative mock-up. The numeric simulation computes the positional response of the full system to the combination of measured interface forces and numerically simulated forces. The transfer system consists of an industrial robot equipped with a 6DOF force/torque sensor at its end effector. As in standard hybrid tests, additional sensors may be fitted directly to the mock-up for the purpose of further data acquisition throughout the study. The fundamental RPsDT architecture then complies with the schematic in Figure 1.

As opposed to standard PsD tests, data from the experimental specimen is not acquired in every time-step but only throughout the contact phase which can be identified based on kinematic constraints in simulation. If in contact, the

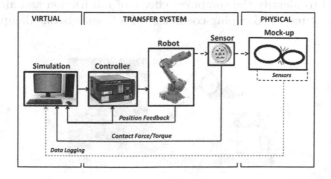

Fig. 1. RPsDT hardware architecture

robot is quasi-statically moved to reproduce the relative position and orientation of the colliding structures. This strains the specimen and allows measurement of the restoring forces and moments which are then fed back into the simulation. In a non-contact phase, the robot is kept stationary and the simulation can advance immediately without prior contact force acquisition.

Upon acquisition of the restoring force, the simulation proceeds by treating the model as an initial value problem: Based on the current states of the contacting structures and physically acquired force measurements from the experimental substructure, the new accelerations are computed and the new system states are obtained with a suitable integration algorithm. The cycle then repeats with the next time-step.

3   RPsDT Drop Test Investigation and Validation

Validation of RPsDT results is difficult because the motivation for RPsDT is the predictive deficit of purely simulated or purely experimental methods. For complex tests, validation approaches must be carefully considered. Here, a simple, reproducible test of a high-speed contact-impact scenario is devised to examine the validity and accuracy of the RPsDT method as a precursor to more complex testing. To this end, the vertical drop of a mass (steel plate) onto a compliant object (tennis ball) is emulated.

3.1   Experimental Setup and Procedure

The basic experimental set-up and test reference frame for RPsDT are illustrated in Figure 2(a). The drop-test rig in Figure 2(b) served the purely experimental reproduction of the contact scenario for validation purposes. Using high speed video capture (1500 frames/sec), plate drops on the experimental rig from an initial height $z_0 = 0.205m$ ($\dot{z}_0 = 0\frac{m}{s}$) were recorded with and without a tennis ball located on the bottom plate. Based on manual frame-by-frame tracking of the dropping plate's lower edge, the true experimental trajectories could be extracted from the video footage. The data from a first drop (without tennis ball) was used to identify the combined effects of rail friction and air resistance and allowed to tune the damping coefficient of a linear viscous damper element

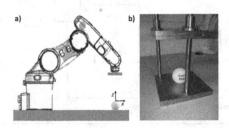

Fig. 2. (a) RPsDT setup and test reference frame. (b) Validation rig.

$(c = 5.18\frac{N}{m/s})$ to give good agreement between the plate trajectories of the experimental drop and a simulated drop prior to contact. The trajectory data from the second drop (with tennis ball located centrally on lower plate) were used to validate the results from a subsequent RPsDT reproduction of the same contact scenario.

The simulated substructure of the RPsDT reproduction featured a point-mass model of the plate ($m = 6.50kg$) which, released from rest in a $1g$ environment and constrained to 1 DOF, drops under the combined influence of rail friction and air resistance as per the previously experimentally identified viscous damping element. The tennis ball and plates from the validation rig (rails removed) were used as specimen in the experimental substructure and the contact force was experimentally measured by the force sensor installed in between robot end effector and 'dropping' plate. As such, the plate's motion was governed by Equation (1).

$$m\ddot{z} = F_c - c\dot{z} - mg \qquad (1)$$

RPsDT was conducted in its simplest from: Based on the newly acquired force measurement F_i at the start of each pseudo-step, the current plate acceleration $\ddot{z}_i$ was computed from Equation (1). The new position z_{i+1} and velocity $\dot{z}_{i+1}$ of the next time-step were found by integration based on the explicit 1st-order Euler method using fixed step sizes of $h_s = 0.01ms$ and $h_c = 0.2ms$ throughout simulation and contact phases respectively.

3.2   Results and Discussion

While the true experimental time for the drop test reproduction using RPsDT amounted to about 90 minutes, RPsDT data presented in this section is plotted against the equivalent 'pseudo-time'. Plate trajectories from both the RPsDT study and drop test on the validation rig are shown in Figure 3(a). Prior to initial impact, both trajectories are in good agreement which emphasis the validity of the model in the virtual substructure throughout non-contact phases. Upon contact, RPsDT and experimental trajectory diverge. In the experimental drop test, the plate loses energy at a much higher rate and settles to rest within 1.5 seconds. The onset of trajectory divergence becomes evident in the initial impact phase. More pronounced asymmetry is apparent for the experimental trajectory, i.e. the experimental trajectory shows a greater difference between rates of compression (faster) and restitution (slower) than the RPsDT trajectory. This is also visible on the corresponding contact force graph in Figure 3(b). Here, force measurements were not available and the experimental contact force was computed as the product of plate mass and plate acceleration (obtained as 2nd derivative of the position trajectory). Despite application of a running-mean filter, noise introduction by double differentiation causes apparent abnormalities in the data, however, general trends remain obvious: (i) compared with RPsDT data the experimental force shows a sharper rise to a higher peak and (ii) the difference in 'sharpness' of contact force increase and contact force decrease is greater in the experimental data, giving a more asymmetric contact force profile.

Both phenomena are attributed to rate dependent damping forces captured as part of the experimental study which during compression act in addition to the restoring forces to decelerate the plate but inhibit plate acceleration in restitution. Due to quasi-static loading, such effects are not observable in RPsDT data and both RPsDT trajectory and contact force graph are consequently more symmetric. Asymmetry that is nonetheless observable in the RPsDT data is attributed to non rate-dependent structural damping which originates from a hysteretic, *i.e.* path dependent stiffness variation that is an inherent property of the tennis ball. This is well-pronounced in Figure 3(c) where contact forces are plotted against tennis ball deformation for all four contact phases of the recorded RPsDT data. The transition from a nominally linear elastic response to a nonlinear response is apparent at around 0.035m deformation, with the deformation from the first impact extending far into the nonlinear region and peaking at about 90% of the ball's original diameter. In addition, it can be noted that RPsDT data shows greater stiffness in the initial compression phase than it does throughout successive contact phases. This stiffness change does not correspond to a true contact phenomenon but is attributed to a time dependent

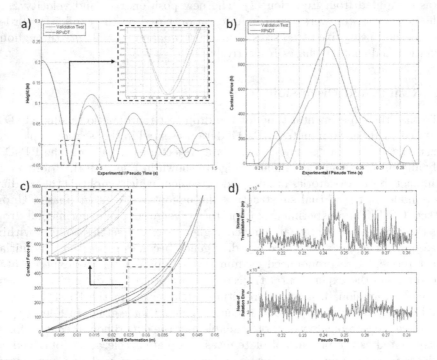

Fig. 3. (a) Experimental and RPsDT plate trajectories. (b) Contact forces on first impact. (c) Contact force vs. tennis ball deformation for RPsDT data. (d) Translation and rotation errors throughout first contact phase expressed as Euclidean and Frobenius norm respectively.

creep caused by sustained stress application over a prolonged period of time in RPsDT.

The extent to which the presented data is afflicted with errors resulting from the robot's positioning accuracy is shown in Figure 3(d). Good position and orientation tracking is suggested with the net translational and fixed frame rotational errors being consistently controlled to within $50\mu m$ and $0.015°$ respectively. It must be noted that this apparent accuracy neglects effects of joint flexure, backlash and deviation from catalogue DH-parameters.

4 Conclusions

This paper has demonstrated the feasibility of studying contact-impact problems in a hybrid test using an off-the-shelf industrial robot in a technique based on the well-established pseudo-dynamic method. This *RPsDT* circumvents robot response time issues for dynamic testing at the expense of disregarding rate dependent effects in the specimen's response. An RPsDT-based investigation of a drop test clearly indicated the ability of the method to account for non-rate dependent effects throughout contact events including the capture of non-linear damping characteristics due to hysteresis. Limitations were identified arising from the neglect of rate- and time-dependent effects, in particular those of viscous damping and creep. Inertial effects are a further concern but had little effect on the observations herein. It is suggested that *a priori* estimates of the time- and rate-dependent effects could be incorporated into the simulation to further improve test fidelity. In conclusion, RPsDT has shown useful potential but a need for careful test design and/or validation has been highlighted. Future work will employ the technique to evaluate contact dynamics in air-to-air refuelling scenarios.This research is sponsored by Cobham Mission Systems.

References

1. du Bois, J.L., Thomas, P.R., Richardson, T.S.: Development of a relative motion facility for simulations of autonomous air to air refuelling. In: 2012 IEEE Aerospace Conference, pp. 1–12. IEEE (2012)
2. Du Bois, J.L., Newell, P., Bullock, S., Thomas, P., Richardson, T.: Vision based closed-loop control systems for satellite rendezvous with model-in-the-loop validation and testing. In: 23rd International Symposium on Space Flight Dynamics. University of Bath (2012)
3. Hogan, N.: Impedance control: an approach to manipulation. In: American Control Conference, pp. 304–313. IEEE (1984)
4. Ortega, R., Van Der Schaft, A.J., Mareels, I., Maschke, B.: Putting energy back in control. IEEE Control Systems **21**(2), 18–33 (2001)
5. De Luca, A.: Feedforward/feedback laws for the control of flexible robots. In: Proceedings of the IEEE International Conference on Robotics and Automation, ICRA 2000, vol. 1, pp. 233–240. IEEE (2000)
6. Williams, M.S., Blakeborough, A.: Laboratory testing of structures under dynamic loads: an introductory review. Philosophical Transactions of the Royal Society of London. Series A: Mathematical, Physical and Engineering Sciences **359**(1786), 1651–1669 (2001)

Hybrid Insect-Inspired Multi-Robot Coverage in Complex Environments

Bastian Broecker[1], Ipek Caliskanelli[1]([✉]), Karl Tuyls[1], Elizabeth I. Sklar[1,2], and Daniel Hennes[3]

[1] Department of Computer Science, University of Liverpool, Liverpool, UK
{bastian.broecker,ipek.caliskanelli,k.tuyls}@liverpool.ac.uk
[2] Department of Informatics, King's College London, London, UK
elizabeth.sklar@kcl.ac.uk
[3] European Space Agency, Noordwijk, The Netherlands
daniel.hennes@esa.int

Abstract. Coordination is one of the most challenging research issues in distributed multi-robot systems (MRS), aiming to improve performance, energy consumption, robustness and reliability of a robotic system in accomplishing complex tasks. Social insect-inspired coordination techniques achieve these goals by applying simple but effective heuristics from which elegant solutions emerge. In our previous research, we demonstrated the effectiveness of a hybrid ant-and-bee inspired approach, HybaCo, designed to provide coordinated multi-robot solutions to area coverage problems in simple environments. In this paper, we extend this work and illustrate the effectiveness of our hybrid ant-and-bee inspired approach (HybaCo) in complex environments with static obstacles. We evaluate both the ant-inspired (StiCo) and bee-inspired (BeePCo) approaches separately, and then compare them according to a number of performance criteria using a high-level simulator. Experimental results indicate that HybaCo improves the area coverage uniformly in complex environments as well as simple environments.

1 Introduction

Recent years have seen a rapidly growing interest in distributed multi-robot systems for automatically exploring and surveilling environments of different sizes, type and complexity. *Multi-robot systems (MRS)* consist of multiple interacting robots, each executing an application-specific control strategy, which is not centrally steered. *Coordination* is a key challenge when deploying teams of distributed multi-robot systems as these systems are often limited in resources (e.g., on-board processors, batteries, controllers). Lightweight interactions among robots (e.g., facilitated by wireless communication) are not only a desired feature for such platforms, but also necessary to overcome practical deployment issues stemming from inconsistent network connectivity. Simple yet effective heuristics that avoid complex, heavy computation and establish lightweight interactions are therefore highly desirable for MRSs. Biologically

© Springer International Publishing Switzerland 2015
C. Dixon and K. Tuyls (Eds.): TAROS 2015, LNAI 9287, pp. 56–68, 2015.
DOI: 10.1007/978-3-319-22416-9_8

inspired, or *bio-inspired*, solutions for the challenging problem of multi-robot coordination are gaining traction.

Swarm algorithms, like ant colony optimisation [10], rely on pheromonal knowledge for communication between agents. Such insect-inspired multi-agent research has also opened the possibility of applying some of these techniques to robotic systems, i.e., swarm robotics [11,16]. Swarm robotic systems are motivated by a wide range of application areas, such as surveillance and patrolling, where mobile guard robots are considered an alternative and improved mechanism over fixed security cameras and even humans. Other application areas include exploration and identification of hazardous environments (e.g., nuclear plants and fire detection), mobile sensor networks, wireless sensor and robot networks, and space exploration.

Previous research, *StiCo* [13], investigated ants' stigmergic behaviour in an attempt to address the coverage issues in multi-robot systems using self-organised coordination techniques. Another approach, *BeePCo* [4], used bee pheromone signalling processes to solve the same multi-robot coverage problem. Both *StiCo* and *BeePCo* rely on pheromone substances for coordination and communication between agents. Based on these approaches, a hybrid ant-and-bee inspired technique was developed, *HybaCo* [2], in an attempt to further improve multi-robot coverage. This included a comparative study of the *StiCo*, *BeePCo* and *HybaCo* algorithms and identified their strengths and weaknesses. The environment used in the experimental scenarios in [2] was a simplistic, obstacle-free square arena. In this paper, we extend this work and introduce more complex environments that contain obstacles, in order to represent more realistic scenarios. We demonstrate the effectiveness of *HybaCo* and compare it with *StiCo* and *BeePCo* on multiple complex environments.

2   Related Work

Many bio-inspired techniques have been examined to address multi-robot coverage by establishing lightweight coordination principles based on the behaviour of social insects. Ants [8,9] and bees [1,12] are the two main families of social insects that have inspired approaches within the fields of robotics and distributed systems as a means to improve self-organisation and autonomous coordination characteristics.

In [13,14], Ranjbar-Sahraei et al. developed their first *stigmergic* approach using a simulation environment to address coordination issues in multi-robot systems. Extensive experimental results on both simple (obstacle free) and complex (with static and dynamic obstacles) environments showed that the *stigmergic* approach (StiCo) applies (almost) uniform coverage. Later in [17], Ranjbar-Sahraei et al. extended their research deployed on a physical robot swarm, which validated the correctness of the simulated results. The mathematical model of this approach is shown in [15] that derives a probabilistic macroscopic model for ant-inspired coverage by multi-robot platform.

In previous work on *pheromone signalling (PS)* based load-balancing, [5,7] presented a dynamic technique for *wireless sensor networks (WSNs)* that is applied at run time in the application layer of a communication network. *PS* is inspired from the pheromone signalling mechanism found in bees and provides distributed WSN control that uses local information only. [6] extend the initial *PS* technique by introducing additional network elements in the form of robotic vehicles for *wireless sensor and robot networks (WSRNs)*. Different subclasses of cyber-physical systems are merged together to increase the area coverage *effectively*, which directly increases the service availability and extend the network lifetime by benefiting from their heterogeneity. The same pheromone signalling principle is applied to multi-robot systems in [2,3] and explained in detail in the next sections.

3   Background: Comparison Between StiCo and BeePCo

This section provides some background information about the *StiCo* and *BeePCo* techniques separately, which is essential for later describing our hybrid *HybaCo* in Section 4.

3.1   StiCo Principle

The *StiCo* approach follows the principle of indirect, stigmergic coordination to establish efficient coverage of an environment by simple means. Classic stigmergic coordination in an *ant system (AS)* is characterised by two properties: (1) agents have a tendency to move straight with minor deviations, and (2) traces (signal trails) act as sources of attraction. In contrast, StiCo robots orbit in circles (instead of moving straight), and their traces have repulsion characteristics (instead of attraction). These two key differences transform the path-finding characteristic of an AS into the efficient area coverage provided by StiCo.

In StiCo, robots are equipped with two simple sensors (pointing in the front-left and front-right directions like ant antennae), capable of detecting immediate traces. Each robot rotates in a circle with a predetermined radius. Based on the circling direction (clockwise, CW, or counter-clockwise, CCW), one sensor is considered as the interior sensor and the other as the exterior one. When the interior sensor detects *pheromone* (virtual substance marking the trace of a robot), the robot changes its circling direction immediately. Otherwise, if the exterior sensor detects pheromone, the robot continues rotating in the same direction until it no longer detects any pheromone. For a more detailed description we refer to [13].

3.2   BeePCo Principle

The *BeePCo* approach follows the principle of pheromone-signalled coordination found in bees to establish direct, lightweight communication between agents and can address multi-robot coverage problems. The *BeePCo* algorithm consists of four parts, which are executed on every robot in the MRS: two parts are time-triggered (differentiation cycle and decay of pheromone), whereas the other two (propagation of pheromone and motion direction and magnitude) occur in parallel invoked by one event-triggered process. During the propagation, robots send pheromone to their neighbours within direct communication range. If a robot receives pheromone, it makes a decision to move and selects a target destination in the opposite direction of the pheromone received. The movement decision is based on vector addition; further description can be found in [4].

3.3 Comparison Between StiCo and BeePCo

Here we characterise the strengths and weaknesses of *StiCo* and *BeePCo*. Three key differences between these two techniques are illustrated in Table 1. The first difference between *StiCo* and *BeePCo* lies in how pheromone is used for *communication*. In the *StiCo* approach, communication between agents is implemented using *indirect* pheromone trails, where pheromone signals are deposited in the robots' environment without knowing which or whether another agent will receive the signal. In contrast, in the *BeePCo* approach, pheromone signalling is implemented by *directly* sending signals to robots within a specific range.

Table 1. Differences between *StiCo* and *BeePCo*

property	StiCo	BeePCo
communication	Indirect	Direct
movement	Circular	Vector-based
speed to converge	Normal	Fast

The second difference between *StiCo* and *BeePCo* lies in the type of *movement* effected by the robots. When robots run *StiCo*, their motion is applied in a circular fashion and only the direction of circling changes. When robots run *BeePCo*, their motion is guided by vectors which influence the straight-line direction and distance for each move.

The difference between indirect and direct communication methods, taken in combination with the different motion methods, have significant impact on the *speed* with which these algorithms converge. The duration from the moment of deployment until the robots reach a stable configuration and until their energy is depleted varies significantly. *BeePCo* produces faster convergence in comparison to *StiCo*, and this is one of the strengths of *BeePCo*. However, because robots in *BeePCo* use direct communication based on transmission range, the robots stop moving once they are not in each other's communication range; and this is the main weakness of this approach.

4   HybaCo: Hybrid Bee-Ant Coverage Algorithm

In [2], we proposed *HybaCo*, which combines the effectiveness of the two pheromone-based approaches (*StiCo* and *BeePCo*) detailed above while overcoming the major weaknesses of each approach taken alone. In this section, we describe *HybaCo* briefly once again before we evaluate it in complex environments. The most important performance bottleneck of the *BeePCo* occurs when the robots move far apart from each other and lose their communication network. This prevents pheromone exchange and, as a result, robots do not move any more. The biggest problems with the *StiCo* approach are the extended time to converge and the lack of coverage redundancy—which is a desirable feature when considering practical deployment. In order to solve these issues, our hybrid approach begins with *BeePCo* but changes dynamically to *StiCo* when the communication network between the robots is lost [2]. Robots apply the *BeePCo* technique when the communication network is still active; but as the robots move further apart from each other, the communication network dies. When the robots lose connectivity (communication links) with all of their neighbours (i.e., others within transmission range), they assume that the *BeePCo* technique is no longer effective and so they switch to *StiCo*. After some time using *StiCo*, the robots will again get close enough to transmit pheromones to each other, at which point they will switch back to *BeePCo*. It is important to underline that the ANT and BEE pheromones are *different*, and are thus declared separately in the algorithm.

5 Experimental Evaluation

We evaluated three main algorithms (*StiCo*, *BeePCo* and *HybaCo*) using custom-built abstract simulators. The original *StiCo* [13,14] was developed in C++, which is extended for the *HybaCo* experiments described here. The set of experiments presented in this section compare three important evaluation metrics:

1. the *area covered* by the robots;
2. the *distribution* of robots in their environment; and
3. the *time* it takes to converge (or stabilise).

In this study, *area coverage* is defined as the maximum of the total non-overlapping area covered by the sensors of the involved robot(s), as defined in [17]. The *distribution* of robots in the arena shows that the robots are moving around without leaving unattended gaps in the environment. This is measured by overlaying a grid of $160,000$ cells ($1cm \times 1cm$) on the environment and calculating how many cells are unattended at any moment in time. For the entire environment, all cells that are "attended" (or covered) by one or more robots are summed and divided by the total number of cells, resulting in the *percentage of coverage*. This metric indicates how evenly the robots are distributed in the arena and what the level of redundancy is for the different algorithms. The result is illustrated in the heatmaps shown in Figures 2, 4 and 6. The lighter the colour of the area's in the arena, the higher is the percentage of the area being covered over the total time of the experiment. The more evenly the total area is coloured, the more uniform is the distribution of the robots' positions over time.

Finally, the time it takes to converge is the duration from the moment of deployment of the robots until they achieve a stable configuration (i.e., the moment the algorithm performs nearly optimally before resources are depleted).

The experiments were carried out with sets robots which allow maximal coverage of $\approx 75\%$ (depending on the environment this number is between 56 to 60 robots), each robot having a sensing and communication radius of $25cm$. The robots' environment (arena) size is $400cm \times 400cm$. Initially, the robots are deployed randomly in the central square region of size $5cm \times 5cm$.

We consider the following five algorithmic variations of *StiCo* and *BeePCo* in our comparisons:

- **StiCo**: The robots execute the stigmergic principle for coordination [14], as described in Section 3.1.
- **BeePCo**: The robots execute the bee-pheromone signalling principle [4] for coordination, as described in Section 3.2.
- **BeePCo-with-rotation**: The *BeePCo* algorithm is extended with a rotational move. This is an intermediate algorithm between *BeePCo* and *HybaCo* in which the robots execute the *BeePCo* approach until they lose communication links with each other. Once communication is lost, the robots apply the rotational move described in [2].
- **HybaCo**: The robots execute the bee-and-ant inspired coordination principle as introduced in [2] and described in Section 4.
- **MaxCo**: This represents the optimal case where the robots' transmission range does not intersect with each other. This scenario is a benchmark for the maximum possible coverage of deployed robots with zero surveillance area overlap. This can also be referred to as *potential* coverage.

These five different algorithms are evaluated over three different environmental setups (topdown, L-shaped, floor plan) as Fig. 1 illustrates. The experimental results presented in this paper is average of five individual runs over each environmental setup for all five algorithms.

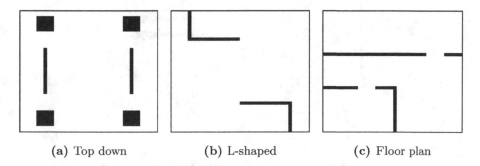

(a) Top down   (b) L-shaped   (c) Floor plan

Fig. 1. Three environment setups used to evaluate the performance

Figures 2 and 3 illustrate the experimental results of an MRS of 56 robots in an arena containing four blocks of square obstacles in the corners and two walls comparing the performance of the *StiCo*, *BeePCo*, *BeePCo-with-rotation* and

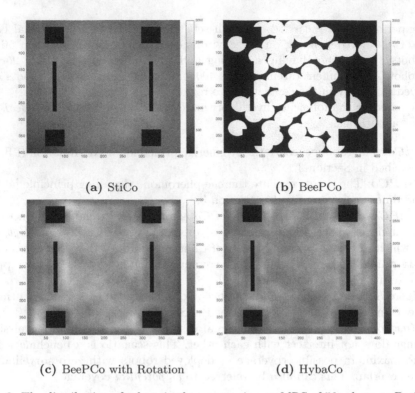

(a) StiCo   (b) BeePCo

(c) BeePCo with Rotation   (d) HybaCo

Fig. 2. The distribution of robots in the arena using an MRS of 56 robots on BeePCo, StiCo, BeePCo-with-rotation and HybaCo

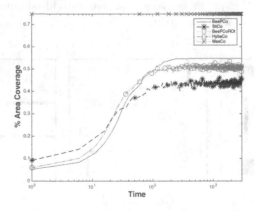

Fig. 3. The percentage of area coverage using MRS of 56 robots on different techniques

HybaCo approaches against each other. Fig. 2 shows the heatmap images of this arena containing a number of static obstacles and our observations are as follows. *StiCo* robots had difficulties passing the walls and the square obstacles, therefore, focused on the middle of the arena, as shown in Fig. 2a with darker corners.

In *BeePCo*, the robots stop spreading after communication links with the other robots are broken, because they are outside of the inter-robot transmission range. Therefore, in Fig. 2b, a white circle represents an area covered by a robot after it stopped moving and did not receive further pheromone from other robots. This results in entirely uncovered areas in the arena, because the robots are not able to move. The *BeePCo-with-rotation* and *HybaCo* approaches provide more even distribution than *BeePCo*, where *BeePCo-with-rotation* distinctively focuses more around the obstacles. This behaviour is less obvious in *HybaCo* and thus makes *HybaCo* more uniformly distributed than *BeePCo-with-rotation*.

Figure 3 shows the percentage of the area coverage with all four techniques. The maximum possible area coverage is shown by *MaxCo*. Among the four techniques, *StiCo* provides the lowest percentage of covered area, whereas *BeePCo* outperforms and achieves the highest percentage of covered area. The difference between *BeePCo-with-rotation* and *HybaCo* is not significant, although *HybaCo* distributes robots more uniformly.

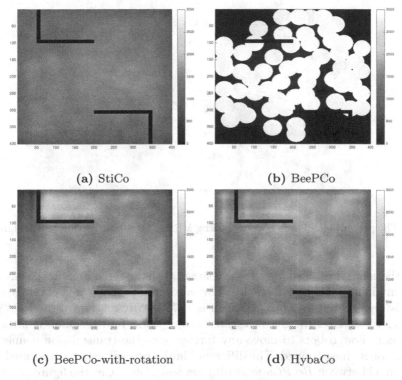

(a) StiCo   (b) BeePCo

(c) BeePCo-with-rotation   (d) HybaCo

Fig. 4. The distribution of robots in the arena using an MRS of 60 robots on BeePCo, StiCo, BeePCo-with-rotation and HybaCo

Figures 4 and 5 show experimental results on an MRS of 60 robots in the arena which contains L-shaped obstacles (e.g., separators or boxes in a room). Figure 4 illustrates the distribution of the robots in this environment on *StiCo, BeePCo, BeePCo-with-rotation* and *HybaCo. StiCo* performs uniformly distributed coverage, although the corners of the arena are slightly less covered, whereas the robots had no issues getting around the L-shaped obstacles, nor providing coverage around the obstacles. In *BeePCo* approach, the robots stop spreading after communication links with the other robots are broken because they are outside of the inter-robot transmission range. The white circles in Fig. 4b show the robots' non-overlapping coverate within the experimental arena. Because the robots do not move for a long time, the percentage of area coverage in *BeePCo* is higher and more unevenly distributed, as opposed to the other three techniques. Figure 4c and 4d illustrate that *BeePCo-with-rotation* and *HybaCo* have a more uniform robot distribution over the environment in comparison to *StiCo* and *BeePCo*. The distribution of the robots is remarkably high around the obstacles in *BeePCo-with-rotation* which shows that robots struggle to get away from the obstacles. This situation is not applicable in *HybaCo*, therefore, *HybaCo* shows more uniformly distributed robots, as Fig. 4d illustrates.

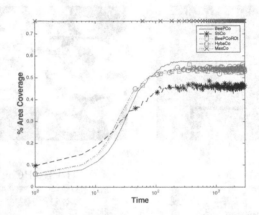

Fig. 5. The percentage of area coverage using MRS of 60 robots on different techniques

Furthermore, Fig. 5 shows the percentage of the covered area in the environment containing L-shaped obstacles. The maximum possible area coverage of 60 robots is represented by *MaxCo*, which is $\approx 75\%$. *StiCo* achieves the least area coverage, whilst *BeePCo* gives the most among these four techniques, although it does not allow robots to move any further once the transmission connection between robots has stopped. The difference in the percentage of the covered area is very small between *BeePCo-with-rotation* and *HybaCo*, as the figure shows. In terms of the time the algorithms take to distribute robots in the period between 10^0 and 10^1 seconds, *StiCo* initially scatters the robots faster than *BeePCo* and converges faster, unlike our expectations, because *StiCo* has a more gradual manner of moving outwards, i.e., circling, whereas in *BeePCo* robots move outwards in a direct line. Later, in the time period between 10^1 and 10^2 seconds,

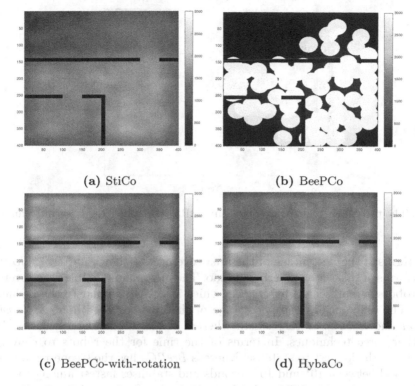

(a) StiCo   (b) BeePCo

(c) BeePCo-with-rotation   (d) HybaCo

Fig. 6. The distribution of robots in the arena using an MRS of 60 robots on BeePCo, StiCo, BeePCo-with-rotation and HybaCo

BeePCo robots benefit from the direct communication exchange and spread out much faster than *StiCo, BeePCo-with-rotation* or *HybaCo*.

Figures 6 and 7 illustrate an MRS of 60 robots on an arena representing floor plans. In Fig. 6, heatmap images represent the distribution of the robots using *StiCo, BeePCo, BeePCo-with-rotation* and *HybaCo* approaches individually. The bottom left corner represents a room where all four sides of the room are surrounded by walls, apart from a small doorway gap close to the middle of the arena. Although *StiCo* and *BeePCo* have some coverage in this room, both *BeePCo-with-rotation* and *HybaCo* provide better coverage and improve the robot distribution as well. The top left corner in the arena is the least covered area for all four approaches due to the long wall stretching across the arena from side-to-side, apart from the little doorway gap. Because the doorway is small, robots struggle to find the gap and pass through the wall, especially covering the upper left corner. *BeePCo* has no coverage on the left corner above the wall, whereas *StiCo* has brief coverage of the same corner. *BeePCo-with-rotation* and *HybaCo* visibly improve coverage of the same corner as a result of the advantages of the combined approaches. Similar to Figs. 2 and 4, *BeePCo-with-rotation* has more coverage around the walls, whereas *HybaCo* uniformly improves the coverage of the arena, mostly beneath the long wall.

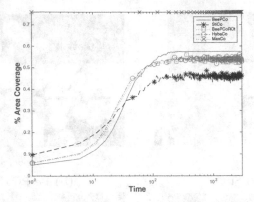

Fig. 7. The percentage of area coverage using MRS of 60 robots on different techniques

In terms of the percentage of the area coverage, as shown in Fig. 7, *BeePCo* outperforms in this set of experiments too. The maximum possible area coverage of 60 robots is represented by *MaxCo*, which is ≈ 75%, similar to the L-shaped arena. The difference in the percentage of the area coverage in *BeePCo-with-rotation* and *HybaCo* is very small, whereas *StiCo* is considerably lower than the other three techniques. In terms of the time for the robots to converge, *StiCo* takes the longest to stabilise, whereas *BeePCo* has the steepest hill in the time period between 10^1 and 10^2 seconds and stabilises fastest among the four techniques compared in this study. A significant performance improvement can be observed in *BeePCo-with-rotation* and *HybaCo* in comparison to *StiCo* and *BeePCo*. Specifically, the experiments show that merging the strengths of both *StiCo* and *BeePCo* leads to superior results with respect to uniform distribution of the robots in the arena.

6   Conclusions

This paper compares four social insect inspired multi-robot coverage approaches, namely stigmergic behaviour of ants (*StiCo*), the pheromone signalling process of bees (*BeePCo*), a derived method based on *BeePCo* (*BeePCo* rotation), and an ant-and-bee inspired hybrid approach (*HybaCo*) in realistic, complex environments. We have shown the performance of all four approaches with respect to a number of criteria, including area coverage, uniformity of distribution and speed of convergence, with a particular focus on our hybrid bee-and-ant inspired approach that merges the strengths of *StiCo* and *BeePCo* into one algorithm. The advantages and disadvantages of these two techniques have been highlighted. In the experimental analysis, we evaluated the effectiveness of the proposed hybrid bee-and-ant inspired approach, i.e. *HybaCo* in MRSs with 56 and 60 robots in a number of different complex environment each containing numerous static objects and reported our observations.

StiCo moves at all times and applies (almost) uniform coverage over the arena. *BeePCo* achieves a higher percentage of area coverage in comparison to *StiCo*; however, it produces non-uniform coverage because the robots stop moving when they step outside of each others' transmission range. The experimental results show that *BeePCo-with-rotation*, which is an extension of *BeePCo*, improves the distribution of the robots, but does not provide the same percentage of area coverage as BeePCo. Finally, our experiments confirm our earlier results [2] and show that *HybaCo* merges the strengths of the *StiCo* and *BeePCo* algorithms.

References

1. Alers, S., Hu, J.: AdMoVeo: a robotic platform for teaching creative programming to designers. In: Chang, M., Kuo, R., Kinshuk, Chen, G.-D., Hirose, M. (eds.) Edutainment 2009. LNCS, vol. 5670, pp. 410–421. Springer, Heidelberg (2009)
2. Broecker, B., Caliskanelli, I., Tuyls, K., Sklar, E., Hennes, D.: Social insect-inspired multi-robot coverage. In: Proc. of Int. Conf. on Autonomous Agents and Multiagent Systems (AAMAS) (2015)
3. Broecker, B., Caliskanelli, I., Tuyls, K., Sklar, E., Hennes, D.: Social insect-inspired multi-robot coverage. In: Proc. of the Workshop on Autonomous Robots and Multirobot Systems (ARMS) at AAMAS (2015)
4. Caliskanelli, I., Broecker, B., Tuyls, K.: Multi-robot coverage: a bee pheromone signalling approach. In: Headleand, C.J., Teahan, W.J., Ap Cenydd, L. (eds.) ALIA 2014. CCIS, vol. 519, pp. 124–140. Springer, Heidelberg (2015)
5. Caliskanelli, I., Harbin, J., Indrusiak, L. Polack, F., Mitchell, P., Chesmore, D.: Runtime optimisation in wsns for load balancing using pheromone signalling. In: 3rd IEEE Int. Conf. on NESEA, December 2012
6. Caliskanelli, I., Indrusiak, L.: Using mobile robotic agents to increase service availability and extend network lifetime on wireless sensor and robot networks. In: 12th IEEE Int. Conf. on INDIN, July 2014
7. Caliskanelli, I., Harbin, J., Soares Indrusiak, L., Mitchell, P., Polack, F., Chesmore, D.: Bio-inspired load balancing in large-scale wsns using pheromone signalling. Int. Journal of Distributed Sensor Networks (2013)
8. Di Caro, G., Dorigo, M.: Antnet: distributed stigmergetic control for communications networks. J. Artif. Int. Res. 9(1), 317–365 (1998)
9. Dorigo, M.: Optimization, Learning and Natural Algorithms. Thesis report, Politecnico di Milano, Italy (1992)
10. Dorigo, M., Birattari, M., Stutzle, T.: Ant colony optimization: Artificial ants as a computational intelligence technique. IEEE Computational Intelligence Magazine 1(4), 28–39 (2006)
11. Dorigo, M., Roosevelt, A.F.: Swarm robotics. In: Special Issue, Autonomous Robots. Citeseer (2004)
12. Lemmens, N., de Jong, S., Tuyls, K., Nowé, A.: Bee behaviour in multi-agent systems. In: Tuyls, K., Nowe, A., Guessoum, Z., Kudenko, D. (eds.) Adaptive Agents and MAS III. LNCS (LNAI), vol. 4865, pp. 145–156. Springer, Heidelberg (2008)
13. Ranjbar-Sahraei, B., Weiss, G., Nakisaee, A.: A multi-robot coverage approach based on stigmergic communication. In: Timm, I.J., Guttmann, C. (eds.) MATES 2012. LNCS, vol. 7598, pp. 126–138. Springer, Heidelberg (2012)

14. Ranjbar-Sahraei, B., Weiss, G., Nakisaee, A.: Stigmergic coverage algorithm for multi-robot systems (demonstration). In: Proceedings of the 11th International Conference on Autonomous Agents and Multiagent Systems, vol. 3, pp. 1497–1498. International Foundation for Autonomous Agents and Multiagent Systems (2012)
15. Ranjbar-Sahraei, B., Weiss, G., Tuyls, K.: A macroscopic model for multi-robot stigmergic coverage. In: Proceedings of the 2013 International Conference on Autonomous Agents and Multi-agent Systems, pp. 1233–1234. International Foundation for Autonomous Agents and Multiagent Systems (2013)
16. Şahin, E.: Swarm robotics: from sources of inspiration to domains of application. In: Şahin, E., Spears, W.M. (eds.) Swarm Robotics WS 2004. LNCS, vol. 3342, pp. 10–20. Springer, Heidelberg (2005)
17. Sahraei, B.R., Alers, S., Tuyls, K., Weiss, G.: Stico in action. In: Int'l. Conf. on Autonomous Agents and Multi-Agent Systems, AAMAS 2013, Saint Paul, MN, USA, May 6–10, 2013, pp. 1403–1404 (2013)

From AgentSpeak to C for Safety Considerations in Unmanned Aerial Vehicles

Samuel Bucheli(✉), Daniel Kroening, Ruben Martins, and Ashutosh Natraj

Department of Computer Science, University of Oxford, Oxford, UK
{samuel.bucheli,daniel.kroening,ruben.martins,
ashutosh.natraj}@cs.ox.ac.uk
http://www.cprover.org

Abstract. Unmanned aerial vehicles (UAV) are becoming increasingly popular for both recreational and industrial applications, leading to growing concerns about safety. Autonomous systems, such as UAVs, are typically hybrid systems consisting of a low-level continuous control part and a high-level discrete decision making part. In this paper, we discuss using the agent programming language AgentSpeak to model the high-level decision making. We present a translation from AgentSpeak to C that bridges the gap between high-level decision making and low-level control code for safety-critical systems. This allows code to be written in a more natural high-level language, thereby reducing its overall complexity and making it easier to maintain, while still conforming to safety guidelines. As an exemplar, we present the code for a UAV autopilot. The generated code is evaluated on a simulator and a Parrot AR.Drone, demonstrating the flexibility and expressiveness of AgentSpeak as a modeling language for UAVs.

1 Introduction

Unmanned Aerial Vehicles (UAVs) have become an area of significant interest in the robotic community. This is mainly due to their capability to operate not only in open outdoor spaces, but also in confined indoor spaces, thus catering to a wide range of applications [27].

Current UAVs are usually remotely controlled by human pilots but partial or even full autonomy is a likely scenario in the foreseeable future, leading to various regulatory questions [21]. An autonomous UAV can be seen as a hybrid system consisting of a low-level continuous control part and a high-level decision making part [9,15]. While, arguably, the low-level control enables autonomy, it is the high-level decision processes that make the system truly autonomous. Owing to their unrestricted movement and usage of non-segregated airspace, safety is a primary concern for UAV use. As outlined in [9,15], questions and procedures relating to the safety of the low-level control are well-established. In this paper, we thus address the issue of the safety of autonomous decision making in UAVs.

Safety certification for aerial vehicles is guided by DO-178C [34]. One particular requirement is coding standards such as MISRA C [23], which are routinely

© Springer International Publishing Switzerland 2015
C. Dixon and K. Tuyls (Eds.): TAROS 2015, LNAI 9287, pp. 69–81, 2015.
DOI: 10.1007/978-3-319-22416-9_9

applied to low-level control code. While it is possible to implement the high-level decision making directly in C, such code would, arguably, be hard to maintain as the language does not provide an adequate level of abstraction, which in turn also complicates establishing the various traceability requirements by DO-178C.

Agent-programming languages [2,39,44], such as AgentSpeak [29], were proposed as a solution for modeling high-level decision making. However, such languages often require an interpreter or similar run-time environment [2], for example Jason [5] in the case of AgentSpeak. Using a setup like [17] allows the high-level decision making to be interfaced with the low-level control code, however, it raises various concerns with regards to safety. The additional complexity of the system not only increases the number of potential sources of failure and required resources, but can also make traceability requirements infeasible, as not only actual agent code is affected, but also the run-time environment.

In order to mitigate these problems, we propose to instead translate the high-level decision making code from the agent-programming language into the language of the low-level control code. This reduces the complexity of the resulting system and enables the reuse of already existing tools and processes for validation of the low-level control code. Furthermore, the traceability of requirements is guaranteed due to the systematic nature of the code translation. This allows the actual development to take place in the more natural setting of the agent-programming language, thus improving code maintainability.

To the best of our knowledge, no such approach has been proposed for AgentSpeak so far. Related work includes the automated code generation from behavior-based agent programming with the aim of facilitating development and code reuse [43], the translation of AgentSpeak to Promela proposed in [3] for the purpose of model checking AgentSpeak, and the translation of AgentSpeak to Erlang in order to improve concurrent performance [11]. In a wider context, the usage and translation of statecharts (e.g. [18,22]) and Stateflow (e.g. [1,38]) for modelling and the relation between statecharts and agent-oriented programming [16] provides further background and inspiration to this work.

This paper makes the following contributions.

- We present an automated translation from AgentSpeak into a fragment of C akin to MISRA C for implementing high-level decision making in UAVs.
- As an exemplar, we present a re-implementation of the autopilot of the `tum_ardrone` package [13] in AgentSpeak and show how the translated code can be used on a Parrot AR.Drone.

2   The Autopilot

Automatic flight control systems are used to maintain given flight dynamic parameters and to augment stability [24].[1] In order to achieve these goals, typically some form of feedback control loop mechanism is employed, e.g., a PID (Proportional-Integral-Derivative) controller. While such a mechanism is clearly

---

[1] While these requirements were originally applied to fixed-wing aircrafts, these high-level aspects translate directly to rotary-wing aircrafts.

a low-level control aspect, one can also discern high-level decision making aspects in an autopilot, in particular the breaking down of the commands and coordinating their execution. One might also imagine a more sophisticated version of an autopilot implementing collision avoidance, or an emergency landing upon low battery status, as even more prominent high-level aspects.

In the following, we will consider a simple autopilot for an unmanned quadrotor system. This autopilot, which we use as an exemplar, is based on the autopilot provided with the tum_ardrone package [13], however, our version only includes essential aspects (see also Section 5). The autopilot supports three basic commands, namely takeoff, goto, and land. The takeoff command engages the rotors and brings the aircraft to a pre-defined altitude before handling any further commands. The goto command allows the specification of a target (in x, y, z and yaw coordinates) and the autopilot then moves the aircraft towards the target until it is reached. The land command lowers the aircraft to the ground and then disengages the rotors. Furthermore, if no commands are given by the user, the autopilot is required to maintain the aircraft's current position.

In the following, we will show how we model the high-level aspects (specifically, the part called the "KI procedures" in the original tum_ardrone package) of this autopilot in AgentSpeak, and how we translate this code in a way that ensures software considerations for safety critical systems.

3   AgentSpeak

AgentSpeak [29] is an agent-oriented programming language [39] based on the BDI (Belief-Desire-Intention) model [30,31]. Thus, in AgentSpeak, the central notion is that of an agent consisting of beliefs and plans. Agents react to events by selecting a plan, and plan selection is guided by the type of event and the current belief state of the agent. Once a plan is selected, it is executed, which can in turn create new events, change an agent's beliefs, or execute basic actions available to the agent. Events can stem from external sources, i.e., the environment, or internal sources, i.e., the agent itself.

In order to illustrate these concepts, let us consider the example given in Figure 1, which gives a (simplified) fragment of an autopilot implemented in AgentSpeak. It gives the plans for an event corresponding to an instruction to go to a given position.

A typical sequence of execution would look as follows, assuming the UAV has already taken off and is airborne. An external source, for example the user, creates a +!goto event with a given target. In this case the first plan is selected, which subsequently updates the agent's beliefs about the last set target and then the agent tries to achieve the goal completeGoto, which in turn creates the (internal) event +!completeGoto.

For this new event, the latter two plans are relevant. If the UAV's current position is close enough to the given target, the third plan is applicable, in which case a notification to the user is issued (using a basic action provided by the environment) and nothing else remains to be done, thus leaving the autopilot ready for handling further events.

```
+!goto(Target) : takenOff
  <- sendHover(); +lastTarget(Target);
     !completeGoto.

+!completeGoto : takenOff & myPosition(Pos) & lastTarget(Target)
                 & not closeEnough(Pos, Target)
  <- Movement = calculateMovement(Pos, Target);
     sendControl(Movement);
     !completeGoto.

+!completeGoto : takenOff & myPosition(Pos) & lastTarget(Target)
                 & closeEnough(Pos, Target)
  <- sendHover(); notifyUser("target reached").
```

Fig. 1. Plans for the goto case of the autopilot in AgentSpeak

```
+!waitForCommand : takenOff & lastTarget(Target)
  <- ?myPosition(Pos);
     Movement = calculateMovement(Pos, Target);
     sendControl(Movement);
     !!waitForCommand. // new focus!
```

Fig. 2. One plan for the waitForCommand case for the autopilot in AgentSpeak

If the current position is not close enough to the target, the second plan is applicable. Here, the agent uses two basic actions in order to calculate the necessary control command and send it to the UAV. Finally, it tries to achieve the goal completeGoto again, which in turn creates a corresponding event and leads to the execution of either of the two plans until the target is reached.

Syntactically, we use a variant similar to AgentSpeak(F) as presented in [3]. For convenience, we add assignments as possible formulae. Another addition is "new focus" goals !!literal, which prove to be useful with the adopted semantics as explained below. The complete grammar is given in Table 1, where atoms are predicates or names of basic actions.

We impose one additional restriction that will become clear in view of the translation target (see Section 4). Roughly speaking, we only allow a restricted form of "recursion". In order to explain this, we briefly need to introduce the following notions. Similar to a call graph, we can create a *trigger graph* of a given set of plans indicating which plans may trigger the execution of which further plans. Note that as in the case of call graphs, this trigger graph is an over-approximation. We call an atom *recursively triggering*, if it occurs as the atom of the triggering event of a plan from which a cycle is reachable in the trigger graph. In the body of a plan, we allow goals and percepts based on recursively triggering atoms only to occur as the last formula; any other goal or percept has to be based on a non-recursively triggering atom.

For an overview of the semantics of AgentSpeak, see [4,42]. The reasoning cycle of an AgentSpeak agent works as follows. First, an event to be handled

Table 1. AgentSpeak syntax

$$
\begin{aligned}
\langle agent \rangle \quad &::= \quad \langle beliefs \rangle \; \langle initial\_goal \rangle \; \langle plans \rangle \\
\langle beliefs \rangle \quad &::= \quad \langle literal \rangle . \; \ldots \langle literal \rangle . \\
\langle initial\_goal \rangle &::= \quad ! \langle literal \rangle \; . \\
\langle plans \rangle \quad &::= \quad \langle plan \rangle \; \ldots \langle plan \rangle \\
\langle plan \rangle \quad &::= \quad \langle triggering\_event \rangle : \langle context \rangle \texttt{<-} \langle body \rangle . \\
\langle triggering\_event \rangle &::= \quad \langle percept \rangle \; | \; \texttt{+}\langle goal \rangle \; | \; \texttt{-}\langle goal \rangle \\
\langle context \rangle \quad &::= \quad \langle condition \rangle \; \texttt{\&} \ldots \texttt{\&} \; \langle condition \rangle \\
\langle condition \rangle &::= \quad \textbf{true} \; | \; \langle literal \rangle \; | \; \textbf{not (} \langle literal \rangle \texttt{)} \\
\langle body \rangle \quad &::= \quad \textbf{true} \; | \; \langle formula \rangle ; \; \ldots ; \; \langle formula \rangle \\
\langle formula \rangle \quad &::= \quad \langle action \rangle \; | \; \langle percept \rangle \; | \; \langle goal \rangle \; | \; \langle assignment \rangle \\
\langle action \rangle \quad &::= \quad \langle literal \rangle \\
\langle percept \rangle \quad &::= \quad \texttt{+}\langle literal \rangle \; | \; \texttt{-}\langle literal \rangle \\
\langle goal \rangle \quad &::= \quad ! \langle literal \rangle \; | \; !! \langle literal \rangle \; | \; ? \langle literal \rangle \\
\langle assignment \rangle &::= \quad \langle variable \rangle = \langle literal \rangle \\
\langle literal \rangle \quad &::= \quad atom(\; \langle term \rangle , \; \ldots , \; \langle term \rangle \;) \\
\langle term \rangle \quad &::= \quad variable \; | \; unnamed\ variable \; | \; number \; | \; string
\end{aligned}
$$

is selected. For this event, all plans that are relevant, i.e., those plans whose triggering event unifies with the selected event, are found. This selection is then narrowed down to applicable plans, i.e., those plans whose context evaluates to true given the agent's current beliefs. From these applicable plans one plan[2] is selected to be actually executed, creating a so-called intention. Then, from all the intentions an agent currently has, one is selected to be executed, which means executing the next formula in the plan body and storing the updated intention accordingly.

The AgentSpeak architecture allows various customization options, see [42]. In particular, belief revision, event selection, and intention selection can all be customized. In view of the translation target, we use the following options.

First, events are handled in "run-to-completion" style, i.e., once a plan for a given event is selected, this plan is run until completed, which also involves any sub-plans triggered. This behavior can be achieved by using customized event and intention selection functions. Note that new focus goals !!literal can be used in order to generate deferred events, i.e., events that do not need to be handled immediately. Consider for example Figure 2, where this is used to implement a default behavior in case no user commands are given.

Second, the belief base of an agent can store at most one instance of a literal, i.e., if an agent holds the belief speed(3), after percept +speed(5), the belief base is updated to speed(5), and will not contain an additional speed(3). This can be achieved using an appropriate belief revision function.

---

[2] The default behavior is to select the first applicable plan in textual order.

```
void next_step(void) {
  updateBeliefs();
  eventt event = get_next_event();
  switch (event.trigger) {
    /* ... */
    case ADD_ACHIEVE_GOTO:
      add_achieve_goto(event.goto_param0); break;
    case ADD_ACHIEVE_COMPLETEGOTO:
      add_achieve_completeGoto(); break;
    /* ... */
  }
}
```

Fig. 3. Selection of relevant plans for the autopilot, translated to C

```
void add_achieve_completeTakeOff(int param0) {
  if (add_achieve_completeTakeOff_plan0(param0)) { return; }
  if (add_achieve_completeTakeOff_plan1(param0)) { return; }
  /* ... handle the case where no plan is applicable ... */
  return;
}
```

Fig. 4. Selection of applicable plans, for the `completeTakeOff` case

The purpose of these restrictions will be further illuminated in the following section. Note that even though we are only considering a fragment of Agent-Speak, it already suffices to model interesting processes, such as the autopilot.

4   Translation

We generally follow the ideas from [3], which present a translation from Agent-Speak to Promela [19] for the purposes of model checking AgentSpeak. As our intention is to run the generated code directly on the platform without any intermediate interpreters, our target language is C. As previously mentioned, software considerations for airborne systems are regulated by DO-178C [34]. While DO-178C does not prescribe the usage of any particular set of coding guidelines, MISRA C [23], or very similar rulesets, has become de-facto standard for safety-critical embedded software. We thus aim for our generated code to comply with the rules imposed by this standard, which prohibits recursion and dynamically allocated memory.

The syntactic restrictions and semantic customizations as introduced in the previous section directly relate to these restrictions. The restriction on "recursion" and the "run-to-completion" semantics give us a limit on the depth of the call stack, eliminate the need for handling and storing multiple intentions, and, furthermore, also limit the number of events generated internally to at most one per deliberation cycle. Thus, we only need to store at most two events, one event

```
bool add_achieve_goto_plan0(positiont param0) {
  positiont Target = param0;
  if (!takenOff_set) { return false; }
  sendHover();
  lastTarget_set = true;
  lastTarget_param0 = Target;
  internal_achieve_completeGoto();
  return true;
}
```

Fig. 5. Translation of the goto plan from Figure 1 into C

```
bool add_achieve_waitForCommand_plan1(void) {
  if (!takenOff_set) { return false; }
  if (!lastTarget_set) { return false; }
  positiont Target = lastTarget_param0;
  if (!myPosition_set) {
    /* ... achieve test goal or handle plan failure ... */
  }
  positiont Pos = myPosition_param0;
  control_commandt Movement = calculateMovement(Pos, Target);
  sendControl(Movement);
  internal_achieve_new_focus_waitForCommand();
  return true;
}
```

Fig. 6. Translation of the waitForCommand plan from Figure 2 into C

that needs to be run to completion,[3] and, possibly, a deferred event. Our custom belief revision allows us to model the belief base using one single variable per literal (plus an additional flag that indicates whether it is set or not), again not requiring dynamic memory. Note also that the restriction on "recursion" gives natural break points to segment long runs to completion into smaller parts, thus ensuring the reactivity of the translated program.

The translation proceeds in three main stages. First of all, the AgentSpeak program and an accompanying configuration file are parsed. Then, the parsed program is analyzed, which includes determining the recursively triggering atoms. Finally, the actual translation takes place, which we will outline below.

Due to space limitations it is not possible to give all details,[4] but the following examples should provide all relevant ideas. At the core of the translated code lies the next_step() function corresponding to one step in the agent's reasoning cycle. First of all, the hook updateBeliefs() allows the environment to update the agent's belief state, if required. Then, the next event is selected. Here, the rtc event is given preference over the deferred event. Then a relevant plan is

---

[3] We will call this the *rtc event* in the following.
[4] The full code is available at [6].

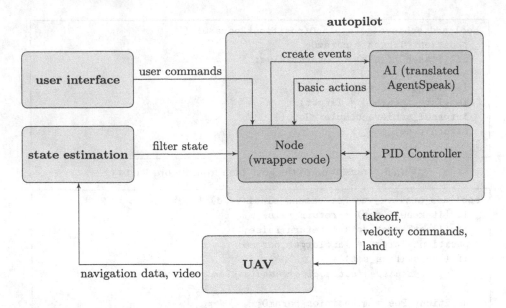

Fig. 7. Experimental setup

selected, as illustrated in Figure 3. Subsequently, the first applicable plan (in order of textual appearance) is selected and executed, see Figure 4 for a concrete instance. Note that this would also allow the inclusion of error handling if no plan is applicable (cf. [4]).

The translation of the plans themselves is illustrated in Figures 5 and 6. First, a plan tests for its own applicability, instantiating variables along the way, if necessary. Basic actions and assignments translate directly to their C counterparts. Percepts set or unset the corresponding variables in the belief base, and test goals read the corresponding variable on belief bases.

Achievement goals require a more elaborate handling, distinguishing three different cases. Non-recursively triggering achievement goals call their relevant plan selection method directly. Recursively triggering achievement goals create a new rtc event, thus guaranteeing it will be handled next. New focus achievement goals create a deferred event, thus allowing for rtc events to be handled before them. Note that external events are created as rtc events and can only be created if there is currently no rtc event set. Otherwise, the creation of the external event has to wait until the current rtc event has finished. However, one might image more elaborate versions where certain external events can pre-empt internal rtc events.

5   Experimental Setup

We evaluated the generated code on the `tum_simulator` package [20] and a Parrot AR.Drone 2.0 using the Robot Operating System (ROS) [28,33] version

Hydro Medusa with the `tum_ardrone` package [13,14]. The availability of the `tum_simulator` facilitated rapid prototyping and development of the translator, as translated code could be easily tested without the need of an extensive experimental setup. Due to the modular nature of ROS, switching to the actual platform was a straightforward operation.

It is important to note that the generated code is not specific to ROS.[5] Thus, it is necessary to provide wrapper code acting as an interface between ROS and the generated code, translating user commands into AgentSpeak events and providing the basic actions required by the AgentSpeak code. The original `tum_ardrone` package already uses a similar arrangement, consisting of three components, namely the autopilot node, a PID controller for low-level control, and so-called "KI procedures" for the high-level control. For our experiments, we replaced these KI procedures with our generated code and we modified the autopilot node accordingly. The complete setup is illustrated in Figure 7. In addition to the autopilot node, a simple user interface for the input of flight plans is provided. Furthermore, a state estimation node is used to track the current position of the UAV based on navigation data and the video feed.

We have tested various flight plans composed of the three basic commands `takeoff`, `goto` and `land`. We compared our autopilot to the original `tum_ardrone` autopilot and have confirmed analogue behavior.

On the technical side, the AgentSpeak code for the autopilot uses about 70 lines of code and generates about 700 lines of C code compared to the about 300 lines of C++ code of the original KI procedures. This shows that AgentSpeak allows a much more compact representation of the high-level behavior of the autopilot, making the code easier to maintain and extend. The modified `tum_ardrone` package is available online [7], where detailed instructions regarding installation are also provided.

6   Conclusions and Future Work

We discuss using AgentSpeak as a modeling language for the high-level decision making of UAVs. This model can then be automatically translated into C, thus bridging the gap between the high-level decision making and the low-level control code. The abstract model in AgentSpeak reduces the complexity of the code and is flexible for maintenance and further extensions. The automatic translation removes possible errors that can be introduced when mixing these high-level aspects directly into the low-level control code and complies with the safety regulations for UAVs. As an example, we show how the autopilot of the `tum_ardrone` package can be easily modeled in AgentSpeak and how the generated C code can be directly used in real world platforms, such as a Parrot AR.Drone. We also remark that this approach is not restricted solely to UAVs, but can be adapted to various other autonomous systems.

---

[5] While it would be possible to generate ROS code directly, we chose a more general solution, such that the generated code can also be used in settings that are not based on ROS.

Future research includes identifying fragments of AgentSpeak that are more expressive while still allowing translation within the given restrictions (cf. [38]). This also includes adequate methods for static analysis and testing, as typically required for safety certification, an essential criteria for the practical usage of UAVs. Formal methods (cf. [35]) complement these approaches and we plan to verify safety properties of the AgentSpeak model by implementing the operational semantics in term rewrite systems like [12,32]. We also plan to validate the translation from AgentSpeak to C code with the assistance of CBMC [8], a bounded model checker for C. Translation validation [26,36,37,40] is a common approach to guarantee that the semantics of the high-level model are preserved in the translated code. Overall, this will guarantee that the safety properties established for the AgentSpeak model can be transferred to the translated code, thereby guaranteeing the required traceability of requirements.

For example, consider [25] stating "[...] satisfying control of the AR.Drone 2.0 is reached by sending the AT-commands every 30 ms [...]".[6] Using [41], one could establish the worst-case execution time of the translated code. On the other hand, the statement *in every deliberation cycle one of the aforementioned basic actions is executed* can be easily formalized in linear temporal logic using BDI primitives. This property could then be checked on the AgentSpeak code using the approach mentioned above. Using the translation validation would then allow us to combine these facts, thus establishing the given requirement.

As a final note, the separation of high-level and low-level concerns allows reusing results on both ends of the translation. For example, one might use Jason [5] (with appropriate custom semantics) as a simulation environment for prototyping. Also, the formal verification on the AgentSpeak is not restricted to BDI properties, but, as outlined in [10] various other options, e.g., probabilistic properties can also be considered, thus allowing even more flexibility.

Acknowledgments. This work was supported by Engineering and Physical Sciences Research Council (EPSRC) grant EP/J012564/1.

References

1. Agrawal, A., Simon, G., Karsai, G.: Semantic Translation of Simulink/Stateflow Models to Hybrid Automata Using Graph Transformations. Electronic Notes in Theoretical Computer Science **109**, 43–56 (2004)
2. Bordini, R.H., Braubach, L., Dastani, M., Fallah-Seghrouchni, A.E., Gómez-Sanz, J., Leite, J., O'Hare, G., Pokahr, A., Ricci, A.: A Survey of Programming Languages and Platforms for Multi-Agent Systems. Informatica **30**(1), 33–44 (2006)
3. Bordini, R.H., Fisher, M., Visser, W., Wooldridge, M.: Verifying Multi-agent Programs by Model Checking. Autonomous Agents and Multi-Agent Systems **12**(2), 239–256 (2006)

---

[6] For our purposes, *AT-commands* can be identified with the basic actions for taking off, landing, and sending movements. Note that hovering is a form of movement.

4. Bordini, R.H., Hübner, J.F.: Semantics for the jason variant of agentspeak (plan failure and some internal actions). In: European Conference on Artificial Intelligence, pp. 635–640. IOS Press (2010)
5. Bordini, R.H., Hübner, J.F., Wooldridge, M.: Programming Multi-Agent Systems in AgentSpeak Using Jason. John Wiley & Sons (2007)
6. Bucheli, S., Kroening, D., Martins, R., Natraj, A.: AgentSpeak Translator. https:// github.com/OxfordUAVAutonomy/AgentSpeakTranslator (last visited June 15, 2015). doi:10.5281/zenodo.18572
7. Bucheli, S., Kroening, D., Martins, R., Natraj, A.: Modified **tum_ardrone** package (last visited 15, June 2015). doi:10.5281/zenodo.18571. https://github.com/ OxfordUAVAutonomy/tum_ardrone
8. Clarke, E., Kroning, D., Lerda, F.: A tool for checking ANSI-C programs. In: Jensen, K., Podelski, A. (eds.) TACAS 2004. LNCS, vol. 2988, pp. 168–176. Springer, Heidelberg (2004)
9. Dennis, L.A., Fisher, M., Lincoln, N., Lisitsa, A., Veres, S.M.: Practical Verification of Decision-Making in Agent-Based Autonomous Systems. Automated Software Engineering Online, 1–55 (2014). doi:10.1007/s10515-014-0168-9
10. Dennis, L.A., Fisher, M., Webster, M.: Two-stage agent program verification. Journal of Logic and Computation Online (2015). doi:10.1093/logcom/exv002
11. Díaz, Á.F., Earle, C.B., Fredlund, L.Å.: eJason: an implementation of jason in erlang. In: Dastani, M., Hübner, J.F., Logan, B. (eds.) ProMAS 2012. LNCS(LNAI), vol. 7837, pp. 1–16. Springer, Heidelberg (2013)
12. Doan, T.T., Yao, Y., Alechina, N., Logan, B.: Verifying heterogeneous multi-agent programs. In: International Joint Conference on Autonomous Agents and Multiagent Systems. IEEE Computer Society (2014)
13. Engel, J., Sturm, J., Cremers, D.: **tum_ardrone**. http://wiki.ros.org/tum_ardrone (last visited June 15, 2015)
14. Engel, J., Sturm, J., Cremers, D.: Scale-Aware Navigation of a Low-Cost Quadrocopter with a Monocular Camera. Robotics and Autonomous Systems **62**(11), 1646–1656 (2014)
15. Fisher, M., Dennis, L., Webster, M.: Verifying autonomous systems. Communications of the ACM **56**(9), 84–93 (2013)
16. Fortino, G., Rango, F., Russo, W., Santoro, C.: Translation of statechart agents into a BDI framework for MAS engineering. Engineering Applications of Artificial Intelligence **41**, 287–297 (2015)
17. Hama, M.T., Allgayer, R.S., Pereira, C.E., Bordini, R.H.: UAVAS: AgentSpeak Agents for Unmanned Aerial Vehicles. In: Workshop on Autonomous Software Systems at CBSoft (Autosoft 2011) (2011)
18. Harel, D., Politi, M.: Modeling Reactive Systems with Statecharts: The Statemate Approach, 1st edn. McGraw-Hill Inc., New York (1998)
19. Holzmann, G.: The Spin Model Checker: Primer and Reference Manual. Addison-Wesley Professional (2003)
20. Huang, H., Sturm, J.: **tum_simulator**. http://wiki.ros.org/tum_simulator (last visited June 15, 2015)
21. ICAO Cir 328: Unmanned Aircraft Systems (UAS). International Civil Aviation Organization (2011)

22. Mikk, E., Lakhnech, Y., Siegel, M., Holzmann, G.: Implementing statecharts in PROMELA/SPIN. In: Workshop on Industrial Strength Formal Specification Techniques, pp. 90–101. IEEE Computer Society Press (1998)
23. Motor Industry Software Reliability Association and Motor Industry Software Reliability Association Staff: MISRA C: 2012: Guidelines for the Use of the C Language in Critical Systems. Motor Industry Research Association (2013)
24. Nelson, R.C.: Flight stability and automatic control. McGraw-Hill (1997)
25. Piskorski, S., Brulez, N., Eline, P., D'Haeyer, F.: AR.Drone Developer Guide Revision SDK 2.0. Parrot S.A. (2012)
26. Pnueli, A., Siegel, M.D., Singerman, E.: Translation validation. In: Steffen, B. (ed.) TACAS 1998. LNCS, vol. 1384, pp. 151–166. Springer, Heidelberg (1998)
27. Puri, A.: A Survey of Unmanned Aerial Vehicles (UAV) for Traffic Surveillance. Department of Computer Science and Engineering, University of South Florida, Tech. rep. (2005)
28. Quigley, M., Conley, K., Gerkey, B., Faust, J., Foote, T., Leibs, J., Wheeler, R., Ng, A.Y.: ROS: an open-source robot operating system. In: ICRA Workshop on Open-Source Software in Robotics (2009)
29. Rao, A.S.: AgentSpeak(L): BDI agents speak out in a logical computable language. In: Perram, J., Van de Velde, W. (eds.) MAAMAW 1996. LNCS, vol. 1038, pp. 42–55. Springer, Heidelberg (1996)
30. Rao, A.S., Georgeff, M.P.: BDI Agents: From theory to practice. In: International Conference on Multi-Agent Systems, pp. 312–319. AAAI press (1995)
31. Rao, A.S., Georgeff, M.P.: Modeling rational agents with a BDI-architecture. In: Readings in Agents, pp. 317–328. Morgan Kaufmann Publishers Inc. (1998)
32. van Riemsdijk, M.B., de Boer, F.S., Dastani, M., Meyer, J.J.C.: Prototyping 3APL in the Maude Term Rewriting Language. In: International Joint Conference on Autonomous Agents and Multiagent Systems, pp. 1279–1281. ACM (2006)
33. ROS: Robot Operating System. http://www.ros.org (last visited June 15, 2015)
34. RTCA: DO-178C, Software Considerations in Airborne Systems and Equipment Certification. Radio Technical Commission for Aeronautics (2011)
35. RTCA: DO-333, Formal Methods Supplement to DO-178C and DO-278A. Radio Technical Commission for Aeronautics (2011)
36. Ryabtsev, M., Strichman, O.: Translation validation: from simulink to C. In: Bouajjani, A., Maler, O. (eds.) CAV 2009. LNCS, vol. 5643, pp. 696–701. Springer, Heidelberg (2009)
37. Sampath, P., Rajeev, A.C., Ramesh, S.: Translation validation for stateflow to C. In: Design Automation Conference on Design Automation Conference, pp. 23:1–23:6. ACM (2014)
38. Scaife, N., Sofronis, C., Caspi, P., Tripakis, S., Maraninchi, F.: Defining and translating a "safe" subset of simulink/stateflow into lustre. In: International Conference on Embedded Software, pp. 259–268. ACM (2004)
39. Shoham, Y.: Agent-oriented Programming. Artificial Intelligence **60**(1), 51–92 (1993)
40. Staats, M., Heimdahl, M.P.E.: Partial translation verification for untrusted code-generators. In: Liu, S., Araki, K. (eds.) ICFEM 2008. LNCS, vol. 5256, pp. 226–237. Springer, Heidelberg (2008)

41. Thesing, S., Souyris, J., Heckmann, R., Randimbivololona, F., Langenbach, M., Wilhelm, R., Ferdinand, C.: An abstract interpretation-based timing validation of hard real-time avionics software. In: International Conference on Dependable Systems and Networks, pp. 625–632. IEEE (2003)
42. Vieira, R., Moreira, A., Wooldridge, M., Bordini, R.H.: On the formal semantics of speech-act based communication in an agent-oriented programming language. Journal of Artificial Intelligence Research 29(1), 221–267 (2007)
43. Vu, T., Veloso, M.: Behavior Programming Language and Automated Code Generation for Agent Behavior Control. School of Computer Science, Carnegie Mellon University, Tech. rep. (2004)
44. Ziafati, P., Dastani, M., Meyer, J.-J., van der Torre, L.: Agent programming languages requirements for programming autonomous robots. In: Dastani, M., Hübner, J.F., Logan, B. (eds.) ProMAS 2012. LNCS(LNAI), vol. 7837, pp. 35–53. Springer, Heidelberg (2013)

Joint Torques and Velocities in a 3-mass Linear Inverted Pendulum Model of Bipedal Gait

Guido Bugmann$^{(\boxtimes)}$

Plymouth University, Plymouth PL4 8AA, UK
gbugmann@plymouth.ac.uk

Abstract. When a humanoid robot walks with pace, different joints face different requirements of speed and torque. This paper shows how to estimate the requirements for three heavily loaded joints: one in the knee and two in the hips. The dynamics of the walking robot is modeled as that of a 3-mass linear inverted pendulum (3-mass LIPM). The 3-mass LIPM is the simplest model that requires considering effects of internal forces and torques dues to swinging masses. The calculations show that the three joints test the characteristics of standard servo motors in different ways, and their requirements can be individually modulated by changing the gait pattern and the robot design.

1 Introduction

How fast can a humanoid robot walk? This question arose from an interest in the Hurosot robot sprint competitions. The answer required the development of a general method for specifying servomotors for walking humanoid robots. During gait, the leg joints of a walking robot are required to rotate at specified speeds and must produce specified torques. Rotation speeds and torques increase with walking speeds and the characteristics of the actuators ultimately limit the walking speed. Three heavily loaded joints are considered in this report: 1) The knee joint of the supporting leg. 2) The roll hip joint of the supporting leg (lateral leg angle control) and 3) the pitch hip joint of the swinging leg (forward-backward motion).

The robot is modeled as a linear inverted pendulum (LIPM) with 3 masses: m_1 represents the upper-body and m_2 and m_3 represent each foot. The use of three masses allows an estimation of the forces on the swinging leg. It generally improves the description of the dynamics of the robot, compared to the standard single-mass LIPM that ignores the effects of moving masses.

As the robot moves in a 3D space, both equation for forward x-motion and lateral y-motion need to be developed and solved. Knowing these motions then allows the calculation of the torques and rotation speeds of any joint. The resulting expressions for the three selected joints were implemented in a spreadsheet, to be made available online, where the user can modify the geometrical parameters of the robot, the servo motor characteristics and the walking speed.

© Springer International Publishing Switzerland 2015
C. Dixon and K. Tuyls (Eds.): TAROS 2015, LNAI 9287, pp. 82–93, 2015.
DOI: 10.1007/978-3-319-22416-9_10

2   Forces and Torques in Multi-mass LIPM Models

The standard inverted pendulum model (IPM) consist of one mass on top of a massless pole (figure 1A). As the robot walks, its weight shifts from one foot to the other. In the IPM model, the short time of double support is ignored, and the contact point of the pole on the ground is instantaneously repositioned at each step. This very simplified model has provided useful insights into human gait (see e.g. [1]) but is mathematically challenging for robot gait control due to the coupled nature of the motion of the pole along the x (forward), y (lateral) and z (vertical).

A significant breakthrough was the Linear Inverted Pendulum where the mass is constrained to move in a plane, e.g. a plane of constant height [4]. This decouples the x and y motion, greatly simplifying the dynamics. In practice, the requirement of keeping the height of the robot constant while walking, or following a constant slope while climbing stairs, was never perceived as a big constraint.

How is a LIPM model used to generate a robot gait? The idea is that, by controlling where and when the pole touches the ground, one can control in what direction the mass is going to fall. By catching a succession of falls with a succession of appropriate foot placements, one can make the robot move in any desired direction. Knowing the dynamics of the pole is very useful as it allows to predict exactly the consequences of a given foot placement. An example of how this is exploited in the case of constant-frequency gaits is given by [3].

One of the problems with this approach is that the robot is not a point-like mass but a multi-body system that changes shape during the phases of the gait. Hence, it only follows approximately the LIPM equations of motion and practical gait design requires the use of fudge factors or of feedback controllers attempting to make the real robot behave like its ideal model. Another problem is that servo-motors have to move objects with masses, e.g. legs. By ignoring such masses, the theory can suggests gait patterns that are unstable because the torques and velocities required are beyond the capabilities of the servo-motors used. Thus, a robot model with a distributed mass system should improve the prediction of its motion as well as the evaluation of the feasibility of a given gait pattern.

2.1   The 1-mass LIPM Equation

To start with, lets consider the one-mass LIPM in figure 1A. The mass m is subject to the gravitational force mg that exerts a torque τ around the pivot point of the pole.

$$\tau = mg\sin(\theta) \cdot L = mg\frac{x}{L} \cdot L = mgx \tag{1}$$

The normal inverted pendulum would fall to the right in the figure. In the LIPM, the pole is provided with a mechanism that extends its length as it falls, to keep the mass at a constant height.

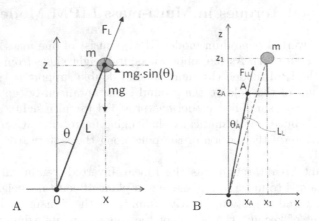

Fig. 1. A. Simple linear inverted pendulum. It is linear because the height z is maintained constant by the axial force F_L. B. A massless hip-like structure supporting the mass m.

The axial force in the pole F_L is such that its its vertical component $F_{L,z}$ compensates for the gravitational downward pull force mg. The resulting horizontal component $F_{L,x}$ accelerates the mass.

$$F_{L,z} = F_L \cdot \cos(\theta) = mg \tag{2}$$

$$F_{L,x} = F_L \cdot \sin(\theta) = m\ddot{x} = \frac{mg}{\cos(\theta)} \sin(\theta) = mg\tan(\theta) = mg\frac{x}{z} \tag{3}$$

$$\rightarrow m\ddot{x} = mg\frac{x}{z} \tag{4}$$

Equation (4) is the standard LIPM equation of motion in the x-direction, when there are no other external forces than gravity. In 3-D, one needs a second equation to describe the projection of the motion in the y-direction. It is the same as (4) but with x replaced by y. By multiplying both sides of equation 4 by z one obtains an equation relating torques:

$$m\ddot{x}z = mgx \tag{5}$$

Equation 5 shows how the rotation torque mgx produced by gravity is converted into a "propulsion torque" $m\ddot{x}z$. Indeed, the axial force in the pole helps to make this happen.

2.2   Force and Torques in a Simple Robot Structure

Lets consider a hip-like structure as in figure 1B where the leg produces an axial force F_{LL} offset from the centre of the mass m.

To maintain joint A at its height z_A, the vertical gravitational force needs to be canceled by the vertical component of F_{LL}, which in turn generates a horizontal force $F_{LL,x}$, as described in the previous section:

$$F_{LL,x} = mg\frac{x_A}{z_A} \tag{6}$$

To maintain the hip structure level, a torque τ needs to be applied at joint A to counteract the torque caused by gravity $\tau_g = mg(x_1 - x_A)$. The torque τ causes a horizontal reaction force at the pivot point (0,0):

$$F_{\tau,x} = \frac{\tau_g}{z_A} = mg\frac{x_1 - x_A}{z_A} \tag{7}$$

The total horizontal force applied at joint A is:

$$F_{A,x} = F_{LL,x} + F_{\tau,x} = mg\frac{x_A}{z_A} + mg\frac{x_1 - x_A}{z_A} = mg\frac{x_1}{z_A} \tag{8}$$

This force is applied to m through a lever-like structure that, in this case, reduces it:

$$F_{m,x} = F_{A,x}\frac{z_A}{z_1} = mg\frac{x_1}{z_A} \cdot \frac{z_A}{z_1}$$
$$\rightarrow m\ddot{x} = mg\frac{x_1}{z_1} \tag{9}$$

This is the same formula as for a simple pole directly linking the pivot point and the mass (4). It is interesting that now we have an extensible leg producing an axial force, and a hip producing roll torque, and none of those needs explicit representation in the equation of motion. Note that the axial force on the pole has also disappeared from equation 4.

2.3   A 2-mass System

Lets add another mass to the 1-mass LIPM, with a link L_2 attached to mass m_1 at the end of which the mass m_2 is located. It is assumed that both masses are in the same plane (d,z) where d is a direction vector defined in the (x,y) plane (figure 2).

Point P_1 is subjected to the weight of m_2 through the axial force in link L_2 and the vertical reaction to the torque exerted by m_2g at the end of L_2. The vertical component of the axial force by the pole L_1 on P_1 now is:

$$F_{L1,z} = (m_1 + m_2)g \tag{10}$$

requiring an axial force F_{L1} with a horizontal component:

$$F_{L1,d} = (m_1 + m_2)g\frac{d_1}{z_1} \tag{11}$$

The joint P_1 applies a torque τ to counteract the torque generated by m_2:

$$\tau_{P_1} = m_2g(d_2 - d_1) \tag{12}$$

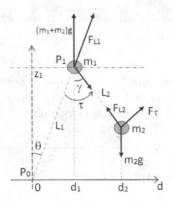

Fig. 2. Side view in the (d,z) plane of a co-planar 2-body inverted pendulum

This torque causes a horizontal reaction force $F_{\tau,P0,d}$ at the pivot point P_0:

$$F_{\tau,P0,d} = \frac{\tau_{P_1}}{z_1} = m_2 g \frac{(d_2 - d_1)}{z_1}, \tag{13}$$

which contributes to the total horizontal force $F_{P1,d}$ at P_1:

$$F_{P1,d} = F_{L1,d} + F_{\tau,P0,d} = (m_1 + m_2)g\frac{d_1}{z_1} + m_2 g\frac{(d_2 - d_1)}{z_1} \tag{14}$$

$$= \frac{1}{z_1}(m_1 d_1 g + m_2 d_2 g) \tag{15}$$

That force drives both m_1 and m_2, the latter via the lever L_2:

$$F_{P1,d} = m_1 \ddot{d}_1 + \frac{z_2}{z_1} m_2 \ddot{d}_2 \tag{16}$$

Replacing $F_{P1,d}$ by (15) and multiplying both sides with z_1 produces:

$$m_1 \ddot{d}_1 z_1 + m_2 \ddot{d}_2 z_2 = m_1 d_1 g + m_2 d_2 g \tag{17}$$

This is an extension of (5) to two masses, showing that the sum of the torques applied will cause the sum of the propulsion torques.

How the total torque applied is divided into accelerating mass m_1 and mass m_2 depends on constraints imposed to the structure, e.g. whether it is rigid or actuated. The section 3 will describe how this question is approached in the case of our walking robot model.

The calculation above can be re-done using only the projection of the forces onto the (x-z) or (y-z) planes and they leads to two equations with the same form as (17) but with d replaced with x or y.

$$m_1 \ddot{x}_1 z_1 + m_2 \ddot{x}_2 z_2 = m_1 x_1 g + m_2 x_2 g$$
$$m_1 \ddot{y}_1 z_1 + m_2 \ddot{y}_2 z_2 = m_1 y_1 g + m_2 y_2 g \tag{18}$$

In the non-coplanar case, e.g. when the swinging foot (m_2) is not in the vertical plane defined by the pivot point P_0 and position P_1 of m_1, the force $m_2 g$ is still conveyed to P_1 by the link L_2. However, the torque on P_1 due to m_2 tends to accelerate the system in a different direction than the torque generated by ($m_1 + m_2$) at P_1. Here again, it can be shown that the x and y motions are described by (18).

Equations (18) can be generalized to any number of masses, leading to the moment equations [2]:

$$\sum_{i=1}^{n} m_i \ddot{x}_i z_i = \sum_{i=1}^{n} g m_i x_i \qquad \sum_{i=1}^{n} m_i \ddot{y}_i z_i = \sum_{i=1}^{n} g m_i y_i \qquad (19)$$

Note that links can be actuated, but the exerted torques or forces disappear from the equation. The actuation is not expressed in terms of torques, but as constraints on the relative positions of the bodies. This is very suitable for robots using servos controlled in position, as will be illustrated in section 3.

3   A Three-Mass Robot Model

Our robot is modeled as a three-mass system for the purpose of analyzing torques on three selected joints (figure 3): 1) The knee joint of the supporting leg. 2) The roll hip joint of the supporting leg (lateral leg angle control) and 3) the pitch hip joint of the swinging leg (forward-backward motion). It could have been modeled with a larger number of masses, with no qualitative gain in information produced.

The mass of the support foot can be ignored, as it constitutes the pivot point. So, in practice, this is a two-mass model, where m_3 is not used.

3.1   Forward Motion: Equation and Solution

Applying equation 19, one obtains

$$m_1 \ddot{x}_1 z_1 + m_2 \ddot{x}_2 z_2 = g m_1 x_1 + g m_2 x_2 \qquad (20)$$

The pivot point of the foot on the ground is located a $x = 0, z = 0$.

I now make the additional assumption that the swinging foot has a height $z_2 = 0$. This assumption is not as odd as it sounds. The pivot of the leg is around 4 cm above ground, determined by the dimensions of the two servomotors in the foot and the thickness of the sole (figure 3). Most of the weight of the foot is located below the pivot point. If we assume that the pivot point has a height $z = 0$, then a height $z_2 = 0$ actually requires the swinging foot to be lifted by several centimeters off the ground. So, assuming $z_2 = 0$ simplifies (20) further:

$$m_1 \ddot{x}_1 z_1 = g m_1 x_1 + g m_2 x_2 \qquad (21)$$

At this point I need to define $x_2(t)$. I assume that the swinging foot is immobile before starting to swing, and is immobile again just before becoming the

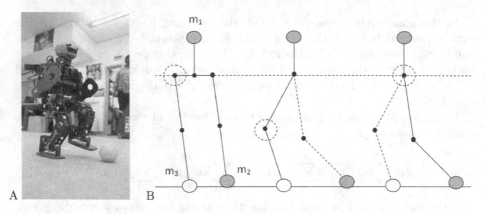

Fig. 3. A. The 50cm tall competition robot "Drake". B. 3-mass robot model showing the joints analyzed in the paper. The mass of the foot on the ground (open circle) is ignored as it constitutes the pivot for the robot. The three joints of interest are: The roll hip joint of the support leg, the knee joint of the support leg, and the tilt hip joint of the swinging leg.

support foot again. There is certainly an optimal way to accelerate and slow down the swinging foot, however, here I will simply assume a sinusoidal function:

$$x_2(t) = -2x_s \cdot \cos(\frac{2\pi}{T}t) \tag{22}$$

where x_s is the half step length and T is the (preset) gait cycle period. A cycle is defined as one foot completing its support and its swing phase once. The swinging foot starts at a point $-2x_s$ behind the support foot and ends up $2x_s$ in front of it (figure 4). Note that in the robot's own reference frame the foot only moves from $-x_s$ to $+x_s$. Foot trajectories are illustrated in figures 6 and 7.

With the above decision on the motion of the swinging foot, the equation of motion (21) becomes:

$$\ddot{x}_1 = x_1 \frac{g}{z_1} - \frac{m_2}{m_1} \frac{g}{z_1} 2x_s \cos(\frac{2\pi}{T}t) \tag{23}$$

$$= x_1 \cdot \alpha_1^2 - \alpha_2 \cdot \cos(wt) \tag{24}$$

Where constants α_1, α_2 and w are introduced to simplify the notation.

Based on a mixture of literature search [2], friendly advice[1] and some intuition, a good solution of this equation of motion is:

$$x_1(t) = C_1 \cosh(\alpha_1 t) + C_2 \sinh(\alpha_1 t) + C_3 \cos(wt) \tag{25}$$

---

[1] Thank you Piotr Bogdan.

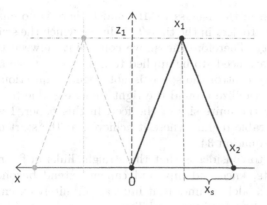

Fig. 4. 3-mass robot model showing mass m_1 and m_2 and the linking virtual poles (in reality, articulated joints are used). The full lines show the configuration at the start of a half gait cycle, with the swinging foot starting a x_2. The robot is moving to the left.

Calculating the first and second derivatives gives:

$$\dot{x}_1(t) = C_1\alpha_1 \sinh(\alpha_1 t) + C_2\alpha_1 \cosh(\alpha_1 t) - C_3 w \sin(wt) \tag{26}$$

$$\ddot{x}_1(t) = C_1\alpha_1^2 \cosh(\alpha_1 t) + C_2\alpha_1^2 \sinh(\alpha_1 t) - C_3 w^2 \cos(wt) \tag{27}$$

$$= \alpha_1^2(x_1(t) - C_3 \cos(wt)) - w^2 C_3 \cos(wt) \tag{28}$$

$$= \alpha_1^2 x_1(t) - (\alpha_1^2 C_3 + w^2 C_3) \cos(wt) \tag{29}$$

Comparing with equation 24 one finds that:

$$\alpha_2 = C_3(\alpha_1^2 + w^2) \tag{30}$$

$$\rightarrow C_3 = \frac{\alpha_2}{\alpha_1^2 + w^2} = \frac{m_2 g 2 x_s}{m_1 z_1(\alpha_1^2 + w^2)} \tag{31}$$

By defining as initial conditions $x_1(t = 0) = x_{1,0}$ and $\dot{x}_1(t = 0) = v_{1,0}$ one finds:

$$C_1 = x_{1,0} - C_3 \qquad C_2 = \frac{v_{1,0}}{\alpha_1} \tag{32}$$

One can see that the effect the mass m_2 is to reduce the distance (25) traveled by m_1 compared to the single-mass LIPM model (defined by $C_3 = 0$):

$$x_1(\frac{T}{2}) = x_{1,0} \cosh(\alpha_1 \frac{T}{2}) + \frac{v_{1,0}}{\alpha_1} \sinh(\alpha_1 \frac{T}{2}) - C_3(1 + \cosh(\alpha_1 \frac{T}{2})) \tag{33}$$

Thus, the 3-mass LIPM requires a higher initial velocity to ensure that m_1 reaches x_s at the same time as the swinging foot reaches $2x_s$. By setting $x_1(\frac{T}{2}) = x_s$ in (33) one finds the required initial velocity:

$$v_{1,0} = \alpha_1 \frac{x_s - (x_{1,0} - C_3) \cosh(\alpha_1 \frac{T}{2}) + C_3}{\sinh(\alpha_1 \frac{T}{2})} \tag{34}$$

Note that, even in the 1-mass LIMP model, there is no guarantee that m_1 will move from $-x_s$ to $+x_s$ in the time $T/2$ during which the swinging leg moves from $-2x_s$ to $+2x_s$. Therefore the above constraint between the stride length and the speed always needs to be applied to generate a stable gait.

How to actually control the speed is not a simple question. Sometime, the robot's speed will stabilize around the right value, e.g. due to mechanical feedback generated by the finite size of the feet. In this paper, I will assume that the robot is in a stable regime where it achieves, at the start of each step, the velocity given by equation 34.

Another point to consider is that the straight links in figure 4 are actually made of two joints, knee and hip, and cannot extend beyond their physical limits. The LIPM model assumes that links are infinitely extensible. Therefore, the solution has to be tested for feasibility.

3.2   Lateral Motion: Equation and Solution

During gait, the body must swing laterally to remove the load on the foot that swings forward. As for the forward motion, the swinging foot is assumed to have a height $z_2 = 0$. Its lateral position does not change during the swing that is a motion of m_1 only (Figure 5). Applying equation 19 the relation between motion induced and torque applied becomes:

$$m_1 \ddot{y}_1 z_1 = m_1 g(y_1 - y_{fs}) - m_2 g \cdot 2 y_{fs} \tag{35}$$

this can be rewritten as:

$$\ddot{y}_1 = y_1 \frac{g}{z_1} - y_{fs} \frac{g}{z_1}(1 + \frac{2m_2}{m_1}) = y_1 \alpha_1^2 - \alpha_y 2 \tag{36}$$

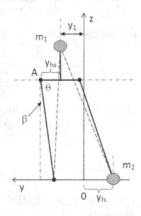

Fig. 5. 3-mass robot model of lateral swing showing masses m_1 and m_2 and the linking virtual poles as doted lines. Mass m_3 is under the contact foot and is ignored. The position of the mass m_1 is relative to the middle point between the feet. The spacing between feet is $2y_{fs}$ and the hip width is $2y_{hs}$.

it solution is:

$$y_1(t) = C_{y1} \cosh(\alpha_1 t) + C_{y2} \sinh(\alpha_1 t) + C_{y3} \tag{37}$$

where:

$$\dot{y}_1(t) = C_{y1}\alpha_1 \sinh(\alpha_1 t) + C_{y2}\alpha_1 \cosh(\alpha_1 t) \tag{38}$$

$$\ddot{y}_1(t) = C_{y1}\alpha_1^2 \cosh(\alpha_1 t) + C_{y2}\alpha_1^2 \sinh(\alpha_1 t) \tag{39}$$

$$= \alpha_1^2[y_1(t) - C_{y3}] \tag{40}$$

and therefore, from (40) and (36):

$$C_{y3} = \frac{\alpha_2}{\alpha_1^2} \tag{41}$$

For the lateral swing, the initial conditions are defined when the robot is centered. This is the configuration where the weight is transferred from one foot to the other. It is the initial configuration in the model of the forward motion. Thus, the initial position is defined as $y_1(0) = 0$ and the position at the end of of a half-cycle as $y_1(\frac{T}{2}) = 0$, leading to:

$$y_1(0) = C_{y1} + C_{y3} = 0 \tag{42}$$

$$\rightarrow C_{y1} = -C_{y3} = -\alpha_{y2} \tag{43}$$

$$y_1(\frac{T}{2}) = C_{y1} \cosh(\alpha_1 \frac{T}{2}) + C_{y2} \sinh(\alpha_1 \frac{T}{2}) + C_{y3} = 0 \tag{44}$$

$$\rightarrow C_{y2} = \frac{-C_{y1} \cosh(\alpha_1 \frac{T}{2}) - C_{y3}}{\sinh(\alpha_1 \frac{T}{2})} \tag{45}$$

The "initial conditions" above correspond to a robot moving laterally at full speed. To initiate the lateral swing, the robot has to lean to one side with both feet on the ground, then lift one foot. The robot then swings back until the foot in the air is placed on the ground again, and the other one is lifted. The maintenance of the oscillation requires some injection of energy, but this is a topic beyond the purpose of this paper.

4 Effects of Gait Parameters on Joint Torques and Velocities

Using methods described in section 2, and applying the solutions of the equations of motion, expressions for the torques and angular velocities of any joint of the hip-leg system can be calculated. The details are not presented here dues to space limitations.

These calculations have then been implemented in a spreadsheet[2] where the parameters of robot geometry and gait can be set. The gait parameters are

---

[2] The spreadsheet is available upon request to the author.

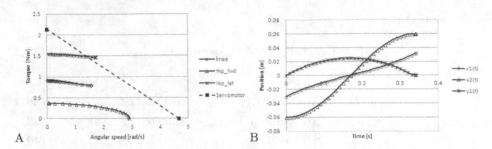

Fig. 6. A. Torque versus angular velocity for three joints and for the MX-28 servomotor. The maximum speed and torque of the servomotor have been reduced by 33% to show the boundary of a 50% controllability margin above requirements. B. Forward displacements of the hip centre x_1 and the swinging foot x_2. Lateral displacement y1 of the hip. The displacements are shown for 1/2 gait cycle. Parameters: walking velocity=18cm/sec, gait frequency=1.5Hz, stride length=3cm.

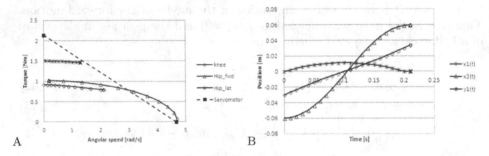

Fig. 7. Same curves as in figure 6 with following parameters: Walking velocity=30cm/sec, gait frequency=2.5Hz, stride length=3cm

the gait frequency and the stride-length. The geometry parameters include the height of the hip that controls the amount of knee bend in a standing position.

The parameters tried are relevant to our competition robot that is using Robotis MX-28 servos [3]. The torque versus velocity for these servos are also drawn on figures 6A and 7A to show when the set gait reaches the controllability limits of the servos.

Figure 7 shows that increasing the gait frequency makes demands on the pitch hip servo beyond its capabilities. On the other hand, a higher frequency reduces the amplitude of the lateral swing y_1 and the requirement of peak hip roll velocity. Through an exploration of gait parameters, the model shows that velocities above 28cm/sec are not achievable with our small-size robot.

5   Conclusion

The 3-mass model is the simplest that allows the representation of the effects of the mass of the swinging foot. A method has been described here to calculate forces and torques in a realistic mechanical structure underlying the idealized 3-mass LIPM. It can be extended to more masses. The tool developed here can now be used to design a faster robot or to optimize servomotors for a given design specification. Future test will show if the three-mass LIPM also constitutes a better model to generate the gait.

References

1. Buczeka, F.L., Cooneya, K.M., Walkera, M.R., Rainbowa, M.J., Conchaa, M.C., Sandersa, J.O.: Performance of an inverted pendulum model directly applied to normal human gait. Clinical Biomechanics **21**(3), 288–296 (2006)
2. Galdeano, D., Chemori, A., Krut, S., Fraisse P.: Optimal Pattern Generator For Dynamic Walking in humanoid Robotics. In: IEEE SSD 2013: 10th International Conference on Systems, Analysis and Automatic Control, vol. 1, pp. 1–6. IEEE, Tunisia (2013)
3. Garton, H., Bugmann, G., Culverhouse, P., Roberts, S., Simpson, C., Santana, A.: Humanoid Robot Gait Generator: Foot Steps Calculation for Trajectory Following. In: Mistry, M., Leonardis, A., Witkowski, M., Melhuish, C. (eds.) TAROS 2014. LNCS, vol. 8717, pp. 251–262. Springer, Heidelberg (2014)
4. Kajita, S., Kanehiro, F., Kaneko, K., Yokoi, K., Hirukawa, H.: The 3D Linear Inverted Pendulum Model: A simple modeling for a biped walking pattern generation. In: Proceedings of the 2001 IROS, Maui, Hawaii, pp. 23–246 (2001)

Limits and Potential of an Experience Metric Space in a Mobile Domestic Robot

Nathan Burke[✉], Joe Saunders, Kerstin Dautenhahn,
and Chrystopher Nehaniv

Adaptive Systems Research Group, University of Hertfordshire,
College Lane, Hatfield AL10 9AB, UK
n.burke@natbur.com, {j.1.Saunders,K.Dautenhahn,C.L.Nehaniv}@herts.ac.uk

Abstract. This paper discusses the concept of an Interaction History Architecture, its use in training a simple task on a mobile domestic robot and some negative results which point to limitations of such an approach. We begin with a brief history its use which motivated the current research. It is based upon Shannon Information Theory and has previously been used in both humanoid and non-humanoid robots. These studies hinted at the ability of using the Interaction History Architecture to classify actions on a broader scale. The experiment outlined is an early test-bed for the use of this on a domestic robot as well as introducing negative rewards to the system. We then present the results from this and discuss and explain some of the difficulties and limitations that were uncovered in this type of approach.

1 Introduction

This paper discusses the adaptation and implementation of the Interaction History Architecture (IHA) on the Sunflower 1-1 robot (Figure 1); a mobile domestic robot built on the Pioneer P3-DX platform. We attempt to use IHA to teach the robot to face an participant. This allowed us to study the first steps of a following task without risking damage to either the robot or the participant. We then provide an analysis of the limits of IHA discovered in this implementation.

Fig. 1. The Sunflower 1-1 Robot (University of Hertfordshire)

© Springer International Publishing Switzerland 2015
C. Dixon and K. Tuyls (Eds.): TAROS 2015, LNAI 9287, pp. 94–99, 2015.
DOI: 10.1007/978-3-319-22416-9_11

2   Motivation

For domestic robots to be successful must capable adaptation and learning. We believe that the history of interactions with their environment provide valuable data from which the robot can learn. In this study we utilize IHA to facilitate learning and evaluate its performance. It has been used to produce desired behaviours from interactions in both humanoid [1] and non-humanoid robots [2].

3   Interaction History Architecture

Beginning with the definition of an interaction history for an embodied agent as

> The temporally extended, dynamically constructed, individual sensorimotor history of an agent situated and acting in its environment, including the social environment, that shapes current and future action [3]

we are able to develop a definition of an experience for an agent. One problem in describing an interaction for an agent is the grounding of the data [4]. By starting with grounded sensorimotor data, and from the agents' perspective, we propose that the experience itself is grounded [3].

IHA has been used to predict the path of a ball [5], play simple turn taking games [3], and proposed as a method to acquire simple language [6]. This flexibility suggests that may be used in a variety of scenarios and robots.

3.1   Information Distance

IHA distinguishes events by using *information distance*, built on Crutchfields' *information metric* and described in terms of bits of Shannon Information[7] between two sources. The sensor data is first categorized into Q bins, which are used to compute the conditional entropy. This reduces computational complexity, at the risk of artificially high entropy at bin boundaries. The measure of uncertainty for value X given Y can be defined as

$$H(X|Y) = \sum_{xX} \sum_{yY} P(x,y) \log_2 P(x|y) \qquad (1)$$

where $P(x,y)$ and $P(x|y)$ are the joint and conditional probability distributions of T most recent values. The information distance between two sources is then

$$d(X,Y) = H(X|Y) + H(Y|X) \qquad (2)$$

a method developed by Mirza[2]. This ignores the time-series of the input stream which can often be an advantage, as when performing mirroring. However, this can become problematic when mirrored actions are expected to be distinct. We cover the impact this had on the study in Section 6 of this paper.

3.2 Metric Space of Experience

We define an experience as sensorimotor information X for a time horizon h. Thus, to find the distance between two experiences, we need to calculate the distance between their sensorimotor information. Defining $X_{t_0,h}$ to be the sensorimotor variable with temporal horizon h starting at time t_0 allows for the experiences of the robot to be defined for temporal horizon h as $E(t, h) = (X_{t,h}^1, \ldots, X_{t,h}^N)$ where N is the number of sensors and $(X_{t,h}^1, \ldots, X_{t,h}^N)$ is the set of all sensorimotor variables available to the robot. Following this, we can define a distance metric for experiences of temporal horizon h as

$$D(E, E') = \sum_{k=1}^{N} d(X_{t,h}^k, X_{t',h}^k) \tag{3}$$

where $E = E(t, h)$ and $E(t', h)$ are two experiences of the robot and d is the information distance [2]. The average of this over the sensors is the final measure used. This was shown to be effective in distinguishing between experiences in [2].

3.3 Action Selection

Each experience in records the action performed and its associated reward. IHA then chooses the next action based on similar past experiences, or at random if no relevant experience is found. To assist in escaping 'local minima' there is an additional chance of choosing a random action, which decreases over time and a limit to the number of times that the same action may repeat. When selecting from similar past experiences, the reward and distance are used as the fitness value in a weighted roulette wheel selection. The action from the selected experience is then performed.

3.4 Rewards

There are three primary reward systems or *motivations M* that contribute to the reward. Each motivation m has a reward m_R and a weight constant m_w. Using human teaching methods has been successfully used in the past with virtual agents [8] and the first motivation tries this with a three button software *clicker interface* that can be used to increase, decrease, and zero its contribution. The *self-preservation* motivation aims to prevent the robot from harm and supplies values in the range $[-1, 0]$ linearly decreasing as the robot moves closer than $0.5m$ to an obstacle. The *closeness* motivation is broken into two parts each with a range of $[0, 1]$. The first part increases as the robot approaches within $1m$ of the participant, reaching $R_{MAX} = 1$ at $0.5m$. The second is based on the direction the robot is facing, with $R_{MAX} = 1$ if the robot is facing directly towards the participant. The total reward value is give by

$$R_{all} = \sum_{m \in M} m_w m_R \tag{4}$$

allowing for total rewards of greater than 1 and less than -1. At each time step the reward for the previous experience is the set to $max(R_{current}, R_{previous})$. Doing so helps the robot discover sequences of actions, as the first action would end up with the maximum reward value that was earned during its execution.

4 Methods

Sunflower is capable of a number of atomic actions A that it can perform. These include basic movements of the base, opening/closing its tray, looking at a specific point and settings its led panel colour. Additionally included is an 'idle' action that does nothing.

The environment consists of typical lounge furnishing in an open floor plan house. The robot utilises IHA to determine the actions to perform, only being interrupted if its reward values fall below -1. The clicker interface is used to reward the robot when it performed a desired action. The experiment was repeated 5 times, with the robots experience space being cleared before each completed run. For this experiment, a bin value Q of 10 was chosen, along with a horizon h of 1 second.

In this experiment $A = [turnleft, turnright, idle]$, a safe subset of all actions. This had the added benefit of reducing the search space during the initial phase of learning. The desired outcome was for the robot to turn to face the researcher, located at a fixed point, then perform 'idle' actions.

5 Results and Conclusions

At the beginning of each iteration, Sunflower performed random movements as expected. The anticipated steady decrease of this behaviour and an associated increase in desired actions did not occur and after a period of ten minutes the iteration was reset and allowed to run again. This was done for five iterations, each with the same result.

A number of incompatibilities between the information distance algorithm and the sensor stream were found which prevented IHA from properly distinguishing experiences. The distance between sensors with little entropy, regardless of explicit value is always zero. The order in which the sensor data is received within the horizon h also does not change the information distance. The impact of these effects were more prominent than was seen in previous experiments[2]. While useful in many scenarios, such as distinguishing walking from turning, the indifference to the time-series of the sensor data when combined with the low-entropy environment caused the information distance calculation for nearly all experiences to be near zero.

5.1 The Importance of Entropy

In IHA, each of the sensors must have entropy in order for the distance to have meaning. As Sunflower was not navigating, much of the sensor data did not vary. In all cases when 'idle' was chosen, the calculated distance 0, as the sensor values were constant over the time horizon. This meant the distance calculation was primarily based on the single computed sensor. For 'noisy' inputs like vision and range sensors this would work but it failed for the computed sensor.With the robot and participant in static locations, only the computed angle changed. While this did have entropy, once normalized and binned bins, this was often lost. The effect of this is that the chosen past experience tended to 22, and the given action 'Turn Left'. This occurred even in experience 25, where the closest other, in terms of raw sensor values, is experience 24 (see Table 1). This action continued until the repeat limit was reached and a random action was chosen.

Table 1. Distances between selected experiences. Distance is a unit-less measure.

–	22	23	24	25	26
22	–	0.587	0.209	0.357	0.056
23	0.587	–	0.414	0.230	0.531
24	0.209	0.414	–	0.185	0.153
25	0.357	0.230	0.185	–	0.302
26	0.056	0.531	0.153	0.302	–

5.2   Time-Series Within an Experience

In [5] IHA had difficulty distinguishing the direction of motion of a ball when its position was halfway through its circular path. During the trials, IHA occasionally predicted the past trajectory as the future path. This effect was far more pronounced during this experiment likely due to lower entropy in the computed sensor.

A further experiment done by Shen, utilized this effect in order to evaluate mirroring between a human and a robot partner. In [9], the robot performed a waving action, IHA was used to evaluate how accurately the human participant was mirroring the action. It was considered to be successful if the action was perfectly synchronous or asynchronous.

6   Discussion and Conclusion

Previous studies have shown the flexibility and potential of IHA. This study has served to highlight an instance in which the information distance metric may not be suited to the given task. Incompatibilities of information distance and the sensors must be addressed prior to IHA being a viable method of comparing experiences in a low-entropy environment or when the order of sensor data is important.

The study presented in this paper illustrates some negative aspects of using classification mechanisms on sensor streams that exhibit low entropy. It is clear that care should be taken when selecting distance metrics appropriate to the application. In this work we focussed on a simple mobile robot task and applied the IHA mechanism to classify actions based on experience histories and using Crutchfield's information distance [10]. Experiences in this framework, based on sensor streams, must have sufficient entropy to allow correct classification. We hope that by describing these negative aspects others can avoid similar paths.

Acknowledgments. The work described in this paper was partially conducted within the EU-FP7 project ACCOMPANY ("Acceptable robotiCs COMPanions for AgeiNg Years") funded by the European Commission under contract number FP7-287624.

References

1. Broz, F., Nehaniv, C.L., Kose-Bagci, H., Dautenhahn, K.: Interaction histories and short term memory: Enactive development of turn-taking behaviors in a childlike humanoid robot (2012). arXiv preprint arXiv:1202.5600
2. Mirza, N.A., Nehaniv, C.L., Dautenhahn, K., te Boekhorst, R.: Interaction histories: From experience to action and back again. In: Proceedings of the 5th IEEE International Conference on Development and Learning (ICDL 2006), Citeseer (2006). ISBN 0-9786456-0-X
3. Mirza, N.A., Nehaniv, C.L., Dautenhahn, K., te Boekhorst, R.: Developing social action capabilities in a humanoid robot using an interaction history architecture. In: 8th IEEE-RAS International Conference on Humanoid Robots, Humanoids 2008, pp. 609–616. IEEE (2008)
4. Harnad, S.: The symbol grounding problem. Physica D: Nonlinear Phenomena **42**(1), 335–346 (1990)
5. Mirza, N.A., Nehaniv, C.L., Dautenhahn, K., te Boekhorst, R.: Anticipating future experience using grounded sensorimotor informational relationships. In: Proceedings of the Eleventh International Conference on the Simulation and Synthesis of Living Systems. MIT Press (2008)
6. Nehaniv, C.L., Forster, F., Saunders, J., Broz, F., Antonova, E., Kose, H., Lyon, C., Lehmann, H., Sato, Y., Dautenhahn, K.: Interaction and experience in enactive intelligence and humanoid robotics. In: IEEE Symposium on Artificial Life (ALIFE), pp. 148–155. IEEE (2013)
7. Shannon, C.E.: A mathematical theory of communication. ACM SIGMOBILE Mobile Computing and Communications Review **5**(1), 3–55 (2001)
8. Thomaz, A.L., Breazeal, C.: Teachable Characters: User Studies, Design Principles, and Learning Performance. In: Gratch, J., Young, M., Aylett, R.S., Ballin, D., Olivier, P. (eds.) IVA 2006. LNCS (LNAI), vol. 4133, pp. 395–406. Springer, Heidelberg (2006)
9. Shen, Q., Saunders, J., Kose-Bagci, H., Dautenhahn, K.: Acting and Interacting Like Me? A Method for Identifying Similarity and Synchronous Behavior between a Human and a Robot. Poster Presentation at IEEE IROS Workshop on "From motor to interaction learning in robots", Nice, France, September 26 (2008)
10. Crutchfield, J.P.: Information and its metric. In: Nonlinear Structures in Physical Systems, pp. 119–130. Springer (1990)

A Novel Approach to Environment Mapping Using Sonar Sensors and Inverse Problems

Eduardo Tondin Ferreira Dias[✉] and Hugo Vieira Neto

Graduate Program in Electrical and Computer Engineering,
Federal University of Technology – Paraná, Curitiba, Brazil
edu.tondin@gmail.com, hvieir@utfpr.edu.br
http://www.cpgei.ct.utfpr.edu.br

Abstract. The traditional approach for environment mapping using sonar sensors in autonomous mobile robots is generally based on time-of-flight, which results in a sparse representation that unfortunately does not make use of the much richer interaction of ultrasonic waves with the obstacles within the environment. In this work, inspiration is taken from techniques used in ultrasound medical imaging, aiming at higher spatial resolution reconstruction of a cross-section of the environment without the need of sweeping it with multiple ultrasonic bursts. A couple of sonar sensors provide raw analogue data that is used to feed an inverse model of the acquisition system and generate an image reconstruction of what can be interpreted as a top view of the environment. Preliminary experiments in a small controlled environment show promising reconstruction results for inverse problem approaches when compared to beamforming techniques normally used in ultrasound medical imaging.

Keywords: Image reconstruction · Sonar sensing · Environment mapping

1 Introduction

Environment mapping is a fundamental task for autonomous mobile robots, which normally use sonar sensors, laser scanners or cameras for obstacle detection and autonomous navigation [1]. Each sensor technology has its own restrictions regarding processing speed, spatial resolution and ways to deal with occlusions – one of the main goals in mobile robotics is to develop methods that enhance mapping results while minimising practical problems.

Sonar sensing is a well-known, widespread method for environment mapping. Sonar sensors are cost-effective and relatively easy to handle when compared to laser scanners or cameras, and several techniques using sonar sensing have been investigated over the years in order to provide robust environmental models. The conventional approach using ultrasound systems to detect objects employs the concept of time-of-flight [2], in which the distances between sonar and obstacles are computed to build the map, providing a sparse representation of the environment and requiring multiple scans in order to achieve better resolution [3].

C. Dixon and K. Tuyls (Eds.): TAROS 2015, LNAI 9287, pp. 100–111, 2015.
DOI: 10.1007/978-3-319-22416-9_12

Time-of-flight techniques have difficulties to deal with the detection of reflections, which commonly hinders the map building, but are still commonly used in probabilistic approaches, such as occupancy grid maps [4]. Other mapping building strategies use geometric primitives, such as edges, planes and corners obtained from time-of-flight sonar data [5].

However, ultrasound systems that use raw analogue data are based on the premise that multiple reflections and wave interferences provide relevant information about the environment – this alternative approach has been used for image reconstruction in the medical field [6] for many years. In the context of ultrasound medical imaging, the image reconstruction algorithms traditionally implemented are based on beamforming techniques [7]. In recent years, the concept of inverse problems has also been used with the purpose of improving the resolution of image reconstruction [8].

In this work, a preliminary study for environment mapping using sonar sensors and image reconstruction methods based in medical imaging technologies is proposed. Differently from time-of-flight approaches, the objective is to build a map of the environment using the concept of inverse problems and raw analogue sonar data. The expected outcome is a reconstructed cross-section image of the environment, which can be interpreted as a top view of the environment. In order to validate this approach, experiments were conducted in a small controlled environment using several inverse problem algorithms available in the literature and the traditional beamforming algorithm as baseline.

The remainder of this paper is structured as follows. Section 2 presents an introduction of image reconstruction algorithms. Then, the experimental setup is presented and discussed in Section 3. Experimental results are presented and discussed in Section 4. Conclusions and future work are finally outlined in Section 5.

2   Image Reconstruction

2.1   Beamforming

Ultrasound medical imaging is traditionally performed using beamforming techniques. These techniques use the principle of delay and sum [7], through which ultrasonic reflections from the Region Of Interest (ROI) are processed for image reconstruction. Beamforming implies fast processing and is easy to implement, but results in low resolution of the reconstructed image.

In this work, the delay and sum model described in [9] was implemented in order to be used as a baseline for comparisons. In this model, the signal from each ultrasonic transducer is delayed in such a way that all signals are temporally aligned, weighted and added. Signal weighting is performed to control the spatial response and focusing of the ultrasonic beam [10].

The image reconstruction maps the output beam according to each specific position within the region of interest, which constitutes a pixel in the reconstructed image.

2.2   Inverse Problems

Ultrasound systems can be modeled mathematically as a well-posed problem:

$$g = Hf + \eta, \tag{1}$$

where g is a vector containing sampled data from ultrasonic reflection signals, H is the model matrix, f is the desired reconstructed image and η is a vector containing system perturbations (e.g. noise and modeling errors).

The matrix H is modeled with ultrasonic reflection signals acquired for small individual objects in each position within the ROI, which are appended to each column of the matrix. Ultrasonic reflection signals are concatenated if they are acquired in the same grid position, composing what is known as the point spread function. This model provides the ultrasonic reflection signal behaviour for each mapped position, theoretically allowing image reconstruction of any arbitrary unknown object within the ROI. The mathematical modeling of ultrasonic reflection signals is described in [11] and the process of construction of H is described in [12].

The initial approach for image reconstruction ($\widehat{f}$) from the ultrasonic reflection signals (g) does not considers perturbations and consists in simply inverting the matrix H, as described below:

$$\widehat{f} = H^{-1}g. \tag{2}$$

Nevertheless, ultrasound systems constitute ill-posed problems, i.e. they do not satisfy the three Hadamard conditions: (i) a solution exists, (ii) the solution is unique, and (iii) the solution depends continuously on the data. In order to solve ill-posed problems, several methods are available in the literature, such as inverse problem and regularization techniques [8].

The presence of perturbations (η) does not allow the simple inversion of H. In other words, there is no g that generates $\widehat{f}$ in Eq. 2. To solve this problem, Least Squares [13] is the method typically used:

$$\widehat{f} = \arg\min_{f} ||g - Hf||_2^2. \tag{3}$$

The Least Squares equation (3) can be solved using the gradient method and results in:

$$\widehat{f} = (H^T H)^{-1} H^T g, \tag{4}$$

where H^T represents the transpose of H and $(H^T H)^{-1} H^T$ is known as the Moore-Penrose pseudo-inverse [14].

However, the model matrix H for ultrasound systems is ill-conditioned and becomes unstable in the presence of noise, resulting in reconstructed images with amplified noise. The solution using the transpose results in reconstructed images similar to the ones obtained with beamforming (delay and sum).

Regularisation is commonly used to stabilise the system, addressing the problem of ill-conditioned matrix modeling and therefore minimising noise amplification [15]. When applying regularisation to Eq. 3, image reconstruction can be done by using [16]:

$$\widehat{f} = \arg\min_{f} ||g - Hf||_2^2 + \lambda \Re(f), \tag{5}$$

where $\Re(\cdot)$ is the regularisation term and λ is the regularisation parameter.

Several methods use the regularisation approach, such as Tikhonov and Total Variation [17]. Iterative algorithms are used to reduce the computational cost and increase the speed of reconstruction. One of the most used algorithms is the Conjugate Gradient. A simple formulation follows [18]:

$$f_{k+1} = f_k + \alpha_k p_k, \tag{6}$$

where f is a solution (reconstructed image), α is a step size coefficient, p is the direction and k is the iteration. The α coefficient is given by:

$$\alpha_k = \frac{p_k^T r_{k-1}}{p_k^T A p_k}, \tag{7}$$

where $A = H^T H$ and r_k is a residual computed as $r_k = H^T g - A f_k$. The direction p is computed as $r_{k+1} = H^T g - A f_k$:

$$p_{k+1} = r_{k+1} + \frac{r_{k+1}^T r_{k+1}}{r_k^T r_k} p_k. \tag{8}$$

Another iterative method is the Fast Iterative Shrinkage-Thresholding Algorithm (FISTA) [19], which is a more efficient version of the Iterative Shrinkage-Thresholding Algorithm (ISTA) [20]. The ISTA algorithm iteration step is:

$$f_{k+1} = S_{\frac{\lambda}{c}} \left(\frac{1}{c} H^T (g - Hf_k) + f_k \right), \tag{9}$$

where S is a shrinkage function and c is the step size parameter. In the FISTA algorithm [19], instead of applying Eq. 9 on f_k, it is applied on a modified vector given by [20]:

$$\widehat{f}_{k+1} = f_k + \frac{t_k - 1}{t_{k+1}} (f_k - f_{k-1}), \tag{10}$$

where $t_k = \left(1 + \sqrt{1 + 4t_{k-1}^2}\right)/2$ with $t_1 = 1$ and $f_0 = 0$.

Other currently investigated approach to improve the resolution of image reconstruction regards changing the modeling of matrix H by assuming complex-valued point spread functions, as opposed to typical models, which are implemented using real-valued point spread functions [21]. In this work, both representation models for ultrasonic reflections were assessed: real-valued and complex-valued. The use of complex-valued models is based on the premise that signal reflections present phase changes and hence are best represented by complex values. For each representation model, two non-iterative (Transpose and Pseudo-inverse) and two iterative (Conjugate Gradient and FISTA) algorithms were investigated, as described in Section 3.

3   Experimental Setup

3.1   Data Collection

A controlled environment was designed to allow experimentation with different reconstruction algorithms and comparisons between their results. This real (physical) environment was composed of a ROI containing 19×19 possible positions for small obstacles. The distance between each grid position was 2.5cm, constituting a square ROI with side length of 45cm, positioned at a distance of 30cm from the sonar sensors. Each position could be filled or not with obstacles, allowing different environmental configurations. Three sonars were positioned in front of the environment, separated from each other by 5cm. Fig. 1 shows a schematic top view of the experimental setup.

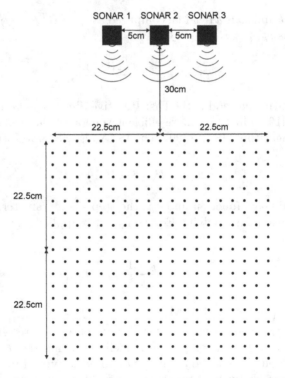

Fig. 1. Experimental setup: three sonars were positioned equidistant from each other by 5cm and at a distance of 30cm from the region of interest, which is composed of an array of 19×19 positions for obstacles, comprising an area of 45×45cm for mapping

The hardware that was used consisted of an array of three Pro-Wave SRM400 sonar ranging modules[1] and a data acquisition board. The choice for these sonar ranging modules was due to the possibility of acquiring raw data, which is fundamental for the development of this research. The SRM400 modules work by sending ultrasonic bursts of 40kHz and detecting the signal of reflected echoes. A Texas Instruments Stellaris LM3S8962 Evaluation Board[2] was used to control the sonar ranging modules and acquire the raw analogue signals from them. Each signal is composed of 776 samples, acquired at a sampling rate of 200kHz.

Acquisition of ultrasonic reflection signals was performed by positioning a cylindrical obstacle with 2cm diameter and 20cm height at every position of the grid within the ROI. Each sonar was individually triggered and used for receiving the ultrasonic reflection signal at each obstacle position, generating a database with $3 \times 19 \times 19 = 1083$ ultrasonic reflection signals. This procedure was performed twice, in order to generate two databases, one that was used for system modeling and another one that was used for validation – in total, 2166 ultrasonic reflection signals were acquired. A single obstacle was present in the environment in each acquisition.

The acquired data was then transferred to a desktop computer for processing. The model matrix H was constructed as follows: (i) the three acquired ultrasonic reflection signal vectors corresponding to each object were concatenated into single column vectors of $3 \times 776 = 2328$ samples; (ii) the column vectors of the $19 \times 19 = 361$ possible object positions in the grid were appended side by side, resulting in a model matrix H of dimensions 2328×361.

4 Experimental Results

4.1 Image Reconstruction

Experiments were performed to assess the ability of each reconstruction method to detect objects in their actual position. The Transpose, Pseudo-inverse, Conjugate Gradient and FISTA algorithms were implemented both using real-valued and complex-valued models. Also, the delay and sum Beamforming algorithm was implemented to serve as a baseline for comparisons. In order to facilitate presentation of results, algorithm names were compressed as follows:

- BMF corresponding to Delay and Sum Beamforming
- TPR corresponding to Transpose using real values
- IVR corresponding to Pseudo-inverse using real values
- CGR corresponding to Conjugate Gradient using real values
- FTR corresponding to FISTA using real values
- TPC corresponding to Transpose using complex values
- IVC corresponding to Pseudo-inverse using complex values

---

[1] http://www.prowave.com.tw/english/products/sr/srm400.htm
[2] http://www.ti.com/lit/ug/spmu032b/spmu032b.pdf

- CGC corresponding to Conjugate Gradient using complex values
- FTC corresponding to FISTA using complex values.

The model matrix H was constructed using data from the first acquired database, whereas the data from the second acquired database was used for image reconstruction validation. In order to double the number of sample results, an additional round of experiments was conducted by reversing the two databases, i.e. the second acquired database was used for the construction of the model matrix H and the first acquired database was used for image reconstruction validation.

A qualitative assessment was conducted to verify if the obstacle present in the environment was detected in its actual position. Image reconstructions were performed for each position of the ROI grid. Fig. 2 shows the image reconstructions of an obstacle positioned exactly in the middle of the region of interest using each of the aforementioned algorithms.

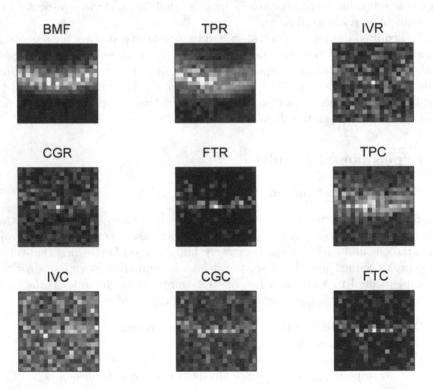

Fig. 2. Image reconstruction: sample reconstructed images for an obstacle presents in the middle of the region of interest using all the algorithms previously discussed. The best results were yielded by iterative algorithms — CGR, FTR, CGC and FTC – and also the non-iterative IVR algorithm.

As can be noticed in Fig. 2, the techniques using the inverse problem formulation provide better reconstruction of the obstacle's original position when compared to delay and sum beamforming. While reconstructions using Beamforming and the Transpose matrix tend to generate distributed artifacts surrounding the actual object to be detected, the iterative Conjugate Gradient and FISTA algorithms, using either real or complex-valued models, tend to reconstruct a single object at the centre of the image, as expected.

In order to achieve more consistent object detection, the resulting reconstructed images of the environment were thresholded at 90% of the maximum pixel intensity. Fig. 3 shows the corresponding results after thresholding for each of the reconstructed images previously shown in Fig. 2. The best results in this case are the ones which identify a single object (white pixel) at the centre of the reconstructed image – the iterative Conjugate Gradient and FISTA algorithms, using either real or complex-valued models, are among the ones that yield the most promising results.

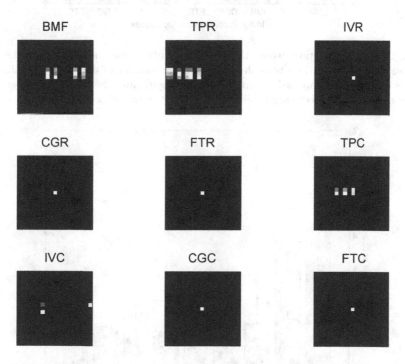

Fig. 3. Detected obstacles: reconstructed images thresholded at 90% of maximum pixel intensity. The best results identify a single object (white pixel) at the centre of the reconstructed image, reinforcing the iterative algorithms CGR, FTR, CGC and FTC and the non-iterative IVR algorithm as the most consistent reconstruction methods.

A first quantitative assessment was also conducted to compare the performance of each reconstruction algorithm concerning the number of detected objects. As there was a single object actually present in different locations of

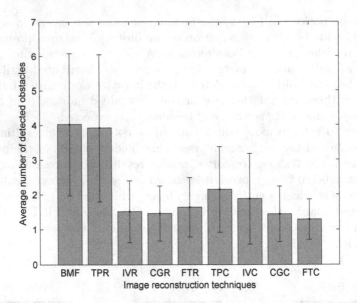

Fig. 4. Obstacle detection using the second database for modeling and the first database for reconstruction: bars show the average number of detected obstacles and the lines indicate the standard deviation for each algorithm. The best results regarding number of detected objects are achieved by IVx, CGx and FTx algorithms.

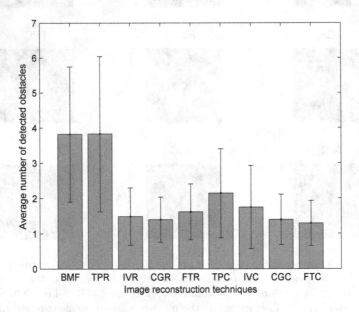

Fig. 5. Obstacle detection using the first database for modeling and the second database for reconstruction: bars show the average number of detected obstacles and the lines indicate the standard deviation for each algorithm. Results are consistent with the ones presented in Fig. 4.

the environment at each moment, the best results are the ones in which a single object is detected. Figs. 4 and 5 show the average number of detected objects for each reconstruction algorithm and each of the two acquired databases, respectively. It can be noticed in Figs. 4 and 5 that the quantitative results are consistent for both databases.

According to Figs. 4 and 5, the iterative algorithms (Conjugate Gradient and FISTA) presented average number of detected objects close to one – this effect was expected, since these methods make use of regularisation, resulting in overall better resolution. Complex-valued models did not show significant improvements over real-valued models.

A second quantitative assessment was computed using the Euclidean distance between the actual obstacle position and the detected obstacle position in the reconstructed image. The closest the results are from zero, the better the accuracy of the algorithm in detecting objects. Fig. 6 shows a boxplot of the distribution of resulting distances for each implemented reconstruction algorithm.

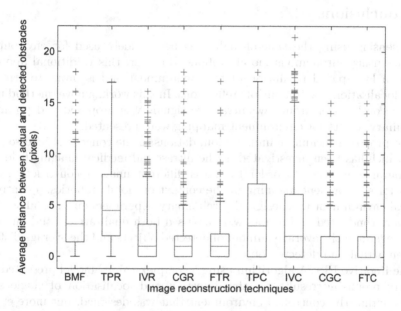

Fig. 6. Boxplot of the distribution of resulting distances (in pixels) between actual and detected objects. A non-parametric Kruskal-Wallis test revealed statistically significant differences between iterative complex-valued algorithms (CGC and FTC) and the other algorithms – the performance yielded by CGC and FTC algorithms regarding distances between actual and detected objects can be considered the best.

The results shown in Fig. 6 were subject to the non-parametric Kruskal-Wallis test, which revealed statistically significant differences between iterative complex-valued algorithms (CGC and FTC) and the other algorithms. Regarding distances between actual and detected objects, the best performance was yielded by CGC and FTC algorithms.

In order to assess the overall quality of image reconstruction achieved using different methods, it is necessary to take into account both the average number of detected objects and the average distance between actual and detected objects. The delay and sum Beamforming algorithm, which is traditionally used in medical imaging, yielded the worst results – its performance was very similar to the performance obtained with the Transpose matrix using the real-valued model, as one could expect. On the other hand, the iterative Conjugate Gradient and FISTA algorithms using the complex-valued model can be considered the best reconstruction algorithms investigated so far.

Optimised implementations of the reconstruction algorithms were out of the scope of this work and therefore a fair and detailed discussion about their processing time can not be made at this time. However, as a general reference, the average processing time for Matlab implementations running in an Intel Core i7 desktop computer with 8GB RAM at 2.93GHz was in the order of 0.01s to 0.1s.

5   Conclusions

Sonar sensing using the time-of-flight has been widely used for environment mapping using autonomous mobile robots. However, this traditional approach results in low spatial resolution of the environment and is prone to errors in object localisation due to multiple reflections. In this work, a novel method using image reconstruction from raw ultrasonic signals was proposed and promising preliminary results for environment mapping were presented.

The proposed technique finds its foundations in the concept of inverse problems, which has been investigated in the ultrasound medical imaging field with meaningful results [8]. In order to assess different image reconstruction algorithms for environment mapping in the context of mobile robotics, a controlled physical environment was designed. Preliminary experiments using real acquired data were conducted and results were assessed both qualitatively and quantitatively, regarding the average number of detected objects and the average distance between actual and detected objects.

The iterative FISTA algorithm using a complex-valued model presented particularly promising results for the detection and localisation of single small objects within the controlled environment that was designed, but more experiments using more complex scenarios still need to be conducted. In future work, experiments will concern multiple objects in more than a single location within the region of interest, as well as the behaviour of the investigated algorithms for obstacles positioned off the modeled grid.

Acknowledgments. This work was partially supported by the Coordination for the Improvement of Higher Education Personnel (CAPES Foundation, Brazil), whose support is gratefully acknowledged. The authors also express their gratitude to the Image Processing and Reconstruction Group at UTFPR, led by Dr. Marcelo V.W. Zibetti, for the many fruitful discussions and help throughout the development of this work.

References

1. Siegwart, R., Nourbakhsh, I.R., Scaramuzza, D.: Introduction to Autonomous Mobile Robots. MIT Press, Massachusetts (2011)
2. Everett, H.R.: Sensors for Mobile Robots: Theory and Application. A.K. Peters, Wellesley (1995)
3. Bank, D., Kämpke, T.: High-resolution ultrasonic environment Imaging. IEEE Trans. on Robotics, 370–381 (2007)
4. Elfes, A., Moravec, H.: High Resolution maps from wide angle sonar. In: Proc. of the IEEE Int. Conf. on Robotics and Automotation, pp. 116–121 (1985)
5. Jimenez, J.A., Urena, J., Mazo, M., Hernandez, A., Santiso, E.: Three-dimensional discrimination between planes, corners and edges using ultrasonic sensors. In: Proc. of Emerging Technologies and Factory Automation, pp. 692–699 (2003)
6. Murino, V., Trucco, A.: Three-dimensional image generation and processing in underwater acoustic vision. In: Proceedings of the IEEE, pp. 1903–1948 (2000)
7. Stergiopoulos, S.: Advanced Signal Processing Handbook: Theory and Implementation for Radar, Sonar, and Medical Imaging Real Time Systems. CRC Press, Florida (2000)
8. Lavarello, R., Kamalabadi, F., O'Brien, W.D.: A regularized inverse approach to ultrasonic pulse-echo imaging. IEEE Trans. on Medical Imaging, 712–722 (2006)
9. Mucci, R.A.: A comparison of efficient beamforming algorithms. IEEE Trans. on Acoustics, Speech and Signal Processing, 548–558 (1984)
10. Synnevag, J.F., Austeng, A., Holm, S.: Adaptive Beamforming Applied to Medical Ultrasound Imaging. IEEE Trans. on Ultrasonics, Ferroelectrics and Frequency Control, pp. 1606–1613 (2007)
11. Jensen, J.A.: A model for the propagation and scattering of ultrasound in tissue. The Journal of the Acoustical Society of America, 182–190 (1991)
12. Viola, F., Ellis, M.A., Walker, W.F.: Time-domain optimized near-field estimator for ultrasound imaging: Initial development and results. IEEE Trans. on Medical Imaging, 99–110 (2008)
13. Björck, A.: Numerical methods for least squares problems. SIAM, Philadelphia (1996)
14. Bertero, M.: Linear inverse and ill-posed problems. Advances in Electronics and Electron Physics, pp. 1–120. Academic Press (1989)
15. Bovik, Al: The Essential Guide to Image Processing. Elsevier, California (2009)
16. Lustig, M., Donoho, D., Pauly, J.M.: Sparse MRI: The application of compressed sensing for rapid MR imaging. Magnetic Resonance in Medicine **58**, 1182–1195 (2007)
17. Karl, W.C. Regularization in image restoration and reconstruction. Handbook of Image and Video Processing, pp. 141–160. Academic Press (2000)
18. Shewchuk, J.R.: An introduction to the conjugate gradient method without the agonizing pain. Technical Report. Carnegie Mellon University (1994)
19. Beck, A., Teboulle, M.: A fast iterative shrinkage-thresholding algorithm for linear inverse problems. SIAM Journal on Imaging Sciences **2**, 183–202 (2009)
20. Zibulevsky, M., Elad, M.: L1-L2 Optimization in signal and image processing. IEEE Signal Processing Magazine, 76–88 (2010)
21. Mu, Z., Plemmons, R.J., Herrington, D.M., Santago, P.: Estimation of complex ultrasonic medium responses by deconvolution. In: IEEE Int. Symposium on Biomedical Imaging, pp. 1047–1050 (2002)

Applying Energy Autonomous Robots
for Dike Inspection

Douwe Dresscher[✉], Theo J.A. de Vries, and Stefano Stramigioli

Robotics and Mechatronics Group, Centre for Telematics and Information
Technology, University of Twente, P.O. Box 217, 7500 AE Enschede, The Netherlands
{D.Dresscher,T.J.A.deVries,S.Stramigioli}@utwente.nl

Abstract. This article presents an exploratory study of an energy-autonomous robot that can be deployed on the Dutch dykes. Based on theory in energy harvesting from sun and wind and the energy-cost of locomotion an analytic expression to determine the feasible daily operational time of such a vehicle is composed. The parameters in this expression are identified using lab results and weather statistics. After an evaluation of the "Energy autonomous robot in the Netherlands" case, the results are generalised by looking at the effects of varying the assumptions. Based on this work, three conclusions can be drawn. Firstly, it is realistic to have an energy-autonomous walking dyke robot in the Netherlands. Secondly, the use of solar panels is probably not feasible if the amount of solar energy that is available is much less than assumed in the study. Finally, in this case study, the inclusion of a wind turbine typically offers a slight benefit. Furthermore, it gives a significant benefit in the months where the incident power of the sun is low, thus allowing a reasonable operational time during the winter.

1 Introduction

The Netherlands - the name itself means "low countries" - is a geographically low lying country. More than 60% of the land lies below sea level, including the densely populated "Randstad" region. Over 7,000,000 people live and work in this region, which is a little more than 40% of the Dutch population. Without effective flood defences, the parts that lie below sea level would frequently (if not permanently) be subjected to flooding. This makes it important for the Dutch to keep their flood defences in good shape and up-to-date. A major breach would be a disaster in many ways.

An important part of the flood defence system consists of dykes. Recent (2003, 2004) dyke failures have shown that knowledge about the flood-defence systems is insufficient to always prevent flooding. The goal of the ROSE project is to develop a 'team' of robotic walkers that will function as an autonomous early-warning system by acquiring data about the composition, consistency and condition of dykes.

This work was conducted as part of the ROSE project, which is financed by the Dutch Technology Foundation STW under grant 10550.

© Springer International Publishing Switzerland 2015
C. Dixon and K. Tuyls (Eds.): TAROS 2015, LNAI 9287, pp. 112–123, 2015.
DOI: 10.1007/978-3-319-22416-9_13

Energy autonomy implies that the robot takes care of its own energy supply. Energy sources available to a robot in an outdoor environment may include unrefined biomass, sun and wind. Of these, robotic energy harvesting from unrefined biomass has not yet progressed beyond the laboratory, although the research is promising [4, 7, 14]. By contrast, equipment to harvest energy from sun and wind has been available COTS (Commercially Off The Shelf) for several decades and is continually improving. For this reason, this work focusses on energy autonomous robot that uses solar and/or wind energy.

This article presents an exploratory study of an energy-autonomous robot that can be deployed on the Dutch dykes. We start by giving some background information on energy harvesting from sun and wind and the energy-cost of locomotion. This information is combined in an analytic expression to determine the feasible daily operational time of such a vehicle. The parameters in this expression are identified for a dyke robot in the Netherlands to which the theory will be applied. Next, the results are generalised by looking at the effects of varying the assumptions. The work closes with a discussion of the results and drawing conclusions.

2   Theoretical Background

To evaluate the energy autonomy of a robot, the achievable operational (working) time per day can be used as a measure. The operational time per day can be determined from the energy that can be harvested in a day and the average power consumption of the robot by applying the following reasoning. Since we do not add/remove energy to/from the system in any other way then by harvesting, the maximum energy that the system can consume, E_{cons}, is equal to the energy that the system harvests, E_{harv}.

$$E_{cons} = E_{harv} \qquad (1)$$

where the energy that the system consumes is equal to the product of the average power consumption, $\bar{P}_{total}$, and the operational time per day, T_{op}: $E_{cons} = T_{op}\bar{P}_{total}$ such that $T_{op}\bar{P}_{total} = E_{harv}$ or:

$$T_{op} = \frac{E_{harv}}{\bar{P}_{total}} \qquad (2)$$

In the following two sections, the derivation of the harvested energy per day and average power consumption is discussed.

2.1   Harvested Energy per Day

For this work, two types of energy harvesting are considered: energy harvesting from solar energy and energy harvesting from wind energy. The total amount of harvested energy is equal to the sum of the harvested solar energy, E_s, and harvested wind energy, E_{wt}:

$$E_{harv} = E_s + E_{wt} \qquad (3)$$

From weather statistics, the average effective incident energy per square metre per day, $\bar{E}_{sun}\{Whm^{-2}\}$, can be obtained. Using this information, the average energy generated by the solar panels, $\bar{E}_s\{Wh\}$, can be calculated based on the surface area $S_s\{m^2\}$ and the efficiency $\gamma_s\{\}$ of the solar panel:

$$\bar{E}_s = S_s\gamma_s\bar{E}_{sun} \tag{4}$$

For the calculation of the energy that can be harvested from the wind, the average wind speed over 24 hours, $\bar{v}_{wind}\{ms^{-1}\}$, is available from weather statistics. Using this information, Betz' law [1] can be used to calculate the maximum average power generation under ideal conditions, using a wind turbine, $\bar{P}_w\{W\}$:

$$\bar{P}_w = 0.5\rho_{air}\bar{v}_{wind}^3 S_w C_p \tag{5}$$

where $\rho_{air}\{kgm^{-3}\}$ is the density of the air, $S_w\{m^2\}$ the effective surface area of the wind turbine (sectional area) and $C_p\{\}$ the power coefficient of the wind turbine. Using this, the amount of wind energy that is harvested in a day can be calculated:

$$E_{wt} = 24\bar{P}_w = 12\rho_{air}\bar{v}_{wind}^3 S_w C_p \tag{6}$$

The total energy that is harvested in a day is now equal to:

$$E_{harv} = E_s + E_{wt} = 24S_s\gamma_s\bar{P}_{sun} + 12\rho_{air}\bar{v}_{wind}^3 S_w C_p \tag{7}$$

2.2 Average Power Consumption

The total power consumption, $\bar{P}_{total}$, can be split into the power consumption of the robot's locomotion system ($\bar{P}_l$) and the power consumption from other equipment ($\bar{P}_o$):

$$\bar{P}_{total} = \bar{P}_l + \bar{P}_o \tag{8}$$

The power consumption of the robot's locomotion system can be calculated using a frequently used measure for the power consumption of a locomotion system, namely the specific resistance as first described by [2]. The specific resistance $\epsilon\{\}$ is defined as the ratio of power used for locomotion $P_l\{W\}$ and the product of the weight $m\{kg\}$, earth's gravitational acceleration $g\{ms^{-2}\}$ and the maximum speed $v_{max}\{ms^{-1}\}$, such that:

$$\epsilon = \frac{P_l}{mgv_{max}} \tag{9}$$

This can be rewritten to:

$$P_l = \epsilon mgv_{max} \tag{10}$$

such that, when given a certain specific resistance of a locomotion system for a certain maximum speed and mass, the power consumption can be calculated.

Splitting the total mass, m, into the mass of the locomotion system, m_l; the solar panels, m_s; the wind turbine, m_w; and the mass of other equipment, m_o, results in:

$$P_l = \epsilon(m_l + m_s + m_w + m_o)gv_{max} \tag{11}$$

Let us assume that, when changing the surface area of a wind turbine, the change in depth is negligible. Then, the mass of the wind turbine, m_w, is approximately linearly dependent on the surface area of the wind turbine, S_w:

$$m_w = \eta_w S_w, \tag{12}$$

where η_w represents the planar density of the wind turbine. It is reasonable to assume that such an approximation also exists for the weight of the solar panels:

$$m_s = \eta_s S_s \tag{13}$$

where η_s is the density of the solar panels. Now:

$$P_l = \epsilon(m_l + \eta_s S_s + \eta_w S_w + m_o)gv_{max} \tag{14}$$

such that the total power consumption is equal to:

$$\bar{P}_{total} = \epsilon(m_l + \eta_s S_s + \eta_w S_w + m_o)gv_{max} + \bar{P}_o \tag{15}$$

2.3   An Analytic Expression for the Operational Time per Day

Combining equations 2, 7 and 15 results in:

$$T_{op} = \frac{S_s \gamma_s \bar{E}_{sun} + 12\rho_{air}\bar{v}_{wind}^3 S_w C_p}{\epsilon(m_l + \eta_s S_s + \eta_w S_w + m_o)gv_{max} + p_o} \tag{16}$$

which enables the relation between solar panel and/or wind turbine surface area and the operational time to be studied.

3   Parameter Identification

Using equation 16, the operational time per day can be evaluated for an area of solar panel and wind turbine surface. The next step is to determine the other parameters used in equation 16 based on the case of a dyke inspection robot in the Netherlands. In this section, weather statistics and experimental results that have been achieved for both energy harvesting and locomotion are used to give a value to the parameters. Equipment other than the locomotion system and harvesting equipment is not considered; this implies $P_o = 0$ and $m_o = 0$.

3.1   Solar Energy Harvesting

As mentioned earlier, the study concerns a robot that will be deployed in the Netherlands and the average incident solar energy is used to obtain a realistic value for $\bar{P}_{sun}$. Fig. 1a shows the average solar energy (Whm^{-2}) that is incident per day, for each month of the year, in blue. Fig. 1b shows the average solar energy (Whm^{-2}) that is incident per day, for several locations in the Netherlands, in blue.

116   D. Dresscher et al.

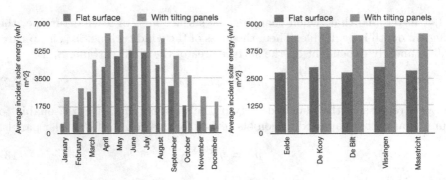

(a) Average incident solar energy per month in the Netherlands

(b) Average incident solar energy for several locations in the Netherlands

Fig. 1. Average incident solar energy in the Netherlands. The figures are shown for a flat surface (source: [9–12]) and an optimally tilted surface.

These numbers represent the incident energy on a flat surface and may be further increased by tilting the panels such that the surface is normal to the sun's rays (for details on the calculations, please refer to [17]). By applying this correction, we obtain incident energy as shown in green. For the case study, the average of these values is used, which is equal to 4334 Whm^{-2}. When the results are generalised later in this paper, monthly and geographical variations are evaluated.

The maximum theoretically achievable efficiency of solar cells is defined by the thermodynamic limit and is equal to 86% [13]. However, current levels are at 37.5% for InGaP/GaAs/InGaAs cells in a lab environment and 28.5% for GaAs (thin film) cells in commercially available modules [3]. Commercially available GaAs (thin film) modules can have a efficiency of 23.5%; this is 82% lower than the cell efficiency. The planar density of current state-of-the-art silicium panels is around $2.2 kgm^{-2}$ [16].

In this study, we look at the opportunities offered by currently commercially available modules ($\gamma_s = 0.235$ and $\eta_s = 2.2$).

3.2   Wind Energy Harvesting

To obtain the average wind speed, average measurements are used as well. Figure 2a shows the average wind speed (ms^{-1}) in the Netherlands, for each month of the year. Fig. 2b shows the average wind speed for various locations in the Netherlands.

For the study, the average of these values is used, which is equal to 4.78 ms^{-1}. When the results are generalised later in this paper, monthly and geographical variations are evaluated. The density of air, ρ_{air}, at $10°C$ is $1.25 kgm^{-3}$.

The maximum theoretically achievable efficiency at which an idealised model of a wind turbine can convert the kinetic energy of wind to useful power is defined

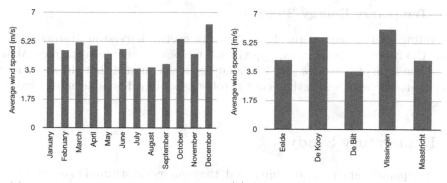

(a) Average wind speed in the Netherlands, for each month of the year

(b) Average wind speed in the Netherlands, for each month of the year

Fig. 2. verage wind speed in the Netherlands (source: [9–12])

by Betz's coefficient and is equal to 0.59 [6]. Currently, the measured power coefficient is in the range 0.4-0.5 [8]. The problem is that currently available wind turbines are large and are not designed for mobile applications. Therefore, it is difficult to say if this number is realistic for a smaller model. However, for this study, we assume that it is realistic to have such a power coefficient for a smaller model. We use $C_p = 0.45$ and look at the implications of applying this assumption when the results are generalised later in this paper.

Currently available wind turbines are not designed for mobile applications and are therefore not optimised for low mass. For mobile applications, it is desirable to have a light turbine with a planar density of (say) 10kg per square metre of surface area. It is assumed that such a wind turbine is available ($\eta_w = 10kg$), and the implications of this assumption will be discussed when the results are generalised later in this paper.

3.3 Locomotion Energy Consumption

Many legged robots have been developed over the years. To identify a realistic set of parameters for the locomotion system, a state-of-the-art example is used. In [5], the specific resistance of a range of vehicles. Among these is Scout II ([15]) which is a state-of-the-art multi-legged robots.

The locomotion system of the Scout II is, with a specific resistance of 1 at a velocity of $1\frac{m}{s}$, one of the most energy-efficient multi-legged walking systems currently available . Although these are experimental results from a lab environment, it is assumed that it is possible to locomote with this efficiency outside the lab, and therefore, the locomotion system of the Scout II robot is used as an example locomotion system in this case study. Therefore, ϵ is equal to 1, v_{max} is equal to $1ms^{-1}$ and m_l is equal to 25kg (including batteries and electronics)[15]. For the gravitational acceleration, g, a value of $9.81ms^{-2}$ is used.

3.4   Temporary Energy Storage

Depending on the short-term fluctuations in energy harvesting during operations, the robot needs to be equipped with temporary energy-storage capabilities in the form of batteries. For this analysis, it is assumed that the amount of battery storage with which the Scout II robot is equipped is sufficient to achieve this goal.

4   Exploratory Study

Once the parameters have been identified, they can by substituted in equation 16. This results in the following expression:

$$T_{op} = \frac{1018.49S_s + 737.2S_w}{245.25 + 21.58S_s + 98.1S_w}, \tag{17}$$

such that the (yearly average) operational time per day can be expressed as a function of the solar panel and wind turbine surface area. When plotting the yearly average operational time per day as a function of the surface area of the solar panel and/or wind turbine, Fig. 3 is obtained; where the horizontal axis represents the area of solar panels/ wind turnbines and the vertical axis the operational time. For this case study, we assume that the robot can carry up to twice its top surface in solar panels; for the Scout II robot, this is $1.637m^2$ of solar panel, and a wind turbine with a radius of the length of the robot body at most; this results in a maximum surface of $0.55m^2$ for the Scout II robot.

From this, it can be seen that it is always beneficial to include solar panels when applying a wind-turbine (Fig. 3b). In addition, it shows that up to a certain amount of solar panel surface area, it is beneficial to include a wind turbine when applying solar panels (Fig. 3a). This boundary is at $2.15m^2$ of solar panel surface area. If the solar panel surface area is larger than $2.15m^2$, the presence of a wind turbine decreases the operational time. A wind turbine offers an improvement in the feasible operational time per day. However, if the maximum amount of solar panel surface is applied, the addition of a wind turbine offers only a small improvement - from 5.96 to 6.21 hours per day - .

5   Generalisation of the Results

In this study, assumptions are made regarding the parameters in equation 16. Since it is likely that a variation to these assumptions will apply in practise, the effect of variations to these assumptions is discussed in this section.

Fig. 4a shows the effect of varying the assumption about the efficiency of the solar panels.

Fig. 4b shows the effect of varying the assumption about the efficiency of wind turbine. It shows that when the efficiency of the wind turbine drops below 0.36, it is no longer beneficial to have a wind turbine when the maximum amount amount solar panels is applied.

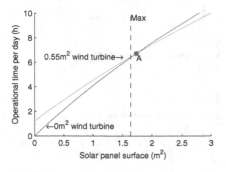

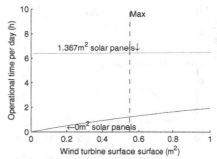

(a) The operational time per day vs the solar panel surface area in the case of no wind-turbine (in blue) and the case of a $0.55m^2$ wind turbine (in green)

(b) The operational time per day vs the wind turbine surface area in the case of no solar panels (in blue) and the case of a $1.637m^2$ wind turbine (in green)

Fig. 3. Case evaluation. Point A in the left figure is the point where there is no change in operational time with the addition of a wind-turbine, the area left of point A is the area where the addition of a wind turbine is beneficial and the area to the right of point A is the area where the addition of a wind turbine is not beneficial. The black, dashed line marks the maximum solar panel surface for this study.

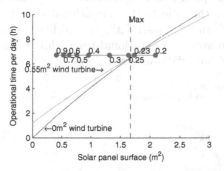

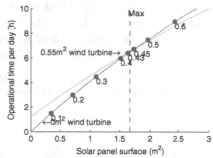

(a) Variations of the solar-panel efficiency γ_s

(b) Variations of the wind-turbine efficiency C_p

Fig. 4. A variation of Fig. 3 with the addition of the red line that shows how point A moves with variations of the solar-panel efficiency γ_s (left) and wind-turbine efficiency C_p (right), the red dots represent the specific values that are given alongside

Figures 5a and 5b show the result of a variation in the amount of incident solar energy and wind speed, respectively.

As shown in Figures 1a and 2a, the wind speed and incident solar energy change over the year; Fig. 6a shows how this affects the operational time per day. The figure shows that the inclusion of a $0.55m^2$ wind turbine reduces the operational time per day for most months. However, it also shows that it improves the operational time per day for the months with the lowest operational time per day. Without a wind turbine, the operational time ranges between 2.6 hours

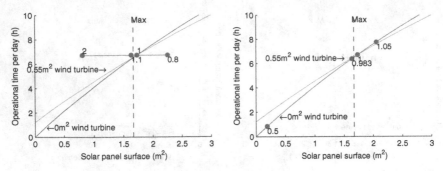

(a) Variations of the incident solar energy $\bar{E}_{sun}$

(b) Variations of the wind speed $\bar{v}_{wind}$

Fig. 5. Variations of Fig. 3 with the addition of the red line that shows how point A moves with variations of the incident solar energy $\bar{E}_{sun}$ and wind speed $\bar{v}_{wind}$. The red points represend the specific values that are given alongside.

and 8.8 hours per day, while it ranges from 3.6 hours and 8.4 hours per day with a $0.55m^2$ wind turbine.

Figures 1b and 2b show that the wind speed and incident solar energy differ for different locations; Fig. 6b shows how this affects the operational time per day. For the regions "Eelde", "De Bilt" and "Maastricht", the addition of a wind turbine has a negative effect to the average operational time while it has a positive effect in the regions "De Kooy" and "Vlissingen".

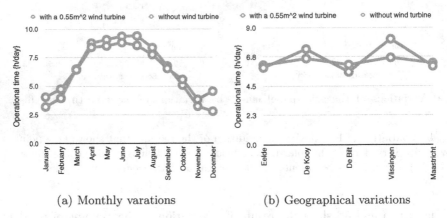

(a) Monthly varations

(b) Geographical variations

Fig. 6. The effect on monthly and geographical variations in incident solar energy and wind speed on the operational time per day for two situations: a $0.55m^2$ wind turbine and no wind turbine. For both situations, the maximum ($1.64m^2$) amount of solar panel surface is assumed.

The effect of applying a different assumption on the weight of the wind turbine and solar panels result in Figures 7a and 7b. Fig. 7a shows that if the

solar panels are heavier, it it has a more significant effect on the operational
time per day than if lighter solar panels are used. Fig. 7b shows that if the wind
turbine planar density is more than $13kg/m^2$, it is no longer beneficial to include
a wind turbine if the maximum area of solar panels is applied.

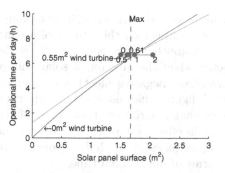 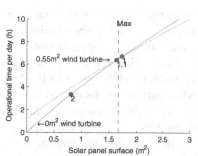

(a) Variations of solar panel weight per square metre η_s

(b) Variations of the wind turbine weight per square metre η_w

Fig. 7. A variation of Fig. 3 with the addition of the red line that shows how point A
moves with variations solar panel weight per square metre η_s and wind turbine weight
per square metre η_w, the red dots represent the specific values that are given alongside
multiplied by the amount of weight per square metre as used in the study

Another parameter that is interesting to vary is the total mass of the locomotion system, including batteries. A different assumption on the battery storage
required or a different mass of the locomotion system affects this parameter.
Figure 8 shows the results of varying this parameter. It shows that a variation
in the weight of the locomotion system significantly affects the operational time
per day.

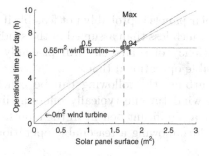

Fig. 8. A variation of Fig. 3 with the addition of the red line that shows how point A
moves with variations of the mass of the locomotion system m_l, the red dots represent
values of 0.5, 0.76 and 1 multiplied by the value used in the study

Furthermore, it is interesting to note the effect of varying the specific resistance (ϵ) and maximum speed (v). Varying these parameters directly scales the operational time.

5.1 Discussion of the Results

From the generalisation of the study, it can be seen that a variation of assumptions for one parameter has significantly more impact on the benefit of a wind turbine and the total operational time than another parameter. Perhaps the most significant differences can be expected when the amount of solar energy turns out to be much lower (which, of course, has a negative effect) than assumed or the wind turbine turns out to be be much lighter than assumed (which has a positive effect). Furthermore, it can be seen that a variation in the mass of the solar panels has - perhaps surprisingly - little effect. Finally, variations in the weight of the locomotion system, including batteries show that a reduction of the weight seems to pay off quite well in terms of operational time.

Based on the results, one could also estimate the distance that can be traversed on a flat terrain in a day by multiplying the operational time per day with the velocity of locomotion, which is $1ms^{-1}$ in this study. Due to space restrictions, this is not done here.

6 Conclusion

In this work, the application of robots as components of an early warning system for the Dutch flood defence system was presented with a exploratory study on an energy-autonomous dyke robot. Based on this work, three conclusions can be drawn.

Firstly, even if only some assumption may prove valid, it is realistic to have an energy-autonomous walking dyke robot in the Netherlands. In much of the world, the amount of energy that can be harvested from the sun is significantly higher than in the Netherlands which would result in an even higher operational time.

Secondly, the use of solar panels is probably not feasible if the amount of solar energy that is available is much less than assumed in the study. However, current developments in the efficiency of solar cells will probably lead to significant improvements in the feasible operational time.

Finally, for a wind turbine, the following can be concluded: in this case study, the inclusion of a wind turbine typically offers a slight benefit. Maybe more importantly, it gives a significant benefit in the months where the incident power of the sun is low, thus allowing a reasonable operational time during the winter.

References

1. Betz, A.: Introduction to the Theory of Flow Machines. Pergamon (1966)
2. Gabrielli, G., von Karman, T.H.: What price speed?: specific power required for propulsion of vehicles. Journal of the American Society for Naval Engineers **63**(1), 188–200 (1951)
3. Green, M., Emery, K., Hishikawa, Y., Warta, W., Dunlop, E.: Solar cell efficiency tables (version 40). Progress in photovoltaics: research and applications, pp. 606–614 (2012)
4. Greenman, J., Holland, O., Kelly, I., Kendall, K., McFarland, D., Melhuish, C.: Towards robot autonomy in the natural world: a robot in predator's clothing. Mechatronics **13**(3), 195–228 (2003)
5. Gregorio, P., Ahmadi, M., Buehler, M.: Design, control, and energetics of an electrically actuated legged robot. IEEE transactions on systems, man, and cybernetics. Part B, Cybernetics : a publication of the IEEE Systems, Man, and Cybernetics Society **27**(4), 626–634 (1997)
6. Huleihil, M.: Maximum windmill efficiency in finite time. Journal of Applied Physics **105**(10), 104908 (2009)
7. Ieropoulos, I., Greenman, J., Melhuish, C., Horsfield, I.: Ecobot-III-a robot with guts. In: ALIFE, pp. 733–740 (2010). https://mitp-web2.mit.edu/sites/default/files/titles/alife/0262290758chap131
8. Inglis, D.R.: A windmills theoretical maximum extraction of power from the wind. American Journal of Physics **47**(5), 416 (1979)
9. KNMI: Jaaroverzicht van het weer in Nederland - 2010. Tech. rep., KNMI (2010)
10. KNMI: Jaaroverzicht van het weer in Nederland - 2011. Tech. rep., KNMI (2011)
11. KNMI: Jaaroverzicht van het weer in Nederland - 2012. Tech. rep., KNMI (2012)
12. KNMI: Jaaroverzicht van het weer in Nederland - 2013. Tech. rep., KNMI (2013)
13. Mart, A., Arajo, G.: Limiting efficiencies for photovoltaic energy conversion in multigap systems. Solar Energy Materials and Solar Cells **43**(1996), 203–222 (1996)
14. Melhuish, C., Ieropoulos, I., Greenman, J., Horsfield, I.: Energetically autonomous robots: Food for thought. Autonomous Robots **21**(3), 187–198 (2006)
15. Poulakakis, I., Smith, J., Buehler, M.: Modeling and Experiments of Untethered Quadrupedal Running with a Bounding Gait: The Scout II Robot. The International Journal of Robotics Research **24**(4), 239–256 (2005)
16. Solbian: Solbian (2013). www.solbian.eu
17. Stine, W.B., Geyer, M.: Power From The Sun (2012). http://www.powerfrom thesun.net/

Advancing Evolutionary Coordination for Fixed-Wing Communications UAVs

Alexandros Giagkos[✉], Elio Tuci, and Myra S. Wilson

Department of Computer Science, Aberystwyth University,
Llandinam Building, Aberystwyth, Ceredigion SY23 3DB, UK
{alg25,elt7,mxw}@aber.ac.uk

Abstract. In this paper we present advances to our previously proposed coordination system for groups of unmanned aerial vehicles that provide a network backbone over mobile ground-based vehicles. Evolutionary algorithms are employed in order to evolve flying manoeuvres that position the aerial vehicles. The updates to the system include obstacle representation, a packing mechanism to permit efficient dynamic allocation of ground-based vehicles to their supporting aerial vehicles within large-scale environments, and changes to time synchronisation. The experimental results presented in this paper show that the system is able to adaptively form sparse formations that cover as many ground-based vehicles as possible, optimising the use of the available power.

Keywords: Evolutionary algorithms · Unmanned Aerial Vehicles · Coordination strategies

1 Introduction

This paper presents advances to the autonomous coordination of multiple unmanned aerial vehicles (UAVs) that provide network coverage for multiple independent ground-based vehicles, initially introduced in [5]. The advances include a way to overcome the known problems of imperfect communication due to the mobility of ground-based vehicles (leading to long distances between them and the aerial vehicles), limited radio frequency power, and other communication failures related to large distances between communicating devices. The aim of the system is to define flying strategies that allow groups of autonomous aerial vehicles to react to topological changes and to ensure that relaying of data between ground-based vehicles is achieved. The use of evolutionary algorithms for similar complex problems has been previously explored, showing great potential ([1,3,7,8]). The system's ultimate objective is to maximise the number of supported ground-based vehicles with respect to the given limited power available to the communication. To achieve such emergent behaviour while staying below the power threshold, a sophisticated method for allocating which ground-based vehicle is to be covered by which unmanned aerial vehicle (referred to as packing) needs to be employed.

C. Dixon and K. Tuyls (Eds.): TAROS 2015, LNAI 9287, pp. 124–135, 2015.
DOI: 10.1007/978-3-319-22416-9_14

In this paper the incorporation of obstacle representation and the use of a packing mechanism that allow simulations of realistic and large-scale scenarios to be achieved, are introduced. A new mechanism for providing coverage considers the limited available power when allocating ground-based vehicles to their supporting aerial vehicles. This mechanism replaces our previous evolutionary algorithm fitness function initially presented in [5]. The original fitness function (equation 1) only considered two objectives; a) maximisation of the network coverage by minimising the overlap between the footprints of the aerial vehicles, and b) minimisation of their average altitudes. The latter ensured that minimum power will be consumed in achieving network coverage.

$$f = \frac{C_{net} - C_{overlap}}{G} \times \left(1 - norm\left(\frac{\sum_{i=0}^{U}(h_i)}{U}\right)\right) \tag{1}$$

Although the equation proved its effectiveness through experiments, it assumes that no power hard limit exists, thus leading to unbalanced situations where some aerial vehicles end up providing network coverage to more ground-based vehicles than others in a stochastic way, independent of the physical positions of the ground-based vehicles. This phenomenon brings to light two important issues. Firstly, in reality, the aerial vehicles have limited power available for communications and thus a method that encourages a balanced way of ground-based vehicles to aerial vehicles allocation is desired. Secondly, using such a fitness function the resulting physical topologies will ultimately lead to communication networks prone to congestion due to bottleneck effects. That is, routing protocols will explore those paths containing overloaded aerial vehicles and thus the traffic will eventually be routed via particular overloaded relays.

The latest system puts emphasis on enhancements that allow large-scale scenarios to be realised (i.e., the obstacle representation and the sophisticated packing mechanism) and a study of analysing the flying formations that emerge.

The structure of the paper is as follows. Section 2 presents the enhancements made to the system. Section 3 is dedicated to the results and their analysis. The paper concludes in section 4, where an overview of the findings is given.

2   Methods

In this section, the key components of the proposed system are briefly discussed with respect to the enhancements made to the new version of the coordination system. Namely, the obstacle representation into the communication model, the chromosome codification with respect to time synchronisation in the system and the packing mechanism that replaces the fitness function of the previous initial work, allowing the system to work efficiently in large-scale dynamic environments.

2.1   Communication Model and Obstacle Representation

As in the initially proposed system, an aerial vehicle is treated as a point object in three-dimensional space with an associated direction vector. At each time step, the position of an aerial vehicle is defined by a latitude, longitude, altitude and heading $(\phi_c, \lambda_c, h_c, \theta_c)$ in the geographic coordination system. Detailed descriptions of the kinematic model responsible for the flying of the fixed-wing aerial vehicles can be found in [6] and in [5].

Networking is achieved by maintaining communication links between the aerial backbone and as many ground-based vehicles as possible. The communication links are treated independently and a transmission is considered successful when the transmitter is able to feed its antenna with enough power, such that it satisfies the desirable quality requirements. In terms of the ground-based vehicles, it is assumed that they have a Global Positioning System (GPS) device and are able to broadcast their own position information at some reasonable interval (the default value is 3 seconds). This data is used by the evolutionary algorithms in order to make appropriate predictions and relocate the aerial vehicles. It is the only way the aerial vehicles can sense the presence of ground-based vehicles and their moving patterns. In terms of the aerial vehicles it is assumed that they are equipped with two radio antennae. One isotropic able to transmit to all directions and a horn-shaped one able to directionally cover an area on the ground. Also, all flying vehicles are equipped by a GPS device and can broadcast information about their current position and directionality at a reasonable interval (default 3 seconds). Here, focus is primarily given to the communication between aerial vehicles and ground-based vehicles using the former horn-shaped antennae, as it dictates the effectiveness of the communication coverage of the mission and the power consumption of a flying mission.

Starting from the initial communication model [5], a link between an aerial vehicle and a ground-based vehicle antennae is considered of a good quality if the ratio of the energy per bit of information E_b to the thermal noise in 1 Hz bandwidth N_0 (normalized signal to noise ratio E_b/N_0) is maintained. The transmitting power P_t that an aerial vehicle is required to feed to its horn-shaped antenna in order to cover a ground-based vehicle in distance d is expressed by the following version of the Friis equation:

$$P_t = p \times d^2 R_b \frac{E_b}{N_0} \frac{1}{G_r G_t} \left(\frac{4\pi f}{c} \right)^2 T_{sys} K \qquad (2)$$

- R_b is the desired data rate on the link (bit/s),
- E_b/N_0 is the target ratio of energy in one bit to the noise in 1 Hz,
- G_r is the receiver's antenna gain (assuming omnidirectional),
- G_t is the transmitter's antenna gain equal to $\frac{2\eta}{1-\cos(HPBW/2)}$, where η and $HPBW$ are the efficiency and the half-power beamwidth angle of the horn-shape antenna,
- T_{sys} is the total system noise temperature,
- K is the Boltzmann K constant,

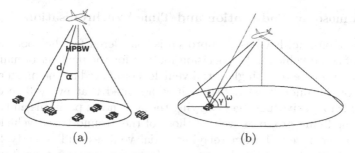

Fig. 1. Slant distance d and angle α of a communication link (a) and angle ϵ defines the area within which links are possible (b). Elevation angle γ is user-defined.

- p is the coverage profile (will be explained below), and
- d is the slant range defining the distance between the aerial vehicle and the ground-based vehicle on the ground.

and has to be reduced from the time step power available threshold P_{max}. In equation 2, most of the terms are known and remain constant during the mission, apart from the slang distance d and coverage profile p.

The latter describes whether a ground-based vehicle is covered by a supporting aerial vehicle. Its value is 1 if the ground-based vehicle lies within its supporting aerial vehicle's footprint, determined using the packing algorithm described in section 2.3. In the new system, there are two criteria that need to be fulfilled so that the coverage profile p of a ground-based vehicle is set to 1. The first criterion is depicted in figure 1(a) and is related to the aerial vehicle's footprint w.r.t. its current altitude as well as its antenna's half-power beamwidth (HPBW) angle. The higher the aerial vehicle flies, the wider its footprint on the ground. The second criterion is related to the existence of obstacles in the system, an enhancement that allows realistic simulations to be conducted. Obstacles are introduced by the use of an elevation angle, as shown in figure 1(b). The two criteria are combined mathematically so as the profile p is written:

$$
L(p) = \begin{cases} 1, & \alpha <^{HPBW}/_2 \quad \text{and} \quad \omega \geq \gamma \\ 0, & \alpha \geq^{HPBW}/_2 \quad \text{and} \quad \omega < \gamma \end{cases} \tag{3}
$$

Finally, at each time step the transmission of 3 UDP datagrams from each aerial vehicle down to any ground-based vehicle it currently supports is simulated, using a downline data rate of 2Mbit/s and frequency of 5GHz to finally ensure a ratio E_b/N_0 of 10db. Ultimately, the number the available communication links that can be accessed depends on the position of the aerial backbone and the number of ground-based vehicles being currently allocated to each of them. In this way, the power consumption per time step is measured and subtracted from the available power to the communication which is set to $P_{max} = $ 50Watts.

2.2   Chromosome Codification and Time Synchronisation

A centralised, on-line, EA-based approach is considered for the coordination of the group of aerial vehicles. The decision making for the next set of manoeuvres for the group is made by a single aerial vehicle, nominated as the master. Taking advantage of the underlying network, it is assumed that every 3 seconds the master is able to receive messages carrying the last known positions and direction vectors of the flying group as well as those of the ground-based vehicles. Data updates may be received from relaying aerial vehicles and directly from the ground-based vehicles within the master's footprint, and are tagged such that the master, and in turn the EA decision unit, are fed with up-to-date knowledge of the topology.

Once the EA has evolved a new set of manoeuvres, the master aerial vehicle is responsible for broadcasting the solutions to the whole group, using the network. As this work mainly focuses on providing network coverage to ground customers, it is assumed that there is no packet loss and that a dynamic routing protocol allows flawless data relaying within the topology.

A flying manoeuvre is described by a Dubins path of 3 segments [5]. Each segment comprises a bank angle and the duration for which the segment's manoeuvre is to be performed. Furthermore, a Dubins path may request a change to the vertical plane, thus require an alteration to the current altitude. The information is stored to the chromosome's genes in a form of "$\beta_1, \delta t_1, \beta_2, \delta t_2, \beta_3, \delta t_3, b, \delta h$" sequences. The first six genes describe the horizontal motion and the duration of each of the 3 segments of the Dubins path and are stored as floating point values. The seventh gene b, as well as the last δh, control the vertical behaviour of the aerial vehicle. When the former is set to 0, the aerial vehicle flies at the same altitude (level flight). If it is set to 1, then the vertical motion is considered and the aerial vehicle is expected to change its altitude by δh within the duration of the Dubins path, $\sum_{i=1}^{3}(\delta t_i)$.

The decision of including two dedicated genes related to the altitude changes instead of just representing the altitude with a single floating point value, is explained when one carefully investigates both the influence and type of parameter each gene represents. When encoding and decoding chromosomes, e.i., constructing flying solutions for the aerial vehicles so they can be evaluated, the genes' values are bounded within pre-defined ranges, according to their nature and meaning. Bank angles drive an aerial vehicle to bank either to the left or to the right, allowing it to gradually change its heading. If β_{min} and β_{max} are the minimum and maximum values, that is -48 and +48 respectively, then the values of the corresponding genes (β_1, β_2 and β_3) are encoded within the range of [0, 1] and decoded after being normalised in the range of $[\beta_{min}, \beta_{max}]$. Similarly, the δh value is decoded after being normalised in the range of $[h_{min}, h_{max}]$, with $h_{min} = -h_{max}$. Clearly, if the 7th gene was not present, the chances of flying level, e.i., $\delta h = 0$, would not be probabilistically fair compared to $\delta h \neq 0$. With the binary gene present, flying level has an equal probability to ascending and descending, thus affecting the power consumption by changing the distance between the transmitting and receiving antennae.

Table 1 depicts an example of chromosome codification, where $C0$ and $C1$ are chromosomes before and after decoding. In this example, the EA-based system proposes that for the next manoeuvre $C1$ will gradually change its altitude by flying -147 metres lower. In this example, the duration of a manoeuvre is set to $\delta t_1 + \delta t_2 + \delta t_2 = 134.164848 + 106.131113 + 71.704039 = 312.0$ seconds.

Table 1. An example of two chromosomes being decoded into meaningful values for the flying of two aerial vehicles

	β_1	δt_1	β_2	δt_2	β_3	δt_3	b	δh
C0:	0.058745	0.345754	0.258386	0.260568	0.997280	0.888298	0	0.587347
C1:	0.197602	0.835057	0.560083	0.565502	0.092618	0.234472	1	0.428889
C0:	-42.360511	88.144913	-23.194961	79.285621	47.738864	144.569466	0	180.807804
C1:	-29.030183	134.164848	5.767955	106.131113	-39.108657	71.704039	1	-147.199089

An evolutionary algorithm using linear ranking is employed to set the parameters of the paths [4]. We consider populations composed of $M = 100$ teams, each consisting of $N = 4$ individuals (the number of aerial vehicles in the flying group). At generation 0 each of the M teams is formed by generating N random chromosomes. For each new generation, the chromosomes of the best team ("the elite") are retained unchanged and copied to the new population. Each of the chromosomes of the other teams is formed by first selecting two old teams using roulette wheel selection. Then, two chromosomes, each randomly selected among the members of the selected teams are recombined with a probability of 0.3 to reproduce one new chromosome. The resulting new chromosome is mutated with a probability of 0.05. This process is repeated to form $M - 1$ new teams of N chromosomes each.

The concept of time and time synchronisation plays a important role in the system. The aerial vehicles are allowed to perform two types of manoeuvres; i) a turn circle manoeuvre of a fixed bank angle, and ii) the resulting evolved manoeuvre generated by the EA decision unit.

As the system does not consider noise at the current, abstract stage (e.g., wind force and variations to cruising speed), an aerial vehicle is expected to perform perfect circles and always reach the same final latitude, longitude, and altitude when completing a revolution. In terms of the manoeuvres generated by the EA, all aerial vehicles are expected to complete their flying at an equal time, due to the Dubins manoeuvres' equal durations. Since the resulting decision that dictates the next move is communicated from the master to the rest of the group using the network, it is understood that there are delays that may affect the transmissions, such that not all of the aerial vehicles will be informed on time. This brings a synchronisation problem which is addressed by increasing the number of revolutions required before a Dubins path is implemented by the aerial vehicles. The significance of this rule is two-fold. Not only it allows all aerial vehicles to successfully receive the next manoeuvre information, depending on the number of repeated revolutions set, but it also ensures that the EA has enough time to reach a solution.

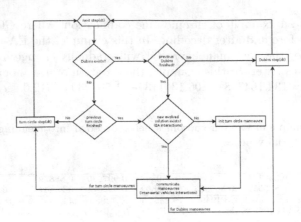

Fig. 2. Process flow for a master aerial vehicle

After solving the time synchronisation issues the EA is able to run in parallel with the rest of the controller, which is responsible for keeping the aerial vehicle flying in order to complete its previous action, either a turn circle or a Dubins manoeuvre. Figure 2 depicts the algorithm flow at the master aerial vehicle. At every time step dt, the algorithm takes a step further to the previous unfinished manoeuvre (as long as it is not completed). If no Dubins manoeuvre is available and a turn circle manoeuvre has been performed, the master interacts with the EA decision unit to receive a freshly evolved solution (set of manoeuvres for all of the aerial vehicles in the group). Notice that in case when the EA has not been successful in evolving a good solution, the master communicates turn circle manoeuvres to the rest group. This allows synchronisation between the aerial vehicles as they are bound to make a revolution, whilst giving more time to the EA algorithm to produce a result.

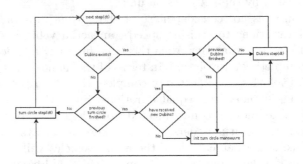

Fig. 3. Process flow for a non-master aerial vehicle

As a non-master aerial vehicle (shown in figure 3) does not interact with an EA decision unit at the current stage of the system, it constructs and performs turn circle manoeuvres only when no Dubins (or turn circle manoeuvre) is received from the master. This feature is rather unusable as the aerial vehicles will always receive a manoeuvre due to the perfect network conditions discussed before. Nevertheless, in reality this will allow a non-master aerial vehicle to continue flying in a controlled attitude and give it time to synchronize with the rest of the group. It is empirically found that a double turn circle manoeuvre of 6 seconds (75 degrees bank angle) gives enough time for the EA decision unit to produce reasonable results for a group of 6 aerial vehicles. Ultimately, the group's flying pattern is as follows: turn circle → turn circle → Dubins → turn circle → turn circle → Dubins → ... and so on.

2.3   Packing Mechanism for Energy Efficiency

Another major enhancement to the proposed system is the use of a packing mechanism which is responsible for allocating ground-based vehicles to supporting aerial vehicles. In order to understand the importance of a packing mechanism and its use in the context of network coverage and coordination of aerial vehicles, it is necessary to deeply understand the internal mechanisms of the evolutionary algorithms while travelling within the search space to converge to an optimal solution. Each population, when generated, has at some point to be evaluated so as the survival of the fittest is ensured. In the context of the proposed system, the evaluation is performed at a group level, where the performance of the group of aerial vehicles is examined in a collective manner. Thus, similarly to the previous fitness function, an optimal solution is a solution that maximises the cumulative network coverage. Still, the influence the altitude of an aerial vehicle has to the amount of power required to achieve communication with ground-based vehicles needs to be incorporated in the system. This is achieved by a packing mechanism.

In this work, a similar packing algorithm used in [2] is adopted. The ability of an aerial vehicle to support ground-based vehicles is described by its packing array, a list of ground-based vehicles that are allocated to this particular flying vehicle. The aim of the packing algorithm is to make the most efficient allocation in terms of the power consumption. In more detail, a general logical map is first built so as to accommodate all aerial vehicles along with all the ground-based vehicles that lie within their footprints, coupled with the power required to support each the intermediate link. Notice that the ground-based vehicles sorted by the power requirement and thus the least expensive lie first. Algorithm 1 is employed ensuring that the number of ground-based vehicles allocated to each aerial vehicle is maximised while the power consumption is minimised, by firstly concentrating on the centre of each aerial vehicle's footprint and gradually expanding to its edges.

Consequently, the group fitness function f is the sum of the packing arrays, i.e., $\sum_{i=1}^{N}(packing[i])$. The bigger that number becomes, the most fitted that particular set of manoeuvres is when implemented. This process allows the EAs to evolve solutions that maximise the network coverage by assigning ground-based vehicles to those aerial vehicles that are able to spend less power to support them. Artificial evolution awards those sets of manoeuvres that achieve a better

---

Algorithm 1. Packing algorithm

1: let G be a sorted logical map
2: initialize packing[N] as empty packing arrays for N aerial vehicles
3: **while** G is not empty of ground-based vehicles **do**
4:      **for each** u in logical map G **do**
5:          let g be the first ground-based vehicle found in u's sorted list
6:          let p be the power required to support g
7:          **if** $powerbudget(u) - p \geq 0$ **then**
8:              $packing[u] \leftarrow g$
9:              $powerbudget(u) = powerbudget(u) - p$
10:         remove remaining instances of g from G

---

score and when all generations complete, the solution is communicated to the flying group for implementation along with the suggested packing arrays used for the fitness score calculations. The receiving aerial vehicles are then forced to fly according to their Dubins path and serve those ground customers that are stored into their packing arrays. It is important to notice that evolution uses information retrieved from frequent broadcast messages sent by all vehicles. In order to ensure that the manoeuvres are generated according to valuable positional information, distances between antennae and in turn power estimation is calculated based on predicted positions.

3   Experiments and Results

The performance of the enhanced proposed system is shown and discussed in this section. The study encompasses an experimental scenario in which groups of three aerial vehicles with limited available communications power are deployed to fulfil the network requirements of several ground-based vehicles. The ultimate aim of the study is to investigate the ability of the system to manage and utilise the aerial vehicles by repositioning them, such that the network coverage is maximised in a power-aware fashion. Also, their performance is evaluated in terms of autonomous flying flexibility, resulting in formations and manoeuvres, in order to challenge its applicability to real world applications. Results related to the aerial vehicles ability to separate and thus allow higher manoeuvring flexibility are included. The configuration settings and simulation conditions are similar in both scenarios and are summarised in table 2.

Results of experiments with three aerial vehicles are presented in figures 4 for both coverage (a, c) and power consumption (b, d). The system is shown to achieve optimal coverage results when covering a small as well as large number of ground-based vehicles. It is also observed that the flexibility in implementing flying manoeuvres, as a result from the design, allows the aerial vehicles quickly find a good formation according to the way their customers are positioned on the ground. This is observed in conjunction with separation results depicted in figure 5. The aerial vehicles fly far away from each other and gradually gain altitude so that more ground-based vehicles lie within their cumulative footprints.

Table 2. Table of parameters and configuration settings

Simulation parameters		
Terrain size	100 km^2	
Duration	6 hours of flying	
Aerial vehicles' parameters		
Init altitude	15000 feet	
Init latitude, longitude and heading	#1 52.8636, -2.6373	270
	#2 52.0512, -1.4219	270
	#3 52.8605, -1.4219	270
Ground-based vehicles's parameters		
Number of units	50 and 200	
Mobility model	Random WayPoint	
Speed	30 mph	

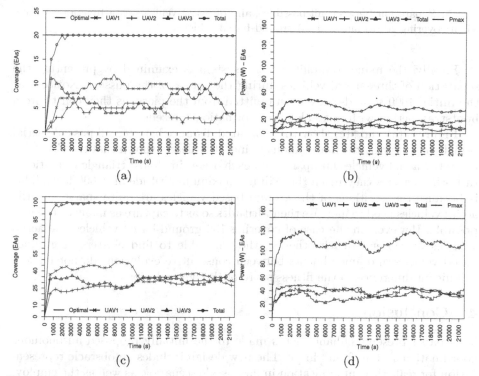

(a) (b)

(c) (d)

Fig. 4. Coverage and power results as a function of time, when covering 20 (a, c) and 100 (b, d) ground-based vehicles

Power consumption is shown in figures 4(b) and 4(d) for small and large number of ground-based vehicles, respectively. In the case of 100 ground-based vehicle participants the system, as expected, consumes more energy yet the packing mechanism balances the number of ground-based vehicles allocated to each aerial vehicle and does not allow aerial vehicles to reach the limit of 50 Watts per time step. The effect is less obvious in the case of supporting only 20 ground-based vehicles as due to the large scenario area, the ground vehicles are sparsely positioned, a physical fact that requires wider footprints to be generated.

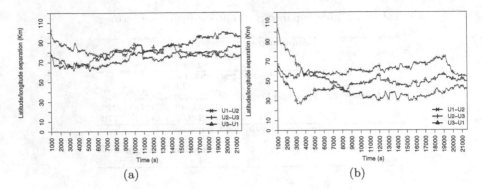

Fig. 5. Separation between all possible pairs of aerial vehicles as a function of time, when covering 20 (a) and 100 (b) ground-based vehicles

Finally, the manoeuvrability of the system is examined by presenting the separation of three aerial vehicles during the course of the mission. Notice that the initial 1000 seconds have been omitted from the plots, as they constitute a bootstrap period with unstable flying formations present.

Figures 5(a) and 5(c) show the separated distance between all possible aerial vehicle pairs during the autonomous flying. It is clearly shown that the EAs tend to fly the aerial vehicles far apart from each other, in almost triangle formations. In both scenarios they reach the ceiling (maximum altitude of 22000 feet). This is because the positions of the ground-based vehicles are quite sparse and the aerial vehicles need to increase their altitudes so as to capture as many or them as possible. However, in the case of covering 100 ground-based vehicles the aerial vehicles fly closer to each other as they are able to find clusters of ground-based vehicles and amendments between consecutive evolutions do not make a significant difference to the fitness score of the new solutions.

4   Conclusion

This paper presents enhancements made to the initially proposed autonomous coordination system found in [5]. The new design includes an obstacle representation for realistic communication in large-scale scenarios, as well as the employment of a packing mechanism that considers limited available power. Emphasis is also given to the design in terms of time synchronisation and chromosome codification.

The experimental results show that the system is able to fulfil the main objective, i.e., to optimise the network coverage while utilising efficiently the power given to the communications. In addition, this paper shows an analysis of the flying behaviour emerging by the use of EAs, in terms of separation in flying. The aerial vehicles are found to quickly adopt sparse formations, a valuable trait for covering large-scale areas.

Acknowledgments. This work is funded by EADS Foundation Wales.

References

1. Agogino, A., HolmesParker, C., Tumer, K.: Evolving large scale UAV communication system. In: Proceedings of the Fourteenth International Conference on Genetic and Evolutionary Computation Conference, GECCO 2012, pp. 1023–1030. ACM, Philadelphia (2012)
2. Charlesworth, P.B.: Simulating missions of a UAV with a communications payload. In: 2013 UKSim 15th International Conference on Computer Modelling and Simulation (UKSim), pp. 650–655, April 2013. doi:10.1109/UKSim.2013.61
3. de la Cruz, J.M., et al.: Evolutionary path planner for UAVs in realistic environments. In: Proceedings of the 10th Annual Conference on Genetic and Evolutionary Computation, GECCO 2008, pp. 1477–1484. ACM, Atlanta (2008)
4. Goldberg, D.E.: Genetic Algorithms in Search, Optimization and Machine Learning. Addison-Wesley, Reading (1989)
5. Giagkos, A., Tuci, E., Wilson, M.S., Charlesworth, P.B.: Evolutionary coordination system for fixed-wing communications unmanned aerial vehicles. In: Mistry, M., Leonardis, A., Witkowski, M., Melhuish, C. (eds.) TAROS 2014. LNCS, vol. 8717, pp. 48–59. Springer, Heidelberg (2014)
6. Giagkos, A., et al.: Evolutionary coordination system for fixed-wing communications unmanned aerial vehicles: supplementary online materials, April 2014. http://www.aber.ac.uk/en/cs/research/ir/projects/nevocab
7. Hasircioglu, I., Topcuoglu, H.R., Ermis, M.: 3D path plan- ning for the navigation of unmanned aerial vehicles by using evolutionary algorithms. In: Proceedings of the 10th Annual Conference on Genetic and Evolutionary Computation, GECCO 2008, pp. 1499–1506. ACM, Atlanta (2008). ISBN: 978-1-60558-130-9
8. O.K. Sahingoz.: Flyable path planning for a multi-UAV system with genetic algorithms and Bézier curves. In: 2013 International Conference on Unmanned Aircraft Systems (ICUAS), pp. 41–48 (2013)

A Generic Approach to Self-localization and Mapping of Mobile Robots Without Using a Kinematic Model

Patrick Kesper[1], Lars Berscheid[1], Florentin Wörgötter[1],
and Poramate Manoonpong[2]($\boxtimes$)

[1] Third Institute of Physics - Biophysics, Georg-August-Universität Göttingen,
Friedrich-Hund-Platz 1, 37077 Göttingen, Germany
{pkesper,worgott}@gwdg.de, lars.berscheid@online.de
[2] CBR Embodied AI and Neurorobotics Lab, The Maersk Mc-Kinney Moller
Institute, University of Southern Denmark, Campusvej 55, 5230 Odense M, Denmark
poma@mmmi.sdu.dk

Abstract. In this paper a generic approach to the SLAM (Simultaneous Localization and Mapping) problem is proposed. The approach is based on a probabilistic SLAM algorithm and employs only two portable sensors, an inertial measurement unit (IMU) and a laser range finder (LRF) to estimate the state and environment of a robot. Scan-matching is applied to compensate for noisy IMU measurements. This approach does not require any robot-specific characteristics, e.g. wheel encoders or kinematic models. In principle, this minimal sensory setup can be mounted on different robot systems without major modifications to the underlying algorithms. The sensory setup with the probabilistic algorithm is tested in real-world experiments on two different kinds of robots: a simple two-wheeled robot and the six-legged hexapod AMOSII. The obtained results indicate a successful implementation of the approach and confirm its generic nature. On both robots, the SLAM problem can be solved with reasonable accuracy.

Keywords: SLAM · Mobile robots · Hexapod robot · Probabilistic robotics · Laser range finder · Inertial measurement unit

1 Introduction

Solving the Simultaneous Localization and Mapping (SLAM) problem is important for a vast variety of different robotic tasks, e.g. performing autonomous navigation [1,2] or completing domestic tasks [3,4]. Probabilistic and other SLAM techniques have been applied to nearly all kinds of robots, e.g. wheeled [5], flying [6,7], walking [8,9] or even underwater robots [10]. In contrast to wheeled and flying robots, the application of SLAM to walking robots is scarce. While all these approaches show impressive results, they typically rely either on visual devices [9,11,12], on leg/body kinematics [13,14] or a multitude of sensors including

© Springer International Publishing Switzerland 2015
C. Dixon and K. Tuyls (Eds.): TAROS 2015, LNAI 9287, pp. 136–142, 2015.
DOI: 10.1007/978-3-319-22416-9_15

robot proprioceptive sensing, e.g wheel encoders. Therefore, they are difficult to transfer to different robotic systems. In this paper we present our Generic SLAM approach. The approach is based on a probabilistic SLAM algorithm and relies only on two portable sensors, an inertial measurement unit (IMU) and a laser range finder (LRF). It can be applied to different robotic systems. We have evaluated the performance of this approach on a wheeled robot and a six-legged walking robot in real-world experiments.

2   Materials and Methods

The Generic SLAM approach is summarized in fig. 1. It receives measurements from IMU and LRF as inputs. Here, two cases are possible. On the one hand the IMU output can be used to compute the translational and rotational velocity of a robot, which is then given to the Velocity Model (VM). This is appropriate, if the IMU data has a low noise level, e.g. when the movement of the robot is mostly linear. On the other hand, a new state estimate can be computed based on the acceleration values in combination with scan matching (SM). The result is then used as an input to the Odometry Model (OM). This approach is especially useful, if the IMU data exhibits a strong background noise. Scan matching is able to compensate for this noise at the cost of increased computational complexity. One of these approaches must be chosen manually. This SLAM/MCL algorithm utilizes the generated control action and the output of the LRF to recursively update the state distribution (belief) of the robot. Based on this belief, the most likely state can be estimated. The individual modules of the approach are explained as follows:

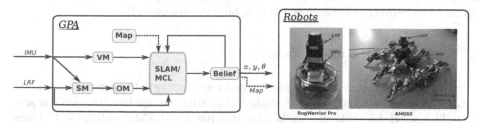

Fig. 1. This figure summarizes the Generic SLAM approach. It receives measurements from an inertial measurement unit (IMU) and a laser range finder (LRF) as an input to the SLAM/Monte-Carlo localization (MCL) algorithm, which updates the belief of the robot. Based on this belief, the most likely state and map can be estimated. The approach is tested on two different robots, the RugWarrior Pro and AMOSII.

Odometry Model (OM): The Odometry Model uses two consecutive states x_{t-1} and x_t to estimate the movement of a robot. The control action is given through $u_t = (\overline{x}_{t-1}, \overline{x}_t)^T$. Based on the difference between both states the rotations δ_{rot1} and δ_{rot2}, as well as the translation δ_{trans} can be computed by applying geometry. To account for model and measurement errors Gaussian noise is added to these variables. The variance of this noise is chosen to reflect the noise characteristics of the utilized IMU. See [15] for more details.

Velocity Model (VM): The Velocity Model follows a different approach compared to the OM. While the OM relies on relative motion information, the VM directly utilizes the translational and rotational velocity of a robot. Thus, the control action is given by $u_t = (v_t, \omega_t)^T$, where v_t denotes the translational velocity of the robot and ω_t the rotational velocity. Again, Gaussian noise is added to these velocities to account for measurement errors. The velocity model assumes these values to be constant over a given time span Δt. In this case the robot moves on the arc of a circle. See [15] for more details.

Scan Matching (SM): It is possible to obtain odometry information with a laser range finder by utilizing a technique called *Scan Matching*. The basic idea of scan matching is to evaluate the relative change between two consecutive scans to obtain an estimate of the corresponding movement of the robot. Here, a technique called *Polar Scan Matching (PSM)* is chosen ([16]). The PSM procedure does not yield any direct information about the velocity of the robot. It only estimates a position. Thus, it has to be used in combination with the OM.

SLAM/MCL: The SLAM/MCL algorithm updates the belief of the robot based on the inputs from OM/VM and the previous belief. Both procedures are implemented using a Particle Filter leading to the so called *FASTSlam* algorithm [15]. To run FASTSlam, a measurement probability $p(z_t|x_t)$ and a state transition probability $p(x_t|u_t, x_{t-1})$ must be known. Possible choices for $p(x_t|u_t, x_{t-1})$ can be derived from the described motion models, *Likelihood Fields* are used to model $p(z_t|x_t)$ [15]. Furthermore, an method to update maps according to the measurement and state of a particle is required. Here, we use *Occupancy Grid Maps* in combination with *Bresenham's Line Algorithm* [17].

3   Experiments and Results

Here, we use two different robot platforms to demonstrate the general use of our approach and to evaluate its performance, the RugWarrior Pro [18] and the six-legged hexapod AMOSII [19]. To be able to solve the SLAM problem the Hokuyo URG-04LX-UG01 Laser Range Finder and the x-IMU are used as a portable sensor modules. For the experiments on the real robots three different courses with varying difficulty and complexity are set up (fig. 2). In all experiments the robots start at the depicted location and traverses the corresponding course until reaching the end. For all experiments the size of one grid cell is 0.05 m x 0.05 m.

RugWarrior Pro: For the RugWarrior Pro the VM was used. The results of all three courses are shown in fig. 2. During the experiment for course a), the robot was steered to the left on purpose to test the functionality of the algorithm. Indeed, the curve can be seen in the computed paths, both for SLAM and MCL. Furthermore, the map created by the SLAM algorithm matches the measured dimensions. However, the computed map appears to be quite noisy.

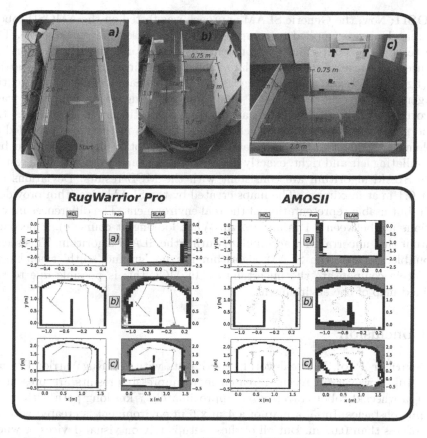

Fig. 2. This figure shows results of the MCL and GridSLAM algorithm for RugWarrior Pro and AMOSII traversing the three different courses a), b), and c). The size of one grid cell is 0.05m x 0.05m. A video illustrating examples of the real robot experiments can be seen at http://manoonpong.com/TAROS2015/supple.mp4

This was due to the unstable mounting of the LRF and IMU on the robot. The MCL algorithm provides a more accurate path than the SLAM procedure due to the higher noise sensitivity of SLAM. Overall, the accuracy of the SLAM algorithm can be estimated with ±0.05 m. The results of course b) are similar to course a). Again, the MCL algorithm provides a slightly less noisy path. But the difference is surprisingly small considering the additional turns of the robot. The achieved accuracy is ±0.1 m. Course c) is the longest and most complex setup. However, the accuracy of the results is similar to course a) and b). The accuracy of the computed map is about ±0.1 m. In summary, regardless of the complexity and length of the course, the SLAM procedure is able to track the path of the robot with reasonable accuracy. The obtained maps match the dimensions of the real environments. Furthermore the MCL and SLAM algorithm are able to successfully deal with erroneous acceleration, velocity and range measurements.

AMOSII: Now, the Generic SLAM approach was tested on the AMOSII robot. Scan matching was applied to compensate the increased movement noise. The same testing procedure and environments as utilized for the RugWarrior Pro were used. The results are very similar to the ones obtained from the RugWarrior Pro (fig. 2). However, the accuracy is slightly decreased. This is due to the stronger noise present in the IMU data. The MCL algorithm is able to track the correct path of the robot on all courses. In particular, it reproduces the erratic movement of AMOSII, which can also be seen in the video provided as supplementary material. The calculated positions do not lie on a straight line, but are oscillating left and right, exactly like the real motion of AMOSII. The results of the SLAM algorithm are consistent with the MCL results. Deviations are about 0.1 m at maximum. The maps created by the SLAM algorithm provide a rough, but usable representation of the real environment. The differences in wall positions are between 0.1 m and 0.2 m. When looking at course c), it becomes apparent that uncertainties accumulate during the SLAM algorithm. This seems reasonable, because the SLAM algorithm has no 'ground truth' available to completely compensate the uncertainties. In MCL this is possible due to the provided map of the environment.

4   Conclusion

RugWarrior Pro: The real-world experiments with the RugWarrior Pro are successful. However, in recent research more impressive results are presented. In [20] a wheeled robot traversed an approximately 7 km long path with many dynamic obstacles. In [5] a map of a 4 m x 6 m environment is created with an error of less than 0.07 m. But all of these setups rely on visual devices or wheel encoders and are generally equipped with more and better hardware. Thus, it is reasonable, that using only two sensory inputs on the inferior RugWarrior Pro sacrifices accuracy.

AMOSII: The accuracy of the obtained results is similar to other approaches. In [8] a humanoid robot maps a 4 m x 7 m environment with a maximum error of 0.1 m. In [11] a humanoid robot walks in a circle with a radius of 0.75 m. The resulting map and trajectory again have a maximum error of 0.1 m. Lastly, in [12] a 0.5 m x 0.5 m environment with rough terrain is mapped with similar accuracy. However, in all of these works the robots used visual information and/or kinematic models. In this paper, we showed that we are able to achieve the same results with our minimal and generic approach.

In large, open environments without any objects the LRF does not return any usable information. Consequently, the accuracy of the MCL and SLAM algorithm decreases drastically. Both algorithms work well in indoor environments due to the abundance of objects and walls. In this paper, a minimal and generic SLAM implementation relying only on a LRF and an IMU is proposed and successfully tested on wheeled and legged robots. Consequently, a possible next step could be experiments with other kinds of robots. In particular, flying robots for indoor navigation are an interesting choice due to their wide availability and versatility.

Acknowledgments. This research was supported by BCCNII Göttingen with grant number 01GQ1005A (project D1).

References

1. Wooden, D., Malchano, M., Blankespoor, K., Howardy, A., Rizzi, A.A., Raibert, M.: Autonomous navigation for bigdog. In: 2010 IEEE International Conference on Robotics and Automation (ICRA), pp. 4736–4741 (May 2010)
2. Burgard, W., Cremers, A.B., Fox, D., Hähnel, D., Lakemeyer, G., Schulz, D., Steiner, W., Thrun, S.: Experiences with an interactive museum tour-guide robot. Artificial Intelligence **114**(1–2), 3–55 (1999)
3. Jones, J.L., Mack, N.E., Nugent, D.M., Sandin, P.E.: Autonomous floor-cleaning robot (August 27, 2013). US Patent 8,516,651
4. Sandin, P.E., Jones, J.L., et al.: Lawn care robot (January 21, 2014). US Patent 8,634,960
5. Wieser, I., Ruiz, A.V., Frassl, M., Angermann, M., Mueller, J., Lichtenstern, M.: Autonomous robotic slam-based indoor navigation for high resolution sampling with complete coverage. In: Position, Location and Navigation Symposium-PLANS 2014, 2014 IEEE/ION, pp. 945–951 (2014)
6. Müller, J.: Autonomous navigation for miniature indoor airships. PhD thesis, Universitätsbibliothek Freiburg (2013)
7. Grzonka, S., Grisetti, G., Burgard, W.: A fully autonomous indoor quadrotor. IEEE Transactions on Robotics **28**(1), 90–100 (2012)
8. Kwak, N., Stasse, O., Foissotte, T., Yokoi, K.: 3d grid and particle based slam for a humanoid robot. In: 9th IEEE-RAS International Conference on Humanoid Robots, Humanoids 2009, pp. 62–67 (2009)
9. Davison, A.J., Reid, I.D., Molton, N.D., Stasse, O.: Monoslam: Real-time single camera slam. IEEE Transactions on Pattern Analysis and Machine Intelligence **29**(6), 1052–1067 (2007)
10. Kim, A., Eustice, R.M.: Real-time visual slam for autonomous underwater hull inspection using visual saliency. IEEE Transactions on Robotics (2013)
11. Stasse, O., Davison, A.J., Sellaouti, R., Yokoi, K.: Real-time 3d slam for humanoid robot considering pattern generator information. In: 2006 IEEE/RSJ International Conference on Intelligent Robots and Systems, pp. 348–355 (2006)
12. Belter, D., Skrzypczynski, P.: Precise self-localization of a walking robot on rough terrain using parallel tracking and mapping. Industrial Robot: An International Journal **40**(3), 229–237 (2013)
13. Bloesch, M., Hutter, M., Hoepflinger, M.A., Leutenegger, S., Gehring, C., Remy, C.D., Siegwart, R.: State estimation for legged robots-consistent fusion of leg kinematics and imu. Robotics 17 (2013)
14. Wawrzyński, P., Możaryn, J., Klimaszewski, J.: Robust estimation of walking robots velocity and tilt using proprioceptive sensors data fusion. Robotics and Autonomous Systems (2014)

15. Thrun, S., Burgard, W., Fox, D.: Probabilistic robotics. MIT Press (2005)
16. Diosi, A., Kleeman, L.: Fast laser scan matching using polar coordinates. The International Journal of Robotics Research **26**(10), 1125–1153 (2007)
17. Bresenham, J.E.: Algorithm for computer control of a digital plotter. IBM Systems Journal **4**(1), 25–30 (1965)
18. Jones, J.L.: RugWarriorPro: Assembly guide. AK Peters Ltd. (1999)
19. Manoonpong, P., Parlitz, U., Wörgötter, F.: Neural control and adaptive neural forward models for insect-like, energy-efficient, and adaptable locomotion of walking machines. Frontiers in Neural Circuits (2013)
20. Kümmerle, R., Ruhnke, M., Steder, B., Stachniss, C., Burgard, W.: Autonomous robot navigation in populated pedestrian zones. Journal of Field Robotics (2014)

Improving Active Vision System Categorization Capability Through Histogram of Oriented Gradients

Olalekan Lanihun$^{(\boxtimes)}$, Bernie Tiddeman, Elio Tuci, and Patricia Shaw

Department of Computer Science,
Aberystwyth University, Aberystwyth SY23 3DB, UK
oal@aber.ac.uk
http://www.aber.ac.uk

Abstract. In the previous work of Mirolli et al. [1], an active vision system controlled by a genetic algorithm evolved neural network was used in simple letter categorization system, using gray-scale average noise filtering of an artificial eye retina. Lanihun et al. [2] further extends on this work by using Uniform Local Binary Patterns (ULBP) [4] as a preprocessing technique, in order to enhance the robustness of the system in categorizing objects in more complex images taken from the camera of a Humanoid (iCub) robot . In this paper we extend on the work in [2], using Histogram of Oriented Gradients (HOG) [5] to improve the performance of this system for the same iCub image problem. We demonstrate this ability by performing comparative experiments among the three methods. Preliminary results show that the proposed HOG method performed better than the ULBP and the gray-scale averaging [1] methods. The approach of better pre-processing with HOG gives a representation that could translate to improve motor responses in enhancing categorization capability for robotic vision control systems.

Keywords: Categorization · Active vision system · Neural network · Genetic algorithm · Histogram of oriented gradients

1 Introduction

An active vision system uses information from sensory-motor coordination in order to intelligently guide the vision mechanism to relevant salient features in a visual scene. The motor control of the visual system can be model by various techniques, but neural network seems to be very appropriate because of its biologically inspired background and suitability for noisy data. However, research in this area is still in its infancy [3]. Most of the work that has been done using this approach was for simple image problems. For Instance, James and Tucker [3] developed an active vision system for categorizing different 2D shapes. The system has the ability to move about in any direction, zoom and rotate; it was also able to categorize different 2D simple shapes irrespective of orientations,

© Springer International Publishing Switzerland 2015
C. Dixon and K. Tuyls (Eds.): TAROS 2015, LNAI 9287, pp. 143–148, 2015.
DOI: 10.1007/978-3-319-22416-9_16

scales and locations. Mirolli et al. [1] developed an active vision system that is based on a genetic algorithm evolved neural network to categorize gray-scale italic alphabet letters in different scales (sizes). The periphery region of the eye was processed by average gray-scale filtering. The movement of the artificial eye was controlled by motor neurons of the output units, which determine the eye location per time step, in order to capture relevant input features for the neural controller. However, Lanihun et al. [2] extended on the work in [1] by using ULBP to process the periphery region, so as to enhance its robustness in categorizing objects in more complex images taken from the camera of a Humanoid robot. They demonstrated the ability of the method to enhance the robustness of this kind of system with improved performance in their proposed system when compared to that of the gray-scale averaging [1] for the same Humanoid image problem. Lastly, in relation to other works listed above, the approach in this paper also involves a visuo-motor coordination of an active vision system based genetic algorithm evolved neural network. We have extended on the work done by Lanihun et al. [2] with the enhancement of the system with HOG to categorize objects in images taken from a Humanoid robot camera.

2   Experimental Details

We have performed three sets of experiments for the categorization of objects in the images taken from the Humanoid camera: (i) the gray-scale averaging experiment in [1], (ii) the ULBP pre-processing as used in [2], (iii) our proposed method that uses HOG for the processing of the retina region. However, we will not explain the details of the gray-scale averaging [1] and ULBP methods as these have been explained in our previous paper in [2]. The artificial agent is in form of a moving eye that explores a visual scene, in order to extract relevant information and process the sensory stimuli. The vision system is controlled by a recurrent neural network evolved by a genetic algorithm, which is similar in approach to [2]. We have also adopted the periphery only architecture of [2] and [1](Fig. 2). The neural network, evolutionary process and the fitness function were the same for the three experiments. The images used in the experiments were coloured of size 320 x 240 pixels (Fig. 1), and were converted to gray-scale images for processing. The objects categorized were: soft toy, remote control set, microphone, board wiper and hammer, which are represented by (ST, RC, MC, BW, H) respectively in the output categorization units of the network (Fig. 2). We have divided our image datasets into two groups as follows:

1. A training set with each object of 3 different sizes with variation of ±15 percent to the intermediate size, and each size with 3 different orientation in the range [4,-4] degrees; and an evaluation set with objects of 2 different sizes, with a variation of 10 and 20 percent to the intermediate size, and each size of 5 different orientations in the range [+3,-3] degrees.
2. A training set of objects of 5 different sizes varied between ±20 percent to the intermediate size, and each size of 3 different orientation in the range [5,-5] degrees; while the evaluation set has 5 varied sizes in the range ±18

Fig. 1. The above figure shows the original coloured images taken from the Humanoid robot camera

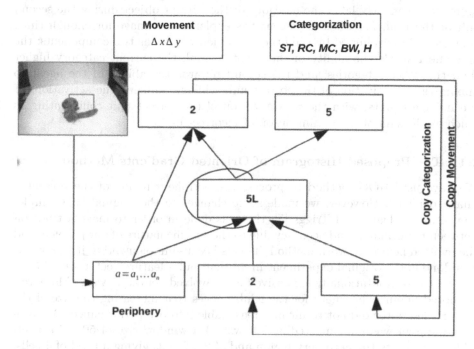

Fig. 2. The neural network architecture and the periphery region scanning the presented soft toy gray image in a trial for categorization

percent to the intermediate size and each size of 3 varied orientation within the range [6,-6] degrees.

The Neural Network and the Evolutionary Task. The recurrent neural network is a 3 layer architecture (Fig. 2). The input units activations were normalized between 0 and 1, and the vector size was determine by the method used in the processing of the periphery region. The input neurons encode the current states of the periphery region and the efferent copies of activations of 2 motor neurons and 5 categorization units (i.e at previous time step $t - 1$). There are 5 hidden units of recurrent activations that depend on the input activations received from the input neurons through the weighted connections and its own activations at the previous time step. In the output layer, the 2 motor neurons determine the movement of the eye per time step (maximal displacement of $[-12, 12]$ pixels in X and Y directions), and the other 5 neurons are the categorization units that labels the category. The agents were evaluated for 150 trials lasting 100 time steps each and for 5000 generations. In each trial the artificial eye is left to freely explore the visual scene (image), however, a trial is ended when the eye can no longer perceive any part of the object in the image through the periphery vision for three consecutive time steps. The task of the agent is to correctly label the category of the current object during the second half of the trial, i.e, when the agent has explored the image for enough time. The agent was evaluated by the fitness function which has two components: the first one rewards the ability of the agent to rank the current category higher than the other categories; and the second rewards the ability of the agent to maximize the activation of the correct unit while minimizing the activations of the incorrect units, with the maximization of the correct unit contributing as much as the sum of the minimization of incorrect units.

2.1   The Proposed Histogram of Oriented Gradients Method

We have used HOG method to process the periphery region of the presented image per trial. However, we made some changes to the original HOG implementation by Dalal and Triggs [5], this was done in order to make a trade-off between performance and accuracy due to the specific nature of our problem and the evolved neural network method. This is because implementing it the way it was done in the original experiment in [5] that had a feature vector size of 3780 will be too computationally expensive for an evolved neural network. Moreover, we noted in our experiment for the iCub images that increasing the size of the feature descriptor did not result into reasonable improve performance. Our proposed system was implemented as follows: (i) a window size of 50 x 50 pixels which represents the periphery region and of 2 x 2 cells giving a total of 4 cells; (ii) gradient orientations were quantized into 9 bins (bin size was 20 degrees for each cell), and orientations were between 0 -180 degrees (i.e signed gradients were ignored), because including signed gradients with orientation of 0 to 360 degrees reduces the performance accuracy of our system for this particular image problem; (iii) gradient magnitudes were truncated into bins and; (iv) histograms were concatenated into feature descriptor of dimension 4 cells x 9 bins giving a feature vector of size 36. This feature vector size is substantially smaller than that of the original implementation by Dalal and Triggs [5] and therefore a more reasonable input vector size for the neural network.

3   Results

We have performed 10 replications of the evolutionary run for each of the 2 training groups for the three set of experiments , and evaluated the performance of the system using the genomes of the best evolved individual from each group. The evaluations were done for 10, 000 trials. The system performance was assessed

Table 1. Experiment One (Gray-scale): Evaluation Test

Current Objects	Average Activation Rates of Objects (Highest Activation Rates in Bold)				
	Soft toy	Remote control	Microphone	Board wiper	Hammer
Soft toy	0.999940	0.499999	0.001440	0.500421	0.504063
Remote control	0.999940	0.499999	0.001441	0.500422	0.504058
Microphone	0.999940	0.499999	0.001441	0.500422	0.504059
Board wiper	0.999940	0.499999	0.001442	0.500422	0.504055
Hammer	0.999940	0.499999	0.001441	0.500422	0.504060

Table 2. Experiment Two (Uniform Local Binary Pattern): Evaluation Test

Current Objects	Average Activation Rates of Objects (Highest Activation Rates in Bold)				
	Soft toy	Remote control	Microphone	Board wiper	Hammer
Soft toy	**0.927989**	0.296624	0.430732	0.400178	0.382848
Remote control	0.303231	**0.851935**	0.222896	0.180930	0.649742
Microphone	0.349986	0.389921	**0.771348**	0.254532	0.529921
Board wiper	0.504821	0.286945	0.000448	**0.770973**	0.002129
Hammer	0.053441	0.894695	0.465550	0.114323	**0.958321**

using the average results of the two evaluation groups based on: the percentage of times during the second half of each trial the categorization unit corresponding to the current object was the most activated, and the average activation rates of current objects in all categorization tasks. In terms of the percentage of times in which current object was the most activated, our proposed HOG method had 81.98 percent accuracy as compared to that of ULBP of 72.98 percent and gray-scale averaging [1] of 20.1 percent (Fig. 3). However, in terms of average activation rates, the gray-scale averaging [1] method was only able to categorize the soft toy (table 1); while the ULBP (table 2) and our proposed HOG method (table 3) were successful for all of the categories. More also, our proposed HOG method had an average activation accuracy rates of 92.42 as compared to ULBP of 85.61 percent and gray-scale averaging [1] method of 50.41 percent of current objects in all categorization tasks (Fig. 4). In respect to the average activation in which our proposed HOG method and the ULBP were both successful in all categorization tasks, the higher differences between the current categories and the other categories in our proposed method as compared to that of the ULBP shows the statistical significant of our method over the ULBP.

Table 3. Experiment Three (Histogram of Oriented Gradients): Evaluation Test

Current Objects	Average Activation Rates of Objects (Highest Activation Rates in Bold)				
	Soft toy	Remote control	Microphone	Board wiper	Hammer
Soft toy	**0.920484**	0.399340	0.024266	0.555876	0.095414
Remote control	0.363643	**0.819481**	0.005429	0.221149	0.612242
Microphone	0.114605	0.066181	**0.921041**	0.072493	0.036939
Board wiper	0.508343	0.067864	0.190685	**0.992130**	0.571213
Hammer	0.038975	0.539647	0.002391	0.481335	**0.967917**

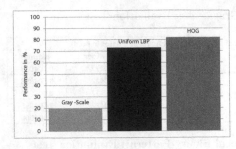
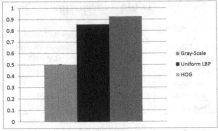

Fig. 3. Accuracy rates in terms of percentage of correct classification

Fig. 4. Accuracy rates in terms of average activations

4   Conclusion

We have done a preliminary investigation using HOG for pre-processing images taken from Humanoid (iCub) robot camera for an active vision system controlled by an evolved neural network. Our proposed method performed better than the ULBP and the gray-scale averaging [1] methods. The results of our investigation show that this kind of approach using pre-processing techniques could be used for solving more complex image analysis of control problems such as in robot vision. Future investigation will be in deep learning techniques as a form of pre-processing so as to give better motor responses in enhancing categorization capabilities.

References

1. Mirolli, M., Ferrauto, T., Nolfi, S.: Categorisation through Evidence Accumulation in an Active Vision System. Connection Science **22**, 331–354 (2010)
2. Lanihun, O., Tiddeman, B., Tuci, E., Shaw, P.: Enhancing active vision system categorization capability through uniform local binary patterns. In: Headleand, C.J., Teahan, W.J., Ap Cenydd, L. (eds.) ALIA 2014. CCIS, vol. 519, pp. 31–43. Springer, Heidelberg (2015)
3. James, D., Tucker, P.: Evolving a neural network active vision system for shape discrimination. In: Genetic and Evolutionary Computation Conference (2005)
4. Ojala, T., Pietikinen, M., Maenpaa, T.: Multi-Resolution Gray-Scale and Rotation Invariant Texture Classification with Local Binary Patterns. IEEE Transactions, Pattern Analysis and Machine Intelligence **29**, 51–59 (2002)
5. Dalal, N., Triggs, B.: Histograms of Oriented Gradients for Human Detection. Computer Vision and Pattern Recognition **1**, 886–893 (2005)

ROBO-GUIDE: Towards Safe, Reliable, Trustworthy, and Natural Behaviours in Robotic Assistants

James Law$^{(\boxtimes)}$, Jonathan M. Aitken, Luke Boorman,
David Cameron, Adriel Chua, Emily C. Collins, Samuel Fernando,
Uriel Martinez-Hernandez, and Owen McAree

Sheffield Robotics, The University of Sheffield, Sheffield, UK
{j.law,jonathan.aitken,l.boorman,d.s.cameron,dxachua,e.c.collins,
s.fernando,uriel.martinez,o.mcaree}@sheffield.ac.uk

Abstract. In this paper we describe a novel scenario, whereby an assistive robot is required to use a lift, and results from a preliminary investigation into floor determination using readily-available information. The aim being to create an assistive robot that can naturally integrate into existing infrastructure.

1 Introduction

For assistive robots (in roles such as carers, guides, companions, and assistants) to be most effective, they will need to seamlessly integrate into the human-centric environments we have designed, interacting through natural communication methods, whilst being safe, reliable, and trustworthy. The ROBO-GUIDE (ROBOtic GUidance and Interaction DEvelopment) project is an interdisciplinary project bringing together engineers and scientists working in computational neuroscience, control systems, formal verification, natural language, and psychology, to address how such a system can be designed and built with a holistic view to deployment. The aim of this project is to develop a guide robot that can navigate inside a large working building in a safe and reliable way, amongst people who are not, on the whole, familiar with robotic technology. A novel scenario, which we are investigating in our current work, is how the robot navigates between floors using a lift.

Prior work in this area has tackled issues relating to identifying and entering a lift [8][1], identifying and pressing buttons within the lift [5][4], or using a wireless interface to control the lift [2]. The novelty of the work in this paper is in addressing the problem of floor determination, and doing so using sensors available to the robot for the purpose of interacting with visitors and navigating corridors. Similar work by Kang et al. [3], demonstrated a mechanism for determining which floor the lift is on by reading the floor indicators, though we go further to describe an approach based on multiple sensor modalities, which adds useful redundancy when measurements from a single source are unreliable.

© Springer International Publishing Switzerland 2015
C. Dixon and K. Tuyls (Eds.): TAROS 2015, LNAI 9287, pp. 149–154, 2015.
DOI: 10.1007/978-3-319-22416-9_17

In the remainder of this paper, we present initial results from an experiment to investigate how confidently the robot can identify a particular floor within a building, using a combination of readily-available navigational information.

2   Floor Determination

For this application, we use the Pioneer LX from Adept MobileRobots[1], which has been extended with a microphone and camera to enable it to interact with building users. For the floor determination task, these sensors along with the inbuilt laser rangefinder allow us to detect and analyse: announcements made within the lift; the floor signage outside the lift; and the local building layout.

We analyse the ability of the robot to identify which floor it is on using each source alone, then combine the measures using a Bayesian filter and assess whether these available measures are sufficient for floor determination.

2.1   Vision

Outside the lift on each floor of the building[2] is an information panel, shown to the left of Fig. 1a, which can be used to visually check which floor the viewer is on. Due to the distance of the panel from the lift, and the relatively low camera resolution, it is not possible for the robot to read the text via an Optical Character Recognition (OCR) technique from within the lift. Instead, the distance between two blue bars, indicating the building and current floor, is correlated with each floor.

Fig. 1a illustrates the process of floor detection based on an image of the information panel. Firstly, the image is thresholded in the Hue-Saturation-Value (HSV) colourspace to isolate the blue areas as a binary image. This is then passed to a blob-analysis routine which calculates the area and centroid position of the two largest blobs. The distance between the blobs is then calculated as

$$d_{scaled} = \frac{d_{centroid}}{\sqrt{A_{largest}}} \tag{1}$$

where $d_{centroid}$ is the Euclidean distance between the centroids and $A_{largest}$ is the area of the largest blob. This scaling ensures the algorithm is insensitive to the size of the information panel in a given image.

Finally, the scaled distance is compared with reference values for each floor to produce a Probability Mass Function (PMF)

$$P_i = \frac{1}{\rho}|d_i - d_{scaled}|^{-1} \tag{2}$$

---

[1] http://www.mobilerobots.com/ResearchRobots/PioneerLX.aspx

[2] Our experiments are based around the lift in the Pam Liversidge Building at The University of Sheffield. Floors are alphabetically labelled, from the ground up, and include floor 'C+' due to a neighbouring mezzanine level.

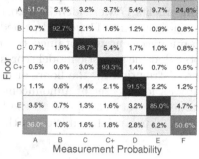

(a) Detection of an information panel   (b) PMFs for each information panel
before and after thresholding

Fig. 1. Visual processing of the information panels

where d_i is the reference distance of the ith floor's information board and ρ is a normalisation term given by

$$\rho = \sum_{i=1}^{N} |d_i - d_{scaled}|^{-1} \tag{3}$$

where N is the number of floors.

Fig. 1b illustrates the PMFs for the information panel on each floor. It can be seen that floors B-E are clearly distinguishable, however floors A and F are easily confused. This occurs as the blue bar highlighting floor A merges with that indicating the building, and the simple algorithm then measures the distance to the bar at the very bottom of the panel. This produces a d_{scaled} value similar to that for floor F.

2.2   Automatic Speech Recognition of Lift Announcements

For the acoustic data we collected 9 minutes and 35 seconds of audio over three sessions. During each session we travelled between all the floors to record the range of announcements that were given by the lift. The data was manually segmented and transcribed to separate each lift utterance from the rest of the background sound and other noises.

The data was analysed using our speech recognition system. We used a cross-validation approach to evaluate the output of the recogniser, using each combination of two sessions as training data and the third as the test. The recognition system was built with the Kaldi toolkit [6], using the SGMM decoding approach [7]. The acoustic models were trained on the WSJ British English spoken corpus. We used the speaker adaptation (SAT) scripts to adapt the models to the acoustic conditions of the lift. The pronunciation dictionary was designed to fit the phrases uttered by the lift, and the language model was created as a constrained grammar to only allow the phrases that are uttered by the lift.

In total there were 20 direction and 20 floor announcements. For the direction announcements 17 were identified correctly, giving an accuracy of 85%. For the floor announcements 11 were identified correctly, giving an accuracy of 55%. For each announcement we obtained the n-best paths output from the recogniser. We looked at the acoustic model scores output for each path to estimate the confidence distribution for the floors or directions, and converted these into PMFs. The PMFs for one set of announcements are shown in Fig. 2a. In many cases the values are very close, due to the level of background noise; this indicates a high level of potential confusion in the floor identification.

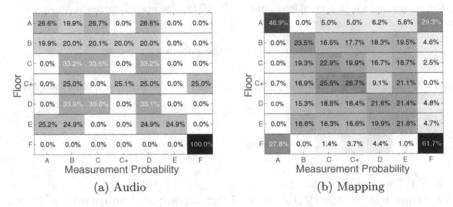

Fig. 2. PMFs for floor announcements and map localisation

2.3   Mapping

Before the experiment, we used the onboard laser scanner and inbuilt software (MobileEyes, Mapper 3, and ARNL Server) to map the 7 floors of the building. The floorplan of the area directly outside the lift is identical on every floor, and so it is necessary for the robot to leave this area before it can detect differences in layout afforded by the various floors.

For our experiment, the robot was driven to a point just outside the lift lobby on each floor, where it was able to detect the layout of the adjacent corridor. A measure of confidence as to which floor the robot was on was then generated. This was achieved by comparing the laser point cloud data from the robot with the pre-recorded maps of each floor, and using MobileEyes to generate a localisation score for each map. These scores were then converted into PMFs, and the results are given in Fig.2b. Floors C, C+, D, and E are similar in their layout.

2.4   Combined Floor Estimation

We now combine the measurements using a Bayesian filter to produce a final estimate of the robot's location. To improve this estimate it is assumed that the robot knows the floor it is starting from precisely and is able to detect the

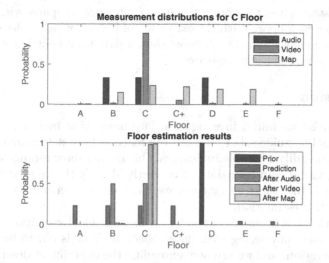

Fig. 3. Prediction of the current floor by the robot. Initially located on D floor, travelling to C floor.

direction of travel of the lift from the announcements with 85% confidence (i.e. there is a 15% chance the robot will think it is going down when it is going up).

Fig. 3 illustrates the execution of the Bayesian filter based on the data presented in the previous sections. In this analysis we assume the robot starts on floor D, the entrance level to our building. It can be seen that the prior distribution indicates the robot is 100% confident in its location (D Floor). After the lift begins to move, the robot's prediction of which floor it will leave on is only based on the direction, leading to an indistinct distribution.

Once the lift stops and announces the floor, the robot is unable to precisely distinguish the announcement between floors B, C and D. Combining this measurement distribution with the prediction, however, discounts D floor as the robot knows this is where it started.

After leaving the lift and capturing an image of the information board, a more distinct measurement is obtained and combined with the estimation. Finally, the robot begins to drive around the floor and assesses the validity of each floor map, increasing its confidence further. Comparison of the individual measurements with the final estimated distribution shows that the incorporation of the additional measures has increased the robot's confidence of being on floor C from a maximum of 89% (based on the video measurement) to over 98%.

Table 1. Comparison of the best confidence measurement for each floor with the estimate achieved by combining measurements

Floor	A	B	C	C+	D	E	F
Best Measurement	51.1%	92.7%	88.7%	93.3%	91.5%	85.0%	100%
Combined Estimate	99.7%	97.2%	98.3%	99.4%	98.3%	99.9%	100%

Table 1 illustrates how the floor estimation confidence compares with the best measurement for each floor. With the exception of F floor, which is detected perfectly by the audio measurement, all floors show a significant confidence improvement by aggregating multiple measures.

3   Conclusion

We have described an initial investigation into navigation between floors in a building by a mobile guide robot, and our approach to floor determination based on the fusion of readily-available indicators. Our results show a range of success in using audio, visual, and laser data to correctly identify the floor on which the robot is situated. In almost every case, we have also shown how the Bayesian filter improves this estimation.

In this work, measurements were taken in near ideal conditions, without other building users interfering with data collection. This is not to be expected in standard operation, and we are now extending the experiment described here to investigate a broad range of more realistic operating conditions.

References

1. Amano, R., Takahashi, K., Yokota, T., Cho, T., Kobayashi, K., Watanabe, K., Kurihara, Y.: Development of automatic elevator navigation algorithm for jaus-compliant mobile robot. In: Proc. SICE, pp. 828–833 (2012)
2. Cavallo, F., Limosani, R., Manzi, A., Bonaccorsi, M., Esposito, R., Di Rocco, M., Pecora, F., Teti, G., Saffiotti, A., Dario, P.: Development of a socially believable multi-robot solution from town to home. Cognitive Comp. **6**(4), 954–967 (2014)
3. Kang, J.G., An, S.Y., Choi, W.S., Oh, S.Y.: Recognition and path planning strategy for autonomous navigation in the elevator environment. International Journal of Control, Automation and Systems **8**(4), 808–821 (2010)
4. Klingbeil, E., Carpenter, B., Russakovsky, O., Ng, A.Y.: Autonomous operation of novel elevators for robot navigation. In: 2010 IEEE International Conference on Robotics and Automation (ICRA), pp. 751–758. IEEE (2010)
5. Miura, J., Iwase, K., Shirai, Y.: Interactive teaching of a mobile robot. In: Proceedings of the 2005 IEEE International Conference on Robotics and Automation, ICRA 2005, pp. 3378–3383 (April 2005)
6. Povey, D., Burget, L., Agarwal, M., Akyazi, P., Feng, K., Ghoshal, A., Glembek, O., Goel, N.K., Karafiát, M., Rastrow, A., et al.: Subspace gaussian mixture models for speech recognition. In: 2010 IEEE International Conference on Acoustics Speech and Signal Processing (ICASSP), pp. 4330–4333. IEEE (2010)
7. Povey, D., Ghoshal, A., Boulianne, G., Burget, L., Glembek, O., Goel, N., Hannemann, M., Motlicek, P., Qian, Y., Schwarz, P., Silovsky, J., Stemmer, G., Vesely, K.: The kaldi speech recognition toolkit. In: IEEE 2011 Workshop on Automatic Speech Recognition and Understanding. IEEE Signal Processing Society (December 2011)
8. Simmons, R., Goldberg, D., Goode, A., Montemerlo, M., Roy, N., Sellner, B., Urmson, C., Bugajska, M., Coblenz, M., Macmahon, M., Perzanowski, D., Horswill, I., Zubek, R., Kortenkamp, D., Wolfe, B., Milam, T., Inc, M., Maxwell, B.: Grace: An autonomous robot for the AAAI robot challenge. AI Magazine **24**, 51–72 (2003)

Mechanical Design of Long Reach Super Thin Discrete Manipulator for Inspections in Fragile Historical Environments

Jason Liu[1](✉), Robert Richardson[1], Rob Hewson[2], and Shaun Whitehead[3]

[1] School of Mechanical Engineering, University of Leeds, Leeds, UK
{mn07jhwl,r.c.richardson}@leeds.ac.uk
[2] Faculty of Engineering, Imperial College, London, UK
r.hewson@imperial.ac.uk
[3] Scoutek Ltd., Saltburn-by-the-Sea, UK
shaun@scoutek.com

Abstract. Long reach and small diameter manipulators are ideal for borehole deployments into search and rescue scenarios and fragile historical environments. Small diameter passageways impose constraints on a snake arm manipulator which severely limit its performance and capabilities. This work investigates the effects of tendon tensions on the maximum working length of a snake arm under tight size constraints and how the maximum length is achieved through an algorithmic approach and consideration of how and when key parts fail.

Keywords: Exploration · Long reach · Discrete backbone · Robot archaeology · Snake arm · Tendon tension · Minimally invasive · Small diameter

1 Introduction

The application of robotic devices has been widely used in exploration and search and rescue (SAR) scenarios [1]. Ideally deployed where human risk is considered too high [2], tools such as the snake arm are often important for examining confined space environments where humans and some robots struggle. These robotic platforms are profoundly influenced by their intended environments and most exploit a single locomotion mechanism to operate in the complex terrains [3-5].

Different environments vary greatly from one to another, and produces a level of uncertainty and challenges for the end user [4]. It can be desirable for small boreholes to be used as means of access. Small boreholes will be faster to drill and reduces secondary collapse hazards; they are also less destructive and aid to preserve a site. Snake arms already have all the necessary locomotion parts anchored to a mobile platform outside the borehole [6-8], this allows the snake arm to fully utilize a boreholes diameter which plays a vital role in the snake arm's length.

The length of a snake arm is representative of the maximum working distance possible with a manipulator. Current small diameter snake arms include the continuous DTRA arm by OC Robotics with a reach of 610 mm and an outer diameter of

© Springer International Publishing Switzerland 2015
C. Dixon and K. Tuyls (Eds.): TAROS 2015, LNAI 9287, pp. 155–160, 2015.
DOI: 10.1007/978-3-319-22416-9_18

12.5 mm [9]. On the other hand long reach snake arms with large diameters already exist where mechanisms to compensate for gravity are possible to achieve unsupported lengths of 6 m with a 100 mm diameter [9].

In this paper the design of a triple jointed snake arm manipulator that conforms to a very restrictive small diameter constraint is firstly introduced, and then an analysis of the snake arm theory is discussed. Lastly, an algorithmic approach is used to determine the maximum working length for the snake arm.

2   Description of the Basic Snake Arm

A discrete backbone snake arm simplifies kinematic formulations and motion control over its continuous backbone counterpart. Formed from a series of links and joints and actuated by a minimum of three tendons per joint, these tendons run through each link and terminate at each joint it is assigned to control. Assuming the boreholes are straight, the snake arm is not expected to maneuver around obstacles until it breaches through the borehole into a target chamber.

A snake arm capable of self-supporting the full length of its own arm in the deployed environment is advantageous for surveying fragile and historically important tomb-like chambers because there would be no need for any contact between the arm and surfaces for any risk of damage to occur. Fig. 1 shows a 12 mm diameter snake arm consisting of a base link (of length B), three two degrees of freedom (DOF) joints (of length J) and three links of identical length (of length L). The diameter places physical constraints on the number of cables controlling the snake arm joints, the thickness of the tubing that makes the links and the diameter of the two DOF joints.

Increasing link length L to create a longer snake arm has the effect of increasing the tendon tensions required to maintain the snake arms horizontal cantilever position. These forces result in greater axial compressive forces acting through the snake arm and possibly leading to joint failure and/or buckling of the links.

As a consequence, a method was required to theoretically calculate the tendon tensions from the snake arms kinematics and analyze the values to determine whether any anticipated failure modes will occur.

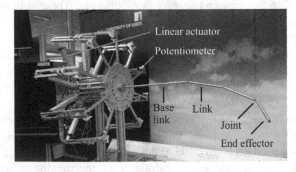

Fig. 1. 12 mm diameter snake arm

3   Discrete Snake Arm Kinematics and Statics

The position of the end effector with respect to the base frame is computed with forward kinematics using transformation matrices produced from joint angles and link lengths. Combined with the Recursive Newton-Euler (RNE) method the toque at each joint is computed and carried over to calculate the cable tensions.

The RNE joint torques reaches a maximum when the snake arm is at a horizontal cantilever position without additional external forces other than gravity acting upon it; therefore at this point it is assumed the associated tendon tensions are also at its maximum. This horizontal state should then be where failure of the snake arm is most likely to occur and is where this analysis is focused on.

Calculating tendon tensions from joint torques can be performed if tensions are assumed constant throughout with negligible friction and all joint angles are known. For a snake arm with two joints and two tendons of link length L, weight W, payload weight P and perpendicular tendon distance of Dy as shown in Fig. 2, multiple tendons cannot occupy the same space for all joints, therefore some tendons are displaced radially about the center. This creates an undesirable lateral load and requires the introduction of additional tendons to counteract the loads.

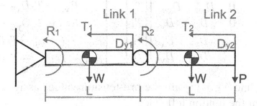

Fig. 2. Simplified snake arm with two joints and two tendons

The RNE method is used to calculate the joint torques R_1 and R_2 and the tendon tensions calculated for the ith joint is expressed as:

$$R_{xy}i = \sum_{j=i}^{N} T_j D_{xy}j \tag{1}$$

Where $R_{xy}i$ is the torque generated at joint i for both yaw (R_x) and pitch (R_y) joint directions. T_j is the tension of each tendon that passes through or terminates at joint j. $D_{xy}j$ is defined as the distance between tendon j and the neutral axis of joint i, it can also be a negative value dependent on the direction the tendon j transmits its force on joint i. Applying equation (1) to the double jointed snake arm as shown on Fig. 2, the equations relating torques to tensions can be produced and solved using matrices.

$$\begin{bmatrix} 0 & D_{y2} \\ D_{y1} & D_{y2} \end{bmatrix} \begin{bmatrix} T_1 \\ T_2 \end{bmatrix} = \begin{bmatrix} R_{y2} \\ R_{y1} \end{bmatrix} \tag{2}$$

Where the tensions T_1 and T_2 can be solved as:

$$\begin{bmatrix} T_1 \\ T_2 \end{bmatrix} = \begin{bmatrix} \dfrac{R_{y1}}{D_{y1}} - \dfrac{R_{y2}}{D_{y1}} \\ \dfrac{R_{p2}}{D_{y2}} \end{bmatrix} \tag{3}$$

For a three jointed snake arm with six DOF and nine control tendons, equation (1) creates a system of linear equations with infinite solutions.

$$
\begin{bmatrix}
0 & 0 & D_{x3} & 0 & 0 & D_{x6} & 0 & 0 & D_{x9} \\
0 & 0 & D_{y3} & 0 & 0 & D_{y6} & 0 & 0 & D_{y9} \\
0 & D_{x2} & D_{x3} & 0 & D_{x5} & D_{x6} & 0 & D_{x8} & D_{x9} \\
0 & D_{y2} & D_{y3} & 0 & D_{y5} & D_{y6} & 0 & D_{y8} & D_{y9} \\
D_{x1} & D_{x2} & D_{x3} & D_{x4} & D_{x5} & D_{x6} & D_{x7} & D_{x8} & D_{x9} \\
D_{y1} & D_{y2} & D_{y3} & D_{y4} & D_{y5} & D_{y6} & D_{y7} & D_{y8} & D_{y9}
\end{bmatrix}
\begin{bmatrix}
T_1 \\ T_2 \\ T_3 \\ T_4 \\ T_5 \\ T_6 \\ T_7 \\ T_8 \\ T_9
\end{bmatrix}
=
\begin{bmatrix}
R_{y3} \\ R_{x3} \\ R_{y2} \\ R_{x2} \\ R_{y1} \\ R_{x1}
\end{bmatrix}
\tag{4}
$$

As the tendons with a negative D_y would not contribute to overcoming the gravity acting on the snake arm, these can be given pre-tension values and this action results in a solvable matrix and the necessary equations for a theoretical tendon tensions.

$$
\begin{bmatrix}
0 & 0 & D_{x3} & 0 & 0 & D_{x9} \\
0 & 0 & D_{y3} & 0 & 0 & D_{y9} \\
0 & D_{x2} & D_{x3} & 0 & D_{x8} & D_{x9} \\
0 & D_{y2} & D_{y3} & 0 & D_{y8} & D_{y9} \\
D_{x1} & D_{x2} & D_{x3} & D_{x4} & D_{x8} & D_{x9} \\
D_{y1} & D_{y2} & D_{y3} & D_{y4} & D_{y8} & D_{y9}
\end{bmatrix}
\begin{bmatrix}
T_1 \\ T_2 \\ T_3 \\ T_4 \\ T_8 \\ T_9
\end{bmatrix}
=
\begin{bmatrix}
R_{y3} - T_6 D_{x6} \\
R_{x3} - T_6 D_{y6} \\
R_{y2} - T_5 D_{x5} - T_6 D_{x6} \\
R_{x2} - T_5 D_{y5} - T_6 D_{y6} \\
R_{y1} - T_5 D_{x5} - T_6 D_{x6} - T_7 D_{x7} \\
R_{x1} - T_5 D_{y5} - T_6 D_{y6} - T_7 D_{y7}
\end{bmatrix}
\tag{5}
$$

The tensions T_5, T_6 and T_7 in (5) are the pre-tension values assigned, where a value of zero would represent the tendon left slack.

4   Cable Tension Experiment

To measure the tension of the tendons in multiple configurations, a test rig modelled after Fig. 2 was assembled with two single DOF joints and links of constant length with capacity for five joints and fifteen tendons at three different diameters.

The methods consisted of holding the arm at a horizontal cantilever position and incrementally attach weights to each tendon. When the arm is released and maintains its position with no change to joint angles the weight is recorded and the process repeated. If inadequate tension is supplied to the tendons the arms would collapse and if too much tension is supplied the arm will rise beyond the horizontal starting position. To alter the torque at each joint only the payload was incremented.

The results shown on Fig. 3 show a close trend between the theoretical equation (3) and the experimental results. Further investigation revealed the zero-shift in the data was the result of friction in the test rig between the tendon, vertebra and pulleys and the theoretical calculations which were assumed to have negligible. To compensate for the error, the relationship between the load acting on the tendons and friction for the test rig was measured with the experimental values adjusted for the additional friction forces dependent on the load on the tendons.

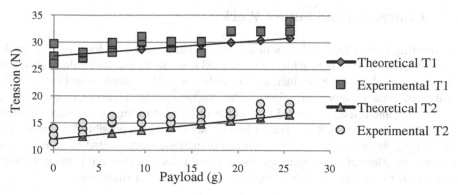

Fig. 3. Comparison between theoretical and experimental tendon tensions for a double jointed snake arm with two tendons

5   Maximum Working Length

MATLAB was used to determine the maximum length of snake arm by algorithmically increasing the length of both links whilst at each step checking against the anticipated failure modes (Table 1). The implementation of factors of safety (FOS) reduces the lengths as a compromise for increased reliability.

The approach was to firstly determine the maximum length of the Links and thereafter determine the length of the Base link. This is because the Base link length does not affect the tendon tensions, the calculations for buckling failure of the three Links or the forces acting onto the 2DOF joints. However the axial stress resulting from the tendon tensions acting through the Base link and weight of the unsupported snake arm does affect the likelihood of buckling for the Base link.

Strongly dependent on the design of the snake arm, the order the failure modes will materialize is difficult to determine and so each mode was analysed individually in the design.

Table 1. Anticipated snake arm failure modes and Factors of Safety

Failure Mode	Check	FOS
Tendon	Each calculated tendon tension was compared to the tendons experimentally found yield stress.	6
Joints	Using Finite element analysis to find the load required for joint failure and compare to the predicted compressive axial loads.	4
Link Buckling	Axial stress through the links was analyzed for buckling using buckling theory for thin walled cylinders in axial compression [10,11].	4

This approach yielded a snake arm of total length 1.011m with a Link length of 0.161m and Base link of 0.478m for the 12mm diameter snake arm.

6   Conclusion and Future Work

The limiting factors for the design of the snake arm were the tendons and link buckling. A length of 1.676 m is achieved if the FOS is set to the point of failure. However a better approach would be to increase the yield stress of the tendons and design links capable of greater compressive loads. The outcome of the algorithmic approach to find the maximum length resulted in a length of 1.011m making this design not ideal for SAR scenarios where much greater lengths are crucial but possible for archeology.

In future, the snake arm will be examined to increase reach. Including an investigation into the effect of axial, bending, and twisting forces through the joints and lastly, further tests by performing fielded experiments in real world situations.

References

1. Jueyao, W., Xiaorui, Z., Fude, T., Tao, Z., Xu, X.: Design of a modular robotic system for archaeological exploration. In: IEEE International Conference on Robotics and Automation, ICRA 2009, pp. 1435–1440, May 12–17, 2009
2. Guizzo, E.: Robots Enter Fukushima Reactors, Detect High Radiation (2011). http://spectrum.ieee.org/automaton/robotics/industrial-robots/robots-enter-fukushima-reactors-detect-high-radiation
3. Daler, L., Lecoeur, J., Hahlen, P.B., Floreano, D.: A flying robot with adaptive morphology for multi-modal locomotion. In: 2013 IEEE/RSJ International Conference on Intelligent Robots and Systems (IROS), pp. 1361–1366, November 3–7, 2013
4. Morris, A., Ferguson, D., Omohundro, Z., Bradley, D., Silver, D., Baker, C., Thayer, S., Whittaker, C., Whittaker, W.: Recent developments in subterranean robotics. Journal of Field Robotics 23(1), 35–57 (2006)
5. Murphy, R.R., Kravitz, J., Stover, S., Shoureshi, R.: Mobile robots in mine rescue and recovery. Robotics & Automation Magazine, IEEE 16(2), 91–103 (2009)
6. Hirose, S.: Biologically inspired robots: snake-like locomotors and manipulators. Oxford University Press (1993)
7. Buckingham, R.O., Graham, A.C.: Dexterous manipulators for nuclear inspection and maintenance - case study. In: 2010 1st International Conference on Applied Robotics for the Power Industry (CARPI), pp. 1–6, October 5–7 2010
8. Junhu, H., Rong, L., Ke, W., Hua, S.: The mechanical design of snake-arm robot. In: 2012 10th IEEE International Conference on Industrial Informatics (INDIN), pp. 758–761, July 25–27, 2012
9. Chalfoun, J., Bidard, C., Keller, D., Perrot, Y., Piolain, G.: Design and flexible modeling of a long reach articulated carrier for inspection. In: IEEE/RSJ International Conference on Intelligent Robots and Systems, IROS 2007, pp. 4013–4019, October 29, 2007–November 2, 2007
10. Howard Allen, P.B.: Cylindrical shell in axial compression. In: Background to Buckling, pp. 515–524. McGraw-Hill Book Company (UK) Limited (1980)
11. Hunter, D.F.: ESDU 88034 avoidance of buckling of some engineering elements (struts, plates and gussets). In: IHS ESDU (1988)

Robot Mapping and Localisation for Feature Sparse Water Pipes Using Voids as Landmarks

Ke Ma[1,2], Juanjuan Zhu[2], Tony J. Dodd[1], Richard Collins[2],
and Sean R. Anderson[1 (✉)]

[1] Department of Automatic Control and Systems Engineering,
University of Sheffield, Sheffield S1 3JD, UK
s.anderson@sheffield.ac.uk
[2] Department of Civil and Structural Engineering,
University of Sheffield, Sheffield S1 3JD, UK

Abstract. Robotic systems for water pipe inspection do not generally include navigation components for mapping the pipe network and locating damage. Such navigation systems would be highly advantageous for water companies because it would allow them to more effectively target maintenance and reduce costs. In water pipes, a major challenge for robot navigation is feature sparsity. In order to address this problem, a novel approach for robot navigation in water pipes is developed here, which uses a new type of landmark feature - voids outside the pipe wall, sensed by ultrasonic scanning. The method was successfully demonstrated in a laboratory environment and showed for the first time the potential of using voids for robot navigation in water pipes.

Keywords: Robot navigation · Mapping · Localisation · Water pipes

1 Introduction

Water, a highly precious resource, is distributed to buildings by networks of pipes. As pipe materials age they are prone to damage, which can cause wastage of water and bacterial infiltration. Water distribution systems therefore require inspection, maintenance and repair [1]. However, water pipes are usually buried and so are difficult to access. Robotic systems have great potential for inspecting these inaccessible pipelines [2]. Whilst there are many techniques for robot pipe inspection itself [3], an as yet unsolved problem is accurately locating damage in pipes once found, to effectively target repair. This problem can be addressed by robot navigation algorithms, specialised to water pipes, which is the focus of this paper.

There are a number of challenges for robot navigation in water pipes. Firstly and most importantly the water pipe is a feature sparse environment. Most current robot navigation systems deal with indoor and outdoor environments, which contain numerous landmark features. However, pipe walls lack features that can be used as landmarks. Secondly, in pipes standard range and bearing sensors can

© Springer International Publishing Switzerland 2015
C. Dixon and K. Tuyls (Eds.): TAROS 2015, LNAI 9287, pp. 161–166, 2015.
DOI: 10.1007/978-3-319-22416-9_19

only detect features that are nearby due to the close and enclosing proximity of the surrounding pipe wall. Thirdly, unlike indoor or outdoor navigation, the in-pipe robot has a very restricted route (moving either forward or backward), which limits the perspective of the robot on landmark features. Therefore, robot navigation in water distribution pipes is a difficult problem.

Robot navigation is often performed using simultaneous localisation and mapping (SLAM), which is where the location of the robot and the map features are represented in the state vector of a state-space model [4]. State estimation techniques are then used to construct the map and localise the robot, using e.g. the extended Kalman filter [5] or the Rao-Blackwellised particle filter [6]. Although these techniques are well-developed, they have rarely been applied in the water pipe environment. Mapping and localisation has been attempted in water pipes based on cameras and inertial measurement units (IMUs) [7,8]. However, the use of cameras is limited by the lack of visual features, and IMUs are subject to drift, meaning that the navigation problem has yet to be adequately solved.

For in-pipe navigation, the feature sparsity problem motivates the development of sensing techniques that can transform the water pipe into a feature-rich environment. The aim of this paper is to improve water pipe navigation by exploiting a novel type of feature - voids that occur outside the pipe wall. Ultrasonic signals can penetrate the wall of plastic pipes to detect the depth between the outside soil and the pipe (and plastic pipes are now typically used by water utilities, especially in the UK). Significant gaps between the pipe wall and soil - voids - can be used as landmark features for navigation. Ultimately, these features can be used to build maps, localise the robot and fuse with standard sensing based on cameras and IMUs.

In this contribution we develop a novel framework for mapping and localisation in water pipes using voids as features, detected by ultrasonic scanning through the pipe wall, and demonstrate the approach experimentally in a laboratory environment. The experimental results show that ultrasonic sensing can be used to successfully build a map based on soil depth outside the pipe wall, and the navigation algorithm can be used to localise using voids as features.

2   Methods

This section describes the navigation algorithm using voids as features, as well as the experimental setup and data used to evaluate the algorithm performance.

2.1   Robot Navigation Algorithm

Problem Statement. The problem that we address here is: (i) to construct a map $g(\mathbf{x}_k)$ for the water pipe environment, that can be used to transform from robot pose $\mathbf{x}_k \in \mathbb{R}^{n_x}$ at time-step k to sensor measurements $\mathbf{y}_k \in \mathbb{R}^{n_y}$, where $g : \mathbf{x}_k \to \tilde{\mathbf{y}}_k$, where typically the robot pose $\mathbf{x}_k = [x\ y\ \theta]^T$, i.e. $\mathbf{x}_k$ contains the spatial location in x-y co-ordinates and heading θ, and $\tilde{\mathbf{y}}_k$ is the noise-free sensor output; and (ii) localise the robot by obtaining the estimate of the pose distribution $p(\mathbf{x}_k|\mathbf{y}_k)$.

Robot Dynamics and Measurement Model. We assume that the dynamics of the water pipe robot can be represented by a state-space model, with state dynamics

$$p(\mathbf{x}_k|\mathbf{x}_{k-1}, \mathbf{u}_{k-1}) \Leftrightarrow \mathbf{x}_k = A\mathbf{x}_{k-1} + B\mathbf{u}_{k-1} + \mathbf{w}_k \tag{1}$$

where $\mathbf{u}_k \in \mathbb{R}^{n_u}$ is the input, assumed to arise from an actuator such as a motor, A is the state transition matrix, B is the input matrix and $\mathbf{w}_k \sim N(0, Q)$ is the state noise. For this investigation we set the state dimension to $n_x = 1$ to only represent the location of the ultrasonic probe in 1-dimension (moving forwards and backwards along a line), with no heading information. The dynamics are described by $A = 1$, $B = [-1\ 1]$, and $\mathbf{u}_{k-1} = [m_{k-1}\ m_k]^T$, where m_k is a motor encoder value - effectively the dynamic model predicts ultrasonic probe location using distance travelled obtained from a motor encoder.

The state-space measurement model is

$$p(\mathbf{y}_k|\mathbf{x}_k) \Leftrightarrow \mathbf{y}_k = g(\mathbf{x}_k) + \mathbf{v}_k \tag{2}$$

where $\mathbf{v}_k \sim N(0, R)$ is the measurement noise. In this case, we use only a single ultrasonic probe, hence $n_y = 1$, although the framework as presented above readily permits the extension to multiple sensors. The nonlinear function $g(.)$ in this case is the mapping from probe location, x_k, to soil depth y_k, which by definition is the map of soil depth over space (see Results, Fig. 2(a)).

Estimation of Robot Location. In order to estimate the location of the probe, i.e. the distribution $p(\mathbf{x}_k|\mathbf{y}_k)$, we used a sequential Monte Carlo algorithm, specifically the bootstrap version of the particle filter, based on sequential importance resampling [9]. Firstly, the particle filter samples are initialised from the prior, which is the starting location of the robot,

$$\mathbf{x}_0^{(i)} \sim p(\mathbf{x}_0), \quad i = 1, \dots, n_s \tag{3}$$

where n_s is the number of samples, and the weights associated with these particles are initialised to $w_0^{(i)} = \frac{1}{n_s}$, for $i = 1, \dots, n_s$. The particle filter algorithm then iterates through the following steps at each sample-time k:

1. The location is predicted by samples drawn from the state equation, Eq. 1,

$$\mathbf{x}_k^{(i)} \sim p(\mathbf{x}_k|\mathbf{x}_{k-1}^{(i)}, \mathbf{u}_{k-1}), \quad i = 1, \dots, n_s \tag{4}$$

   where we make the standard assumption that the state equation can be used as the importance distribution of the particle filter [9].
2. The weights are updated as

$$w_k^{(i)} \propto p(\mathbf{y}_k|\mathbf{x}_k^{(i)}), \quad i = 1, \dots, n_s \tag{5}$$

   where from the assumption of Gaussian noise $\mathbf{v}_k$ on the sensor output,

$$w_k^{(i)} = \exp\left(-\frac{1}{2}\left(\mathbf{y}_k - \hat{\mathbf{y}}_k^{(i)}\right)^T R^{-1} \left(\mathbf{y}_k - \hat{\mathbf{y}}_k^{(i)}\right)\right), \quad i = 1, \dots, n_s \tag{6}$$

where the map $g(.)$ is used to predict the soil depth from the sensor location, $\hat{\mathbf{y}}_k^{(i)} = g\left(\mathbf{x}_k^{(i)}\right)$. The weights are then normalised to sum to unity.

3. The final step is a check for particle degeneracy by counting the effective number of particles - if less than some threshold γ resampling is performed using the stratified resampling algorithm [9].

To implement the localisation algorithm the following parameters were used: $n_s = 300$, $\sqrt{Q} = 7$, $\sqrt{R} = 120$, and $\gamma = 0.6 n_s$.

In the absence of any detected void, i.e. when the observed soil depth is below some small threshold ρ (here $\rho = 100$), the samples are predicted at each time step k with no correction. This results in an increasing spread of samples until a void feature is encountered, which then corrects the location estimate.

2.2   Experimental Setup

In the laboratory setup an ultrasonic transducer was moved through a water bath over plastic pipe material to emulate the water pipe environment. The base of the water bath was covered in soil, with the plastic pipe material resting on top (Fig. 1). At certain locations in the soil, voids were inserted to create landmark features for evaluating the navigation algorithm.

The ultrasonic transducer had a central excitation frequency of 10 MHz and focal distance of 75 mm, mounted to the gantry of a stepper motor driven scanning table. The transducer was pulsed at a rate of 160 pulse/s using PC mounted pulser-receiver and digitisation cards. The location of the transducer was recorded for each pulse. The reflected ultrasound was windowed such that the reflections extending from the upper pipe surface to approximately 80 mm past the lower surface of the pipe could be observed and digitised at a rate of 100 MSamples/s.

3   Results

To evaluate the use of voids as features for navigation, an ultrasonic probe was used to scan soil beyond plastic pipe material, in a setup that was designed to emulate the water pipe environment (Fig. 1). When the sensor was moved from left-to-right, the soil depth was measured and the void map was created (Fig. 2(a)). After the sensor reached the right end point, the map construction was finished, including voids, and was then used for localisation when moving from right-to-left.

To evaluate results, the motor location from the x-y table was used as the ground truth because it is highly accurate. To illustrate the effectiveness of localisation using voids a degraded version of the motor signal was used in the state equation to predict location, where the motor position was corrupted by white noise and a sinusoidal term.

The key result on the localisation is that when there is no void, the samples increasingly spread, but once a void is present, the location estimate is corrected,

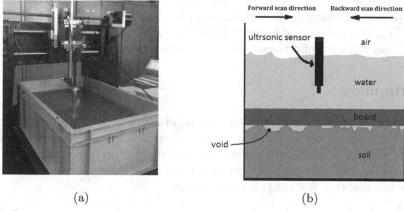

(a)   (b)

Fig. 1. Experimental setup in the laboratory environment. (a) Ultrasonic sensing probe, mounted on an x-y motorised arm in a water bath. At the base of the water bath is a layer of soil, over which is a plastic board of similar width and material to water pipe. (b) Diagram of the lab setup shown in panel (a).

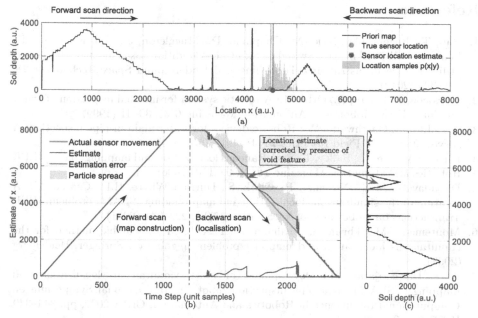

Fig. 2. Mapping and localisation experimental results. (a) Map of the soil depth outside the pipe wall, which corresponds to the observation function $g(.)$. The location estimate in blue corresponds to the sample with the maximum weight. (b) Mapping and localisation results where the map is constructed on the forward pass and localisation is performed on the backward pass. The location estimate in blue corresponds to the sample with the maximum weight. (c) Map of soil depth as shown in panel (a) but orientated to correspond to panel (b).

e.g. see time step $k = 1650$ in Fig. 2(b). Beyond time step $k = 1700$, when the void is passed, the samples diverge again. At $k = 2100$, the probe reaches another void, and all sample estimates once more converge on the true location. Crucially, this demonstrates the effectiveness of using voids as features for robot navigation.

4   Summary

The aim of this paper was to develop the use of a new type of feature, voids, for use in water pipe robot navigation. The navigation framework was successfully demonstrated in a laboratory experimental setup. In the future, this approach could be combined with multiple sensors, in order to transform water pipes from a feature sparse to a feature rich environment.

Acknowledgments. The authors gratefully acknowledge that this work was supported by the EPSRC, UK, grant Assessing the Underworld (EP/K021699/1) and a PhD scholarship award to Ke Ma by the University of Sheffield.

References

1. Hao, T., Rogers, C., Metje, N., Chapman, D., Muggleton, J., Foo, K., Wang, P., Pennock, S.R., Atkins, P., Swingler, S., et al.: Condition assessment of the buried utility service infrastructure. Tunnelling and Underground Space Technology **28**, 331–344 (2012)
2. Moraleda, J., Ollero, A., Orte, M.: A robotic system for internal inspection of water pipelines. IEEE Robotics & Automation Magazine **6**(3), 30–41 (1999)
3. Tur, J.M.M., Garthwaite, W.: Robotic devices for water main in-pipe inspection: A survey. Journal of Field Robotics **27**(4), 491–508 (2010)
4. Durrant-Whyte, H., Bailey, T.: Simultaneous localization and mapping: part I. IEEE Robotics & Automation Magazine **13**(2), 99–110 (2006)
5. Dissanayake, M.G., Newman, P., Clark, S., Durrant-Whyte, H.F., Csorba, M.: A solution to the simultaneous localization and map building (SLAM) problem. IEEE Transactions on Robotics and Automation **17**(3), 229–241 (2001)
6. Montemerlo, M., Thrun, S., Siciliano, B.: FastSLAM: A scalable method for the simultaneous localization and mapping problem in robotics. Springer, Heidelberg (2007)
7. Krys, D., Najjaran, H.: Development of visual simultaneous localization and mapping (VSLAM) for a pipe inspection robot. In: International Symposium on Computational Intelligence in Robotics and Automation, CIRA 2007, pp. 344–349. IEEE (2007)
8. Lim, H., Choi, J.Y., Kwon, Y.S., Jung, E.-J., Yi, B.-J.: SLAM in indoor pipelines with 15mm diameter. In: IEEE International Conference on Robotics and Automation, ICRA 2008, pp. 4005–4011 (2008)
9. Sarkka, S.: Bayesian Filtering and Smoothing. Cambridge University Press, Cambridge (2013)

Standardized Field Testing of Assistant Robots in a Mars-Like Environment

Graham Mann[1]([✉]), Nicolas Small[1], Kevin Lee[2], Jonathan Clarke[3], and Raymond Sheh[4]

[1] Murdoch University, Murdoch, WA, Australia
{g.mann,n.small}@murdoch.edu.au
[2] Nottingham Trent University, Nottingham, UK
kevin.lee@ntu.ac.uk
[3] Mars Society Australia, Clifton Hill, VIC, Australia
president@marssociety.org.au
[4] Curtin University, Bentley, WA, Australia
raymond.sheh@curtin.edu.au

Abstract. Controlled testing on standard tasks and within standard environments can provide meaningful performance comparisons between robots of heterogeneous design. But because they must perform practical tasks in unstructured, and therefore non-standard, environments, the benefits of this approach have barely begun to accrue for field robots. This work describes a desert trial of six student prototypes of astronaut-support robots using a set of standardized engineering tests developed by the US National Institute of Standards and Technology (NIST), along with three operational tests in natural Mars-like terrain. The results suggest that standards developed for emergency response robots are also applicable to the astronaut support domain, yielding useful insights into the differences in capabilities between robots and real design improvements. The exercise shows the value of combining repeatable engineering tests with task-specific application-testing in the field.

Keywords: Test methods · Field testing · Astronaut assistant robots

1 Introduction

By their nature, field robots are difficult to evaluate objectively. Apart from the complexity of the machines themselves, they must operate in natural, unstructured environments, which cannot be easily characterized or measured. The kinds of tasks they must undertake can be uncommon and poorly described. We expect robots to be behaviorally flexible, which means that describing a typical task will generally underspecify usage. Worse still, machine design, task and environment are not independent factors, since they might interact in complex ways. Another complication is that most field robots in real applications are still teleoperated, which adds the attendant problems of evaluating the human controller and interface. Published work in this area tends to focus on demonstrating the robot's

© Springer International Publishing Switzerland 2015
C. Dixon and K. Tuyls (Eds.): TAROS 2015, LNAI 9287, pp. 167–179, 2015.
DOI: 10.1007/978-3-319-22416-9_20

Fig. 1. Miner during sustained speed test (left). Corobot on pitch/roll ramps (right).

fitness for purpose based on specific requirements, often according to the contingencies of practical funding. That commits the studies of performance to tasks which are not necessarily standard, or even particularly well-described, and to measurements within environments that cannot easily be duplicated.

In recent years good progress has been made towards widely-accepted standard benchmarks [1]. By now robot competitions that try to hold constant the task, environment and behaviors, such as the Robocup events [2] are well-established. In some cases much of the robot hardware is also fixed, leaving only software solutions and some details of sensors or manipulators as the key design differences to be compared. This approach has proven quite productive, but it has limitations. For some specialized robots, suitable competitions events may not be held often enough or locally enough. Competition between rival teams might tend to suppress sharing of solutions, especially in the commercial arena. Still another concern is that competitions tend to be held indoors, under controlled lighting, weather and surface conditions, as well as clearly marked task setups - far from ideal for field robot testing. Best practice can now be found in the DHS-NIST-ASTM International Standard Test Methods for Response Robots [3]. This is a comprehensive program of tests consisting of elemental tasks, in elemental test rigs. The results are then combined in different ways to represent the expected performance of robots in a wide variety of applications. The testing procedures also allow for purpose-built operational tests, which put the robots into realistic scenarios according to their special functions. For instance, bomb disposal robots can be tested by hiding suspicious packages on a bus, which the robots must remove.

An opportunity to see if these tests can be adapted to a different purpose arose in the Arkaroola Mars Robot Challenge. Four student teams brought six field robots to a test site in Arkaroola, a remote desert station in central Australia. The machines represented the students' design concepts for robots capable of assisting astronauts performing tasks on the Martian surface. Both a selection of standard engineering benchmarks, and operational tests representing specific astronaut assistance operations in harsh Mars-like terrain were made. The intention was to encourage innovation among the students in friendly, low-competition field trials, make useful measurements on their prototypes, compare the performances of the various designs, and gain experience with these relatively new tests.

A few caveats are in order. First, unlike the original NIST evaluation exercise [4], most of the participating robots were not commercial products, but were built by students with limited budgets. Second, none of the robots were built specifically to score highly on these tests, though details of tests were circulated weeks before the event. This was the first robotics competition held by MSA; it is hoped that in future versions, advance knowledge of the event will allow teams time to build more competitive robots. Third, the tests themselves are a work in progress: not all DHS-NIST-ASTM specifications and procedures are yet fully developed. Fourth, in consideration of transport logistics, local conditions and resource limits prevailing at the remote test site, it was necessary to adapt some of the standard tests. However, every effort was made to preserve the essential standards. In the end, useful data was gathered in a difficult outdoor environment. This event represents one of the first examples of the use of DHS-NIST-ASTM testing outside the emergency response domain.

2   The Test Program

The participating robots were not envisaged as autonomous explorers that are sent to Mars to accomplish scientific missions, such as the Mars Science Laboratory, Curiosity. Instead they were designed to support human science activity on the surface. Although they might be used for remote science sampling or photographic surveys, they may equally be required to perform routine maintenance tasks or to fetch-and-carry tools. As such any level of human control (ideally, kept to a minimum to free up the small human crew) would be provided locally, rather than over planetary distances. Therefore issues of long, variable-length radio propagation delays were not considered for the purposes of these tests. Even under the most optimistic exploration scenarios, human attention and working hours on Mars will be at a premium, so the need for autonomous behavior is clear. We believe that a high level of automation is desirable for this application, but could not find it among the participating robots in these tests. Possibly direct human teleoperation will be at least one operational mode for deployed astronaut support robots though.

A well-designed robot for Mars would have much in common with other field robots for terrestrial use. The selection of suitable tests for this event began with a basic list of desirable attributes: all terrain capability, reliable navigation, suitable payload-carrying capability, endurance sufficient for a working day, ability to locate and image objects, ability to manipulate small objects and wireless vision and sound and control links over kilometers of range. These are not specific enough to be called requirements, but more refinement should be an outcome of this exercise. Current concepts for astronaut-assistance robots are organized around collaborative networks involving one or more robots that follow behind the astronauts during sorties, but independent navigation is a must. At present there is no GPS-like satellite location system at Mars, but it is likely that an equivalent guidance service would be operating on the surface by the time human explorers arrive, operating either by one of the proposed microsatellite constellation systems e.g.[5] or by a network of ground based radio beacons [6]. If not, autonomous navigation using visual methods of relating stored elevation maps to LiDAR suface features, such as multi-frame odometry-compensated global alignment (MOGA) [7] would be practical. Some notes about on-board cameras will be made in Section 4.

Tested prototypes for such machines range from small NASA K-10 four-wheelers capable of carrying 13.6kg of science equipment [8] to the golf-cart sized RAVEN, able to carry an injured person in an emergency [9]. Under this latter requirement, the useful payload should extend to the weight of an average person wearing a Z-series spacesuit (approximately 130kg on Earth) - beyond the reach of most of our test prototypes, but their load-bearing capacity should be assessed. Not many field robots today would have sufficient endurance for an 8-hour working period, though we can expect robots to enjoy the benefits of the expected future increase in energy density in batteries [10]. Tests favoring long endurance will foster development in this direction.

2.1   Selected DHS-NIST-ASTM Tests

Based on these considerations, the following tests were selected from the DHS-NIST-ASTM standards. Note that none of these tests need approximate Mars surface conditions, since they are engineering benchmarks only.

1. Logistics: Robot Test Config. and Cache Packing. The process required the completion of forms for every participating machine to capture details of the physical properties, equipment specifications, configurations, toolkit, packing and transport logistics. The information includes specific photographs of a robot, in different poses and from various angles, against a calibrated background. The information is particularly important for managing the configuration of robots from one test to another.

Fig. 2. (Left to right) Continuous 15° pitch/roll ramps; Far-field acuity test; Rover traversing irregular terrain

2. Energy/Power: Endurance : Terrains: Pitch/Roll Ramps. A test rig consisting of 24 15° wooden ramps measuring 1200 x 600mm was laid out to repeatably measure the robots' performance on discontinuous terrain. Participants guided the robots around a 15m figure-eight path on the ramps around two suspended pylons. Distance d_i and time t_i from full battery charge to inoperability are measured. Because this test had to be conducted in the field, it was necessary to eliminate the side walls to save weight and reduce wind stresses (Fig. 2, left). We found it to be impractical to bench test sets of batteries through multiple charge-recharge cycles in the field.

3. Mobility: Terrains: Flat/Paved Surfaces (100m). Two pylons were placed 50m apart on a flat surface. The ground around each was marked with a circle 2m in diameter. The robots were to make 10 timed figure-of-eight laps around this course, without deviating from the circumscribed path. Table 1 reports their average speeds in meters per second.

4. Mobility: Towing: Grasped Sleds (100m). The robots dragged an aluminum sled, carrying an operator-designated payload, around 10 figure-of-eight laps on the 100m course specified in test 3. Average velocities v_{av} and maximum achieved weights m were recorded. Ideally, the test should be conducted on a concrete paved surface, but this was not available at the test site, so a flat roadway of limestone gravel had to serve. To compare these performances to those of any test on concrete, the different coefficients of sliding friction μ_k between the metal sled and the two surfaces must be taken into account. The two coefficients were experimentally measured, yielding averages of μ_k (Al-concrete) of 0.70 and μ_k (Al-gravel) of 0.42. Thus for a given mass, 40% more applied force would be required to achieve the same performance on concrete.

5. Radio Comms:Line-Of-Sight Environments. The robots were tested for navigation control and video feed on a straight course at 50m, then stations every 100m thereafter. The robot circumnavigated each station at a radius of 2m, reading a 35 x 35mm bold letter and identifying a standard 100 x 100 mm hazardous material label on the four vertical faces of a box atop a pylon. The last station at which both navigation control and video were perfectly reliable (complete circle and all four visual tests correct) was reported.

6. Sensors:Video:Acuity Charts and Field of View Measures. The robots were placed on a 15° ramp 6m from a far-field Landolt-C vision chart (Fig. 2, center). The operator viewed the chart at their control station via the robot's camera and read down the chart to the smallest line at which the orientations of the C shapes were discernible. No more than two errors were permitted on a line. This is reported as l_{far} a percentage of the 6-6 (20:20) vision standard. The same procedure was used for the near-field Landolt-C chart, except that the distance was then 40cm. The horizontal field of view fov_h was calculated by measuring the distance between the far-field chart and the camera at the point where the long sides of the chart are at the edges of the video screen.

2.2   Operational Tests

The following three operational tests were designed to evaluate the robots in tasks approximating their intended purpose on Mars. Some of these were inspired by the University Rover Challenge [11], others from our own prototyping of such a system in the Mascot hexapod [12].

1. Irregular Terrain Traversal. A 106m course consisting of four gates (1.2m pylons spaced 2m apart) was arranged over natural Mars-like terrain. It included a slopes of between approximately 20° to 40°, loose sand, and large irregular stones (Fig. 2, right). The robots were video recorded and timed during their traversal of the course.

2. Context Imaging. A small, brightly painted 100g target object was placed at a random locations on roughly level ground at distances of between 43 and 76m from the starting point. The operator was given the object's GPS coordinates. The operator was to locate the object as quickly as possible, then photograph it in context. Time to locate the target t_{loc} and distance to target d_t were recorded. Each operator chose his best four images to be rated for quality. Each image was later examined by three expert field geologists who rated each according to five criteria: object in context, image composition, brightness and contrast, sharpness of focus and image resolution. The mean rating over all images, experts and criteria was then computed and expressed as a percentage of the perfect score q_{av}.

3. Sample Return. Operators of robots equipped with a manipulator had the option to use it in a variation of the Context Imaging task. The robots had to carry a small geologist's scale, place it alongside the located target object, photograph the object in context, collect the object then return it to the starting location. Time to return t_{ret} was reported.

3   Results

The tests were conducted in July, 2014, in three locations: a flat camping area near the Arkaroola station facilities, a gravel airstrip used by the neighboring Wooltana station (Fig. 1, left) and a disused quarry with a variety of Mars-like ground conditions including a curved, gullied slope for the operational tests. Conditions were generally favourable but wind, dust and, on one occasion, rain created problems for the test program. In particular, very fine dust combined with dry air (relative humidity range 21-44%) caused a number of failures of robot and test electronics. The most serious of these were a malfunctioning Arduino board on the Miner which took it out of service and the catastrophic failure of a compact laptop in the Mascot hexapod which eliminated it from all but the visual acuity tests.

    To gain the full benefit of comparisons between the robot designs, it is important to describe the design features of the competing machines. One limitation of the published results of the Emergency Response Robot Evaluation Exercise [4] was that for commercial confidentiality, and because the NIST must not

Table 1. Summary of NIST tests

	Endurance $d_i(m)$ $t_i(s)$		Mobility $v_{av}(m/s)$	Sled Towing $m(kg)$ $v_{av}(m/s)$		Comm Range $d_s(m)$	Visual Acuity $l_{far}(\%)$ $l_{near}(\%)$		Horiz.FoV $fov_h(deg)$
Little Blue	105	2363	0.47	abstained		200	15	25	39.3
Miner	abstained		0.34	6.47	0.33	200	15	20	68.6
UNSW Rover	abstained		1.59	31.47	1.27	500	20	25	66.9
Corobot	150	2032	0.36	3.47	0.27	50	<10	5	120
Mascot	-	-	-	-	-	-	12	15	63.3
Phantom 2	>1720*	-	0.58	abstained		>860*	<10	<5	125.3

*result of non-standard test

appear to endorse or disendorse products or companies, the names and design details of machines were not disclosed. Instead, individual machines were represented by nominal codes. The Arkaroola Mars Robot Challenge was under no such constraints, and so it was possible to disclose these details (Fig. 3) to study how specific design features contributed to the performances of the individual entries. Table 1 briefly summarizes the results. Other details are provided in the discussion section as needed.

Fig. 3. Summary design features of participating robots

Table 1 gives the results of the selected DHS-NIST-ASTM tests. As specified in NIST documents and as required by our human research ethics permit, teams could abstain from any test without prejudice. The data reported here represent the target metrics specified by NIST, but the units have been altered to the conventional MKS system in some cases. In the event, the teams found the pitch/roll ramps a particularly challenging environment, either because it was hard on robots with small wheels and no suspension, or because some robots seemed too large to be compatible with the apparatus. The latter may be a shortcoming of this test selection. Only two robots participated, both requiring multiple operator interventions (from jamming, toppling, or loss of wheels) and

Table 2. Summary of operational tests

	Irreg. Terrain $v_{av}(m/sec)$	Context Imaging $t_{loc}(s)$	$d_t(m)$	$q_{av}(\%)$	Sample Return $t_{ret}(s)$
Little Blue	0.2	417	43.1	65.0	N/A
Miner	0.11	-	-	-	N/A
UNSW Rover	0.27	789	47.1	82.5	1569
Corobot	0.08	1980	52.1	79.7	-
Mascot	-	-	-	-	N/A
Phantom 2	0.74	163	76.0	76.7	311

only one (Corobot) completed the entire 150m course. All of the wheeled ground robots were able to participate in the other tests, but during days with many tests a shortage of spare fully-charged battery packs sometimes limited what could be done. Little Blue was the least massive (7.5 kg) and had no tow point for sled dragging, so the operator abstained from that task.

The UNSW rover lacked any axle suspension and had a clearance of less than 10cm, yet was able to outperform the other vehicles in speed, load-carrying capacity, radio range and visual acuity. Miner initially suffered poor traction due to inadequate tread on its eight plastic tires until the operators improvised a repair from strips of rubber fixed axially to the wheels to overcome the slippage. Results on the visual acuity tests were generally poor (but see Section 4). As an unmanned aerial vehicle (UAV), the Phantom 2 defied many of the test protocols, so some roughly equivalent, though non-standard, test results were included for informal comparison. It had sufficient range to exhaust the expanse of flat ground available for formal range testing, but the operator flew the robot to a peak approximately 860m from the launch point and circumnavigated a cairn there several times with full video feedback before returning. Several countries including Australia, the UK and the USA impose civil flight regulations on UAVs operated for commercial or research purposes, which would now prohibit such long range demonstrations. Only range testing should, however, generally be affected.

Even after the wheel slippage was corrected, the Miner experienced difficulties in the operational tests because not all the wheels were powered, so that traction was still erratic. Close study of the video shot during the Irregular Terrain Traversal task suggested that the rocker-bogie suspension might perform better on uneven surfaces if i) the center bearings of the rocker arms were improved to allow them to rotate more freely and ii) all eight wheels were driven. If the driven wheel of a rocker left the ground, the other tended to stay in contact, but because it was not powered, that corner of vehicle lost control and the vehicle tended to yaw. Driving all eight wheels might increase power consumption, but lower-powered motors could be used since there would be more torque available at the ground. The Phantom 2 was opportunistically tested on the operational tasks, and displayed a high level of performance on tests where speed and maneuverability were important. It was able to successfully complete the Sample Return task, by suspending a rare-earth magnet on a 190cm line to collect the steel target object. A special-purpose standard test, Aerial:sUAS (Group 1) VTOL Station-Keeping, was attempted for this robot (Fig. 4, right). The operator kept the machine hovering 2m from five pairs of visual targets arranged at the corners and center of a 5 x 5m square as they were identified. Each target combined a 35 x 35mm bold letter, a standard 100 x 100 mm Hazmat diamond and concentric Landolt-C figures representing feature sizes of 20, 8, 3.2, 1.3 and 0.52mm. The operator visited each pair twice while identifying the letter, Hazmat sign and the orientation of the smallest possible C figure on one target each visit, using successive 12 MP still images. The results show a mean repetition rate of 75 seconds, and average feature size of 3.2mm identifiable at the 2m specified altitude.

Fig. 4. UNSW Rover collecting the target object during Sample Return task (left). Phantom 2 quadrotor above horizontal visual targets (right).

4 Analysis and Design Implications

Experiment results were compiled into machine-specific reports for the participating teams for their use. In the case of the Mascot and Corobot robots, two outcomes have so far emerged.

Implications of the FoV/Acuity Tradeoff. Fig. 4 plots measured acuity (lowest readable lines on the Landolt-C near-field chart) against the camera's horizontal field of view (FoV). The relationship is non-linear, but considering that this test actually records the judgments of different human operators viewing the outputs of different cameras on screens of various sizes and quality, it nevertheless suggests a fall in acuity as the field of view increases. The FoV/resolution tradeoff is well known [13], but the variable measured here is not simple image resolution. The acuity achievable by a human operator viewing the scene transmitted from the robot also depends on the sharpness and contrast available at the screen as well as their age-related quality of vision [14]. Importantly, we observed that

FoV mattered for locating targets. This variable is even less simply related to image resolution: the operator's skill at pointing the camera and the speed of the robot will affect target acquisition times. All the ground vehicle operators reported depending heavily on visual imaging, rather than GPS location, during the context imaging task. The error radii of the GPS equipment used was too great to be of much assistance at this scale. The location times t_{loc} were divided by distance from the starting point d_t to normalise them. Although there are only three observations (the Phantom 2 is excluded here because of its relatively high speed, its advantageous aerial viewpoint and because its operator primarily used GPS localisation rather than visual imaging), this is consistent with the above observation. Yet a narrow field of view is likely to be detrimental to situational awareness [15]. The Corobot's 120° Genius camera allowed its operator to avoid numerous obstacles and snares that bedeviled Little Blue. Providing multiple cameras with suitable FoVs and mounting points is the most practical solution for field robots. The Corobot's operator has now added a small 640 x 480 pixel EXOO camera with a 30° FoV to the arm (Fig. 6, right).

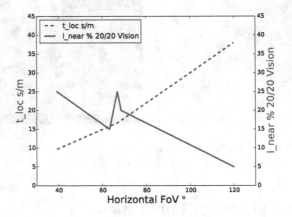

Fig. 5. Near-field visual acuity and target locating ability as functions of camera FoV

Reliability of Radio Link. Two kinds of radio communication links between operator control unit (OCU) and robot were on display: FM analog radio control (RC) transmitter/receiver pairs (Little Blue, Miner, Mascot and Phantom 2), and orthodox WiFi between a laptop and a wireless modem (UNSW Rover and Corobot). During the trials, the reliability of WiFi links was observed to be poor, with long setup times and numerous interruptions to control and telemetry services, while the analog RC suffered no such problems. The best of today's analog RC hardware can operate reliably at ranges of over 15 km (e.g. DragonLink). A disadvantage for teleoperation systems has traditionally been a lack of a data transfer capability, but bidirectional telemetry links are now available. For example, the Mascot had poor OCU control before its computer failed, so

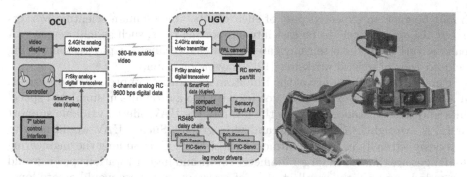

Fig. 6. Design modifications to two test robots following field tests. The Mascot hexapod requires a better RC link which can carry digital data between the robot's laptop and a tablet computer with touch screen command interface (left). Addition of a 30° FoV camera on the Corobot's manipulator arm after the trials (right).

its designers plan to upgrade to an eight-channel RC link using a Taranis X9D transmitter and a FrSky D8R-XP receiver. As well as the servo channels, this pair offers duplex RS-232C data transfer at up to 9600bps, which is enough to be used to form a command link between the Mascot's on-board computer and a tablet at the control station running a bespoke touch-screen interface (Fig. 6, left).

5 Conclusions

In the light of this experience, we offer some recommendations for future tests of astronaut-assist robots and of field robots in general. First, although there is much room for improvement in our test regimen, we claim that there is value in the basic plan of performing both standard engineering tests and specialized operational tests for many kinds of field robots. Because they control for task and environment, standard engineering tests starkly reveal design as the key variable affecting performance. They can allow a profitable comparison of diverse designs, even if conducted at different times and places, but only if the test standards are clearly specified, widely adopted, and only if details of the robot designs are made public. Though Mars-capable robots are a special case, still lessons can be learned from robots designed for another purpose and vice versa. Second, more and better standard tests could be imagined. In particular, a larger irregular terrain environment is needed to accommodate machines, and more practical ways of measuring endurance, radio range and radio occlusion than those of the DHS-NIST-ASTM would be useful. Third, some specific requirements for this application should be dealt with in the operational tests. These should include robot target-finding under more Mars-like lighting conditions, perhaps by fitting filters that both reduce the available light (by about 43%) and shift the color balance toward the red end of the spectrum. A surface dust removal test would stimulate innovation in brushing or sweeping tools. Tests of autonomous behavior should become an expected part of such events. Once human scientific workers arrive, robots are likely to be cast into supporting roles, such as maintenance. A realistic test might be a self-navigated maintenance photography task involving a closed tour of several worksites [12], drawing on a map.

Finally, the performance of the Phantom 2 quadrotor impressed all observers. In a number of short exploratory sorties with two astronauts in simulated space-suits and one assistance robot on the ground, the UAV offered valuable support for EVA oversight, target finding and videography. Such a UAV would need to be redesigned to fly in the thin atmosphere of Mars. Chief among the engineering challenges are that much larger diameter rotors and higher rotation speeds would be needed, even for the smallest class of assistant and these would create long-lasting dust clouds at altitudes of a few meters [16]. Furthermore, the estimated energy demands for a useful payload carrying and range would exceed the performance of the best electrical batteries, while liquid-fueled (such as hydrazine) engines would be limited by the available supply of fuel. Nevertheless, the potential utility is so high that a design study of a small Mars quadrotor has begun at Murdoch University and the MSA Board is already discussing an automated, balloon-borne high-altitude flight test for such a prototype.

Acknowledgments. This research was supported by the Australia-India Council, the Commonwealth Scientific and Industrial Research Organization and the School of Engineering and Information Technology, Murdoch University.

References

1. Moon, S., Rhim, S., Cho, Y.-J., Park, K.-H., Virk, G.S.: Summary of Recent Standardization Activities in the Field of Robotics. Robotica **31**(2), 217–224 (2013)
2. Birk, A.: The True Spirit of Robocup [Education]. Robotics Automation Magazine **17**(4), 108 (2010)
3. Jacoff, A., Downs, A., Huang, H., Messina, E., Saidi, K., Sheh, R., Virts, A.: Standard Test Methods for Response Robots. ASTM International Committee on Homeland Security Applications: Operational Equipment; Robots (E54.08.01). NIST (2014)
4. Jacoff, A., Huang, H., Virts, A., Downs, A., Sheh, R.: Emergency response robot evaluation exercise. In: Proc. of the Workshop on Performance Metrics for Intelligent Systems, pp. 145–154. ACM (2012)
5. Pirondini, F., Fernandez, A.J.: A new approach to the design of navigation constellations around mars: the marco polo evolutionary system. In: AIAA 57th International Astronautical Congress, vol. 7, pp. 4692–4700. IAC (2006)
6. Matsuoka, M., Rock, S.M., Bualat, M.G.: Autonomous deployment of a self-calibrating pseudolite array for mars rover navigation. In: Position Location and Navigation Symposium, PLANS 2004, pp. 733–739. IEEE Press (2004)
7. Carle, P.J.F., Furgale, P.T., Barfoot, T.D.: Long Range Rover Localization by Matching LIDAR Scans to Orbital Elevation Maps. J. of Field Robotics **27**(3), 344–370 (2010)
8. Fong, T., Kunz, C., Hiatt, L.M., Bugajska, M.: The human-robot interaction operating system. In: Proc. of the 1st ACM SIGCHI/SIGART Conf. on Human-Robot Interaction, pp. 41–48. ACM (2006)

9. Akin, D.L., Bowden, M.L., Saripalli, S., Hodges, K.: Developing technologies and techniques for robot-augmented human surface science. In: AIAA Space 2010 Conf. and Exhibition. AIAA, Anaheim (2010)

10. Gao, X.-P., Yang, H.X.: Multi-Electron Reaction Materials for High Energy Density Batteries. Energy & Environmental Science **3**(2), 174–189 (2010)

11. Post, M.A., Lee, R.: Lessons Learned from the York University Rover Team (YURT) at the University Rover Challenge 2008–2009. Acta Astronautica **68**(7), 1343–1352 (2011)

12. Mann, G.A., Baumik, A.: A hexapodal robot for maintenance operations at a future mars base. In: 11th Australian Mars Exploration Conf. MSA, Perth (2011)

13. Lai, J.S., Ford, J.J., Mejias, L., Wainwright, A.L., O'Shea, P.J., Walker, R.A.: Field-of-view, detection range, and false alarm trade-offs in vision-based aircraft detection. In: Int. Cong. of the Aeronautical Sciences. ICAS, Brisbane (2012)

14. Barten, P.G.J.: Contrast Sensitivity of the Human Eye and its Effects on Image Quality, vol. 72. SPIE Press (1999)

15. Hughes, S., Manojlovich, S., Lewis, M., Gennari, J.: Control and decoupled motion for teleoperation. In: International Conference on Systems, Man and Cybernetics 2003, vol. 2, pp. 1339–1344. IEEE (2003)

16. Young, L.A., Aiken, E., Lee, P., Briggs, G.: Mars rotorcraft: possibilities, limitations, and implications for human/robotic exploration. In: Aerospace Conf. 2005, pp. 300–318. IEEE (2005)

Towards Intelligent Lower Limb Prostheses with Activity Recognition

Hafiz Farhan Maqbool$^{(\boxtimes)}$, Pouyan Mehryar, Muhammad Afif B. Husman,
Mohammed I. Awad, Alireza Abouhossein, and Abbas A. Dehghani-Sanij

Institute of Design, Robotics and Optimisation (iDRO),
University of Leeds, Leeds LS2 9JT, UK
{mnhfm,mnpm,mnmabh,m.i.awad,a.abouhossein,
a.a.dehghani-sanij}@leeds.ac.uk

Abstract. User's volitional control of lower limb prostheses is still challenging task despite technological advancements. There is still a need for amputees to impose their will upon the prosthesis to drive in an accurate and interactive fashion. This study represents a brief review on control strategies using different sensor modalities for the purpose of phases/events detection and activity recognition. The preliminary work that is associated with middle-level control shows a simple and reliable method for event detection in real-time using a single inertial measurement unit. The outcome shows promising results.

Keywords: Intent recognition · Lower limb prostheses · Pattern recognition · Electromyography (EMG) · Mechanical sensors · Multi sensor fusion

1   Introduction

One of the most physically and mentally devastating events that can occur to a person is limb loss. There are more than 32 million amputees all around the world in which 75% accounts for lower limb amputees [1]. In England, the number of amputees and limb deficient people reach about 45,000 [2].

The use of prosthetic devices after amputation is one of the interventions to improve the amputees' quality of life. The commercially available prostheses related to lower limb extremity is divided into three types: mechanically passive, microprocessor-controlled passive and powered devices. The mechanically passive and microprocessor devices perform relatively well during simple activities (e.g. level ground walking). However, their inability to produce positive energy, when is needed in many activities (e.g. during stair ascent), is a serious limitation. Powered prostheses use active actuators to generate joint torque which result in powering the knee and ankle joints. Therefore, improved performance has been perceived in complex activities, such as stair ascent, compared to passive devices [3]. Pattern recognition (PR) is the most commonly used control strategy for powered prostheses. Young et al. used supervised PR algorithms to infer the user's intent in real-time [4]. High classification accuracies can be achieved by this approach, however; it requires an extensive

C. Dixon and K. Tuyls (Eds.): TAROS 2015, LNAI 9287, pp. 180–185, 2015.
DOI: 10.1007/978-3-319-22416-9_21

collection of data for training the classifier [4]. The main challenge in the powered devices is the lack of direct control by amputees [5]. Therefore, the need to control the prostheses intuitively has brought the ideas of using surface electromyography (sEMG), mechanical sensors or a fusion-based control. One of the major sources of biological signals in neural control is electromyographic signal (EMG). Surface EMG (sEMG) electrodes have been used to record muscle activities signals from amputees wearing passive prostheses and powered prostheses [6]. Several studies investigated EMG PR to identify the user intent in different activities [5,7,8] for smooth, intuitive and natural control of prostheses. A number of studies have reported the use of mechanical sensors (inertial measurement units (IMUs)), load sensors and pressure-sensitive insoles) for lower limb activity recognition [6,9,10]. All these techniques have achieved reasonable recognition accuracies in steady-state, while the accuracy is much lower in transition between activities [6]. Sensor fusion-based PR for identifying different activities to improve the accuracy and responsiveness have been discussed in [6,11] .

Researchers have segmented the gait cycle in various ways to impose controlling strategy over the prosthesis. Many control algorithms have been implemented using machine learning techniques and simple rule-based approaches [3,12,13] to identify gait phases/events. However, none of the previous studies have dealt with transfemoral amputees (TFA). The aim of this study was to carry out a preliminary work for detecting events including initial contact (IC) and toe off (TO) in real-time using a single IMU. The idea of using multi-sensory system for further improvement in control of lower limb prostheses will remain to be investigated.

2   Control Architecture for Lower Limb Prostheses

The generalized control scheme for the lower limb prostheses consists of three level hierarchy as shown in Figure 1 adapted from [14]. The high level deals with the perception of user's intent based on the signals from prosthesis, environment and the user. The middle level controller translates the perceived user's intent to the desired output state (e.g. desired torque) after implementing detected phases and events. The low level control scheme deals with the feedback control of actuator dynamics for the desired movements (e.g. torque) related to the prosthesis.

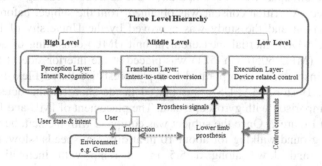

Fig. 1. Generalized control scheme for lower limb prosthesis

2.1   High Level Control

In the high level control, several machine learning techniques were used for accurate identification of locomotive modes. These machine learning techniques require a series of steps including signal processing (filtration and segmentation), feature extraction (time domain, frequency domain and time-frequency domain), feature selection (filter, wrapper method) and classifiers based on unsupervised and supervised learning methods [10].

2.2   Middle Level Control

Middle level control converts the estimated intent from a high level controller to a desired device state by dividing the gait into phases/events. A combination of temporal information, user or device states which are used to identify the gait phases/events, is the main difference between middle level and high level control [14]. The gait cycle (GC) is generally divided into two main phases: stance and swing phases. Some of the sub-phases include mid-stance (MSt), terminal stance (TSt), pre-swing (PSw) and terminal swing (TSw). Furthermore, GC can also be categorized in terms of events such as IC and TO. IC and TO mark the beginning of stance and swing phases, respectively. They are considered important events to objectively assess the gait progress. Accurate identification of gait phases or events is important for controlling lower limb prostheses. The C-leg for instance, is equipped with different sensors (strain gauges, angle sensor) for measuring bending moment, flexion angle and angular velocity of the knee joint. All these measurements detect the gait phases/events and provide necessary damping resistances for user's ambulation.

3   Preliminary Work

3.1   Subjects and Experimental Protocol

One TFA (age: 52 years old; height: 166.1 cm; weight: 66.7 Kg) with two different types of prosthesis A: Ottobock 3R80 (knee) with College Park Venture (foot) and B: C-Leg (knee) with Ottobock 1E56 Axtion (foot) participated in this study. A & B refers to type of prosthesis in Table 1. The amputee had no other neurological or pathological problem apart from his amputation due to trauma leading to chronic infection of the knee. A written consent was obtained from the subject before proceeding for the experiment and the study was approved by the University of Leeds Ethics Committee. A 6-DOF inertial measurement unit (IMU) consisting of accelerometer and gyroscope (MPU 6050, GY-521) was placed at the interior side of the shank. A foot pressure insole with incorporated four piezoresistive based Flexi-Force sensors (Tekscan Inc., Boston, MA, US), was placed inside shoe for the detection of gait events and comparison with gyroscope data. The placement of IMU and foot switches can be seen in Figure 2. Once the subject was equipped with the suit, he was asked to perform level ground walking for about 10 m at different speeds (slow, normal, fast) and walk up and down along a 5.5 m on ramp with inclination of $5°$ at self-selected speed. 10 minutes break was provided in between these activities.

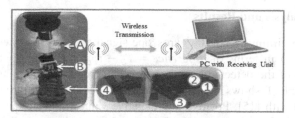

Fig. 2. Experimental Setup: Placement of Sensors; A: IMU, B: Base unit and Footswitches 1: Heel, 2: 1st Metatarsal, 3: 5th metatarsal, 4: Toe

3.2   Real-Time Gait Event Detection Algorithm Description and Validation

Preliminary trials of two healthy subjects were conducted to develop event detection algorithm based on using signals from a gyroscope attached on the shank. The shank angular velocity signal shows distinct characteristic of positive peak (maxima) followed by two negative peaks (minima). Positive peak is known as Mid-Swing (MSw) and two negative peaks on either side of positive peak are known as TO and IC. The proposed algorithm is based on simple heuristic rules and evaluates each sample sequentially, hence facilitates in real-time implementation. The data was captured at a sampling rate of 100 Hz and then filtered out using 2nd order Butterworth low-pass filter at cut-off frequency of 10 Hz. In the start, the algorithm searches the maximum positive value and marks it as MSw after a threshold value greater than 100 degree/sec. Once MSw is marked, it searches for the immediate negative peak and marks it as IC. After IC detection, it waits for 300 ms, then searches the second minima and marks this as TO, provided that the angular velocity is smaller than -20 degree/sec. The threshold values were selected empirically based on the preliminary data. A sample of real-time event detection is shown in Figure 3.

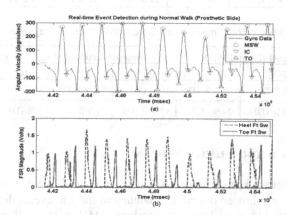

Fig. 3. Sample of real-time gait event detection using signals of (a) IMU (b) heel and toe off foot switches for validation Note: MSw: Mid-Swing, IC: Initial Contact, TO: Toe Off, Ft Sw: Foot Switch

3.3   Data Analysis and Results

The difference in timings from the gyroscope signal and two foot switches (heal and toe) was evaluated in terms of $difference = T_G - T_{FtSw}$, where T_G and T_{FtSw} indicate the timings of the detected events (IC or TO) from gyroscope and footswitches respectively. Table 1 shows the mean differences (MD) for different activities and comparing them with [15]. Positive and negative values indicate the delay and early detection respectively, when compared against footswitch approach. No other work has been carried out with TFA so direct comparison cannot be made with prosthetic side. However; this work is compared with healthy subjects reported in [15]. The MD and percentage increase/decrease (% I/D) of IC for prosthetic side was found to be slightly higher for level ground and ramp ascending whereas for TO it was significantly reduced for all activities when compared with [15] shown in Table 1 and Table 2. The significant improvements were obtained for intact side in terms of MD and % I/D.

Table 1. Mean Difference ± Standard Deviation, all expressed in milliseconds during detection of IC and TO between gyroscope and foot switches

		Level Ground Walk		Ramp Ascending		Ramp Descending	
	Prosthesis	IC	TO	IC	TO	IC	TO
Prosthetic Side	A	13 ± 34	13 ± 10	37 ± 28	23 ± 7.7	-13 ± 15	17 ± 11
	B	34.5 ± 30	-11.7 ± 13	18 ± 12	-34 ± 10	10 ± 25	-122 ± 44
	Total	**24.8 ± 33**	**-0.8 ± 17**	**28 ± 23**	**-5.5 ± 31**	**-1.2 ± 24**	**-53 ± 77**
Intact Side	A	11 ± 13	-44.6 ± 12	13 ± 13	-40.6 ± 6	11.5 ± 12	-41.5 ± 7
	B	2.5 ± 30	-32 ± 15	15 ± 7.7	-20 ± 11	5.6 ± 14	-32.5 ± 14
	Total	**6.4 ± 24**	**-38 ± 15**	**14 ± 11**	**-30.4 ± 13**	**8.5 ± 13**	**-37 ± 12**
[15]		**8 ± 9**	**-50 ± 14**	**21 ± 15**	**-43 ± 10**	**9 ± 20**	**-73 ± 12**

Table 2. % I/D of average mean error between this study and previous work [15]

		Level Ground Walk		Ramp Ascending		Ramp Descending	
		IC	TO	IC	TO	IC	TO
% I/D	Prosthetic	67.7 % (I)	98.4 % (D)	25 % (I)	87.2 % (D)	86.6 % (D)	27.4 (D)
	Intact	20 % (D)	24 % (D)	33.3 % (D)	29.3 % (D)	5 % (D)	49.3 % (D)

4   Conclusion and Future Works

In this study, a brief background was conducted on control of lower limb prostheses. The preliminary work showed overall low latency of the gait events detection for both prosthetic & intact side in real-time using single IMU at shank. Further work will include detection of phases/events with higher number of subjects and other locomotive modes. In addition, implementation of different classifiers to recognize various activities and user intent based on multi-sensor fusion for application of lower limb prostheses will be investigated.

Acknowledgement. This work is linked to a current research sponsored by EPSRC (EP/K020462/1).

References

1. Zhang, X., Liu, Y., Zhang, F., Ren, J., Sun, Y.L., Yang, Q., Huang, H.: On Design and Implementation of Neural-Machine Interface for Artificial Legs. Ieee Transactions on Industrial Informatics **8**(2), 418–429 (2012). doi:10.1109/tii.2011.2166770
2. N. H. S. Commissioning: NHS Commissioning - D01. Complex Disbaility Equipment. England.nhs.uk   (2015).   http://www.england.nhs.uk/commissioning/spec-services/ npc-crg/group-d/d01/
3. Goršič, M., Kamnik, R., Ambrožič, L., Vitiello, N., Lefeber, D., Pasquini, G., Munih, M.: Online Phase Detection Using Wearable Sensors for Walking with a Robotic Prosthesis. Sensors **14**(2), 2776–2794 (2014)
4. Young, A., Hargrove, L.: A Classification Method for User-Independent Intent Recognition for Transfemoral Amputees Using Powered Lower Limb Prostheses (2015)
5. Huang, H., Kuiken, T.A., Lipschutz, R.D.: A strategy for identifying locomotion modes using surface electromyography. IEEE Transactions on Biomedical Engineering **56**(1), 65–73 (2009)
6. Young, A., Kuiken, T., Hargrove, L.: Analysis of using EMG and mechanical sensors to enhance intent recognition in powered lower limb prostheses. Journal of Neural Engineering **11**(5), 056021 (2014)
7. Englehart, K., Hudgins, B.: A robust, real-time control scheme for multifunction myoelectric control. IEEE Transactions on Biomedical Engineering **50**(7), 848–854 (2003)
8. Ajiboye, A.B., Weir, R.F.: A heuristic fuzzy logic approach to EMG pattern recognition for multifunctional prosthesis control. IEEE Transactions on Neural Systems and Rehabilitation Engineering **13**(3), 280–291 (2005)
9. Varol, H.A., Sup, F., Goldfarb, M.: Multiclass real-time intent recognition of a powered lower limb prosthesis. IEEE Transactions on Biomedical Engineering **57**(3), 542–551 (2010)
10. Gupta, P., Dallas, T.: Feature Selection and Activity Recognition System using a Single Tri-axial Accelerometer (2014)
11. Huang, H., Zhang, F., Hargrove, L.J., Dou, Z., Rogers, D.R., Englehart, K.B.: Continuous locomotion-mode identification for prosthetic legs based on neuromuscular–mechanical fusion. IEEE Transactions on Biomedical Engineering **58**(10), 2867–2875 (2011)
12. González, R.C., López, A.M., Rodriguez-Uría, J., Álvarez, D., Alvarez, J.C.: Real-time gait event detection for normal subjects from lower trunk accelerations. Gait & Posture **31**(3), 322–325 (2010)
13. Hanlon, M., Anderson, R.: Real-time gait event detection using wearable sensors. Gait & Posture **30**(4), 523–527 (2009)
14. Tucker, M.R., Olivier, J., Pagel, A., Bleuler, H., Bouri, M., Lambercy, O., Millan, JdR, Riener, R., Vallery, H., Gassert, R.: Control strategies for active lower extremity prosthetics and orthotics: a review. Journal of Neuroengineering and Rehabilitation **12**(1), 1 (2015)
15. Catalfamo, P., Ghoussayni, S., Ewins, D.: Gait event detection on level ground and incline walking using a rate gyroscope. Sensors **10**(6), 5683–5702 (2010)

Discretizing the State Space of Multiple Moving Robots to Verify Visibility Properties

Ali Narenji Sheshkalani[✉], Ramtin Khosravi, and Mohammad K. Fallah

School of Electrical and Computer Engineering, University of Tehran, Tehran, Iran
{narenji,r.khosravi,mk.fallah}@ut.ac.ir

Abstract. In a multi-robot system, a number of autonomous robots sense, communicate, and decide to move within a given domain to achieve a common goal. To prove such a system satisfies certain properties, one must either provide a manual proof, or use an automated verification method. To enable the second approach, we propose a method to automatically generate a discrete state space of a given robot system. This allows using existing tools and algorithms for model checking a system against temporal logic properties. We construct the state space such that properties regarding the visibility of the robots moving along the boundaries of a simple polygon can be model checked. Using our method, there is no need to manually prove that the properties are preserved with every change in the motion algorithms of the robots.

Keywords: Moving robots · Model checking · Visibility

1 Introduction

Mobile robots are able to sense, communicate, and interact with the physical world, and are able to collaboratively solve problems in a wide range of applications (see, for example, [2,3,5,14]).

There has been a close relationship between robot motion planning and computational geometry in the applications where the robots are constrained to move within a geometric domain, like extensions of art-gallery problems [13] to the cases where a number of moving robots must guard a polygonal domain. Efrat et al. [7] considered the problem of sweeping simple polygons with a chain of guards. They developed an algorithm to compute the minimum number of guards needed to sweep an n-vertex polygon that runs in $O(n^3)$ time and $O(n^2)$ working space.

Traditionally, the correctness of robot motion planning algorithms within the context of computational geometry is investigated by manual proofs. It may be hard for certain types of planning algorithms to prove they correctly satisfy the problem's constraints. On the other hand, when it comes to practical applications of motion planning algorithms, the designer may heuristically adjust the algorithm's parameters or even the whole strategy in order to find the best solution that fits both the problem constraints and practical restrictions. In these cases, manually proving the algorithm with every change may be impractical.

© Springer International Publishing Switzerland 2015
C. Dixon and K. Tuyls (Eds.): TAROS 2015, LNAI 9287, pp. 186–191, 2015.
DOI: 10.1007/978-3-319-22416-9_22

An alternative and more reliable approach to examine the correctness of the planning algorithms is formal verification, and more specifically, model-checking [4]. In a few existing works such as [8,9], model checking has been employed in motion planning algorithms. Another related area to which model-checking techniques have been applied are robot swarms. In [12] a swarm of foraging robots is analyzed using the probabilistic model-checker PRISM [11]. Dixon et al. [6] used model-checking techniques to check whether temporal properties are satisfied in order to analyze emergent behaviors of robotic swarms.

In this paper, we focus on discretizing the state space to verify certain properties (*Connectivity* and *Covering*) on a multi-robot system where each robot is programmed with an arbitrary navigation algorithm.

2 Problem Definition

A simple polygon P is defined as a closed region in the plane bounded by a finite set of line segments such that there exists a path between any two points inside P which intersects no edge of P [10]. For simplicity, the boundary of P is denoted by $\beta(P)$. Two points p and q in P are said to be *visible* if the line segment joining p and q contains no point on the exterior of P.

For a simple polygon P, we use the notation V_p for the visibility polygon of a point $p \in P$. Removing V_p from P may result in a number of disconnected regions we call *invisible regions*. Any invisible region has exactly one edge in common with V_p, called a *window* of p, which is characterized by a reflex vertex of P visible from p, like p'. The window is defined as the extension of the (directed) segment pp' from p' to $\beta(P)$. We denote such a window by $w(p, p')$.

Consider a simple polygon P whose boundary is specified by the sequence of n vertices $< p_1, p_2, \ldots, p_n >$ including the set of reflex vertices P_{ref} and convex vertices P_{conv}, and a set $R = \{r_1, r_2, \ldots, r_k\}$ of k robots. The set of robot navigation algorithms $Alg = \{a_1, a_2, \ldots, a_k\}$ (a_i is the navigation algorithm of robot r_i) is given with the following properties:

1. The robots only move on $\beta(P)$,
2. Each step in the movement of each robot is specified by two parameters: direction (clockwise or counterclockwise) and distance (real positive number).

To discretize the state space of the problem, we assume that the robots have turn-based movements. It means that during the movement of a robot, the position of other robots is fixed. By decreasing the time units, we can get closer to a more realistic movement of robots. Having the state space of the robot system in terms of a transition system, we can apply existing model-checking algorithms to verify the correctness of the desired properties.

The correctness properties may be built using temporal logics which are formalisms to describe temporal properties of reactive systems [1]. We define the following two atomic propositions to be used in temporal logic formulas:

Definition 1 (Connectivity). *The set of robots are connected if the graph induced by the visibility relation between pairs of robots is connected.*

Definition 2 (Covering). *The robots cover P if the union of the visibility polygons of all robots covers the whole P.*

Since we do not deal with the details of model-checking algorithms directly in this paper, we refer the reader for a detailed description of temporal logics to [1]. However, to bring an example, the LTL formula $\Diamond(Covering \wedge \neg Connectivity)$ describes the property that eventually (represented by $\Diamond$) the system reaches a state in which P is covered by the robots, but the robots are not connected.

We define a robot system RS as the triple $(P, init, Alg)$ in which P indicates the environment of the robots to navigate, $init$ specifies the initial positions of robots along $\beta(P)$, and Alg is the set of navigation algorithms of robots. Our goal is to define a transition system equivalent of RS, over which temporal logic formulas may be model-checked.

3   Constructing the Discrete State Space

With the ultimate goal of verifying a temporal logic formula over a robot system $RS = (P, init, Alg)$, we must first construct the equivalent transition system of RS. As mentioned before, the states are labeled with the atomic propositions. Hence, the transition system is called a labeled transition system (LTS) [1].

We define the LTS of RS as the tuple $(S, Act, \hookrightarrow, s_0, AP, L)$ where

- S is the set of states (defined bellow),
- $Act = \{\overrightarrow{move}_r, \overleftarrow{move}_r | r \in R\}$ is the set of actions denoting the movement of robot r clockwise or counterclockwise respectively,
- $\hookrightarrow \subseteq S \times Act \times S$ is the transition relation, (we use the notation $s \overset{\alpha}{\hookrightarrow} s'$ whenever $(s, \alpha, s') \in \hookrightarrow$),
- $s_0 \in S$ is the initial state (determined based on $init$),
- $AP = \{Connectivity, Covering\}$ is the set of atomic propositions,
- $L : S \longrightarrow 2^{AP}$ is the labeling function.

3.1   System States

The satisfiability of AP depends on the distribution of robots on $\beta(P)$. We model each state of the system based on the topology of robots and vertices of P.

Definition 3. *Let the window $w(p, p')$ has two endpoints p' and p''. We call the endpoint p'' the projection of p on $\beta(P)$ with respect to p', and denote it by $\pi_p(p')$. Also, we lift the notation to $\pi_p = \underset{p' \in P_{ref}}{\cup} \pi_p(p')$, and further to $\Pi = \underset{p \in P \cup R}{\cup} \pi_p$.*

We define a state as the sequence $s =< q_1, q_2, \ldots, q_m >$ of all points in $\Pi \cup P \cup R$ on $\beta(P)$ in the clockwise order. Without loss of generality, we consider $q_1 = p_1 \in P$ as the starting point in s. Obviously, there exist some sequences which they do not present any feasible state.

Since model checking algorithms assume each atomic proposition is either true or false in a state, the following lemma states that moving of the robots does not change the validity of the propositions $Covering$ and $Connectivity$, as long as the sequence defined above remains the same.

Lemma 1. *Each state s can be uniquely labeled with the atomic propositions $AP = \{Connectivity, Covering\}$.*

Proof. Assume that the labeling $L(s) \in 2^{AP}$ is satisfied by the current state s. It is sufficient to prove that by moving the robots, $L(s)$ is valid while s does not change. We discuss two atomic propositions separately.

Connectivity. The change in connectivity of the robots may happen only when the visible set (of robots) of at least one robot is changed. The visible set of a robot r changes if r crosses a window of another robot r'. Note that in this case the visible set of r' changes too. Since the endpoints of all windows are included in the set $\Pi \cup P_{ref}$, the sequence of the points in s is changed in this case.

Covering. Assume that robot r is going to move. The covering of P may change only in the following two cases, in both there is a change in the current state.

(a) The number of vertices of P visible by r changes. In this case, the point $r \in R$ crosses some points in the set $\underset{p \in P}{\cup} \pi_p$.

(b) There exists a robot r' such that one of the endpoints of the visible part of $\beta(P)$ to r crosses an endpoint of the visible part of $\beta(P)$ to r'. In this case, some point in the set π_r crosses some point in the set $\pi_{r'} \subseteq \Pi$.   □

Corollary 1. *The satisfiability of each element of AP is decidable for all states, and each state can be uniquely labeled with respect to Lemma 1.*

3.2 Transitions Events

We define $\overrightarrow{move}_r$ to be the tuple (CW, δ), where δ is the smallest distance robot r can move in clockwise direction which causes a change in state. we define $\overleftarrow{move}_r$ for the counterclockwise (CCW) direction similarly. We define $\hookrightarrow$ as the smallest relation containing the triples (s, α, s'), where $s \in S$, $\alpha \in \underset{r \in R}{\cup} \{\overrightarrow{move}_r, \overleftarrow{move}_r\}$, and s' is the state obtained from s by taking the action α.

A transition $s \overset{\alpha}{\hookrightarrow} s'$ can occur for two reasons:

(a) Robot r crosses a point in $\Pi \cup P \cup R$,
(b) One of the points in π_r crosses a point in Π.

Assume that robot r moves in some direction (actions $\overrightarrow{move}_r$ or $\overleftarrow{move}_r$), and the current state s changes s'. Based on the definition of transitions, one of the following cases happens:

1. r crosses a point $p \in P \cup R$: in this case, the order of r and p are swapped with each other in the sequence of s, and for all $p' \in P_{ref}$ such that $rp', pp' \in \Pi$ the order of rp' is swapped with pp'.
2. r crosses a point in Π: in this case, there exists a point $p' \in P_{ref}$ such that $pp' \in \Pi$. So, r is swapped with pp', and rp' is swapped with p in s.
3. r crosses a point in Π': in this case, there exist points $p \in P \cup R$ and $p', p'' \in P_{ref}$ such that $pp' \in \Pi$, and r crosses the point $pp'p'' \in \Pi'$. So, the point pp' is swapped with rp'' in the sequence of s.

4   Analysis

Lemma 2 demonstrates an upper bound on the maximum number of states for the given robot system $RS = (P, init, Alg)$. The upper bound obtained in the lemma is not tight. In other words, the geometrical properties of polygon P and the geometry of the projections between the sets R and $\bigcup_{p \in R} \pi_p$ highly affects on the size of the state space.

Lemma 2. *The maximum number of states in order to verify the given robot system $RS = (P, init, Alg)$ has the complexity of $O(n^{2^{k+2}})$ in which n indicates the number of vertices of P, and k specifies the number of robots on $\beta(P)$.*

Proof. Let $\Pi_P = \bigcup_{p \in P} \pi_p$. We define the set of event points $I = \bigcup_{p \in \Pi_P} \pi_p \cup \bigcup_{p \in P} \pi_p \cup P$ as the set of events whose placements are fixed on $\beta(P)$. The following recursive relation $T(k) \in O(n^{2^{k+2}})$ formulates an upper bound on the maximum number of states with presence of k robots on $\beta(P)$:

$$T(k) = \begin{cases} |I| & \text{if } k = 1 \\ T(k-1) \times \left(T(k-1) + O(|P_{ref}|^2)\right) & \text{if } k \geq 2 \end{cases} \qquad \square$$

We have implemented a program to enumerate the states for a given robot system. Table 1 indicates the maximum number of possible states with respect to the polygons shown in Fig. 1 and the number of robots on the boundary of the polygon.

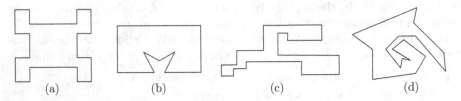

| (a) | (b) | (c) | (d) |

Fig. 1. The polygons used for experimental results

Table 1. The maximum number of states computed with respect to type of the polygons in Fig. 1 and the number of robots

Polygon	Complexity			
	$k = 1$	$k = 2$	$k = 3$	$k = 4$
Polygon (a)	133	3,389	233,683	363,719,760
Polygon (b)	23	406	9,674	482,640
Polygon (c)	135	4,254	308,706	418,697,132
Polygon (d)	84	2,481	137,806	61,352,369

5   Conclusion

We presented a method to construct a discrete state space for a multi-robot system with the aim of verifying correctness properties expressed in temporal logic formulas. The notion of state has been defined in such a way that each state can be uniquely labeled with the atomic propositions *Connectivity* and *Covering*. This way, the modeler can provide the navigation algorithms and verify temporal formulas constructed over the mentioned propositions using existing model checking algorithms. This eliminates the need for manually proving the algorithm(s) each time a change in the algorithm is made.

References

1. Baier, C., Katoen, J.P.: Principles of model checking. MIT Press (2008)
2. Beard, R.W., McLain, T.W., Nelson, D.B., Kingston, D., Johanson, D.: Decentralized cooperative aerial surveillance using fixed-wing miniature uavs. Proceedings of the IEEE **94**(7), 1306–1324 (2006)
3. Cannell, C.J., Stilwell, D.J.: A comparison of two approaches for adaptive sampling of environmental processes using autonomous underwater vehicles. In: Proceedings of the MTS/IEEE OCEANS, Washington, DC, pp. 1514–1521 (2005)
4. Clarke, E., Grumberg, O., Peled, D.: Model Checking. MIT Press (1999)
5. Davison, A., Kita, N.: Active visual localisation for cooperating inspection robots. In: Proceedings of the IEEE/RSJ International Conference on Intelligent Robots and Systems, Takamatsu, Japan, pp. 1709–1715 (2000)
6. Dixon, C., Winfield, A.F.T., Fisher, M., Zeng, C.: Towards temporal verification of swarm robotic systems. Robotics and Autonomous Systems **60**(2), 1429–1441 (2012)
7. Efrat, A., Leonidas, J.G., Har-Peled, S., Lin, D.C., Mitchell, J.S.B., Murali, T.M.: Sweeping simple polygons with a chain of guards. In: SODA 2000, pp. 927–936 (2000)
8. Fainekos, G.E., Girard, A., Kress-Gazit, H., Pappas, G.J.: Temporal logic motion planning for dynamic robots. Automatica **45**(2), 343–352 (2009)
9. Fainekos, G.E., Kress-Gazit, H., Pappas, G.: Temporal logic motion planning for mobile robots. In: Proceedings of the IEEE International Conference on Robotics and Automation (ICRA), pp. 2020–2025. IEEE Computer Society Press (2005)
10. Ghosh, S.K.: Visibility algorithms in the plane. Cambridge University Press (2007)
11. Hinton, A., Kwiatkowska, M., Norman, G., Parker, D.: PRISM: a tool for automatic verification of probabilistic systems. In: Hermanns, H., Palsberg, J. (eds.) TACAS 2006. LNCS, vol. 3920, pp. 441–444. Springer, Heidelberg (2006)
12. Konur, S., Dixon, C., Fisher, M.: Analysis robot swarm behaviour via probabilistic model checking. Robotics and Autonomous Systems **60**(2), 199–213 (2012)
13. O'rourke, J.: Art gallery theorems and algorithms. Oxford University Press (1987)
14. Sugiyama, H., Tsujioka, T., Murata, M.: Collaborative movement of rescue robots for reliable and effective networking in disaster area. In: Proceedings of the International Conference on Collaborative Computing: Networking, Applications and Worksharing, San Jose, CA (2005)

Design of a Continuously Varying Electro-Permanent Magnet Adhesion Mechanism for Climbing Robots

Francisco Ochoa-Cardenas[(✉)] and Tony J. Dodd

University of Sheffield, Sheffield, UK
{cop12fo,t.j.dodd}@sheffield.ac.uk

Abstract. Magnetic adhesion is the most common mechanism employed in Mobile Robots (MRs) used for inspection tasks in ferromagnetic structures. Both Permanent Magnets (PM) and Electro-Magnets (EM) present inherent constraints: constant magnetic force, PMs; or a continuous electric power supply, EMs. These constraints impact significantly on the performance of the MR when it comes to manoeuvring in complex structures. However, this paper presents a novel approach by implementing Electro-Permanent Magnet (EPM) technology. A single short electric pulse is enough to switch On and Off the adhesion force, and by controlling its amplitude the magnetic force can be driven to a desired value, enabling the continuously varying of the magnetic adhesive force. A simple wheel design is proposed and a set of simulations and related experiments performed.

1 Introduction

Adhesion mechanisms, for the particular case of ferric structures, have been constrained due to the lack of an efficient, continually adaptable adhesion mechanism, which allows Unmanned Vehicles (UV) to manoeuvre in complex structures. Even though different types of adhesion technologies can be employed for attaching to complex structures [1–4], these present several disadvantages when compared to magnetic technologies.

Passive mechanisms use Permanent Magnets (PM) as source of adhesion force, resulting in a constant magnetic attaching force [5, 6]. This constraint makes it difficult for the mobile robot to move in a complex environment.

On the other hand, Electro Magnets (EM), active mechanism, allow continuous variation of the magnetic force [7, 8]. The down side is the constant power supply needed to keep the mobile robot adhered to any surface, limiting the manoeuvrability and range of the UV.

In this paper we propose Electro-Permanent Magnet (EPM) technology in the magnetic adhesion mechanism [9], improving the efficiency, performance and size of the mechanism. The main contribution of this paper is the introduction of a novel active, continuously varying, control of the adhesive force for complex ferromagnetic structures, which is achieved by just a single short electrical pulse, minimising the power consumption, and, depending on the amplitude of the electric pulse, a specific value between the minimum and maximum adhesive force values can be selected, thus enhancing manoeuvrability.

© Springer International Publishing Switzerland 2015
C. Dixon and K. Tuyls (Eds.): TAROS 2015, LNAI 9287, pp. 192–197, 2015.
DOI: 10.1007/978-3-319-22416-9_23

2   Theory and Principles of Electro-Permanent Magnets

The basic design of an EPM device consists of three main components: (1) two different PMs, one magnetically hard material (e.g. Neodymium -NdFeB-) and the other a semi-hard material (e.g. Alnico –AlNiCo5-); (2) two keepers made of a soft material (e.g. iron); and (3) copper enamelled wire. Both PMs are placed with their magnetic fields parallel. The semi-hard PM is coiled with the enamelled wire. Finally, the steel keepers are placed at both ends of the PMs, to act as the poles of the EPM.

A short electric pulse is applied through the windings, in order to generate an external magnetic field. This external magnetic field has to be strong enough to alter the magnetic field of the semi-hard PM. Resulting in a change in the magnetic flux direction of the semi-hard PM. The direction of the semi-hard PM's magnetic flux, with respect to the hard PM, will determine the final magnetic attracting force against the target. Fig. 1 shows the On-Off switching cycle of an EPM.

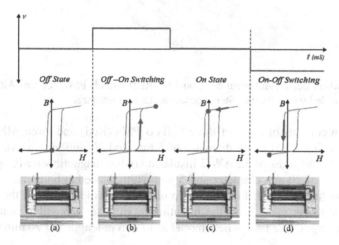

Fig. 1. (Top) EPM On-Off switching pulse cycle. (Middle) Magnetic hysteresis loop of the EPM. (Bottom) Magnetic flux interactions of both PMS, red arrows indicate direction of the magnetic flux. (Left to right) Steps of the Off-On cycle.

The main characteristics of EPM are:

- Power Consumption: EPMs have the advantage of requiring just a fraction of the instantaneous power that an EM [10].
- Scalability: EPMs are scalable in two ways, volume versus power and area versus attracting force [10].
- Working Period (Switching): EPMs are suitable for applications where the switching frequency is small (<10ms) which will derive energy savings [10].
- Low Temperature Rise: In contrast to EM, EPMs have a low temperature rise [10].

3   Design, Simulations and Experiments

3.1   Proposed Wheel Design

A simple wheel design was implemented in order to demonstrate the viability of using EPMs as adhesion mechanism in a mobile robot. The design, Fig. 2, was constrained by the requirement to validate the simulations results using readily available off-the-shelf components.

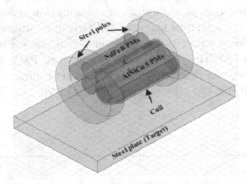

Fig. 2. EPM wheel design and main components, (Red) NdFeB PMs, (Green) AlNiCo5 PMs, (Copper) Enamelled wire, (Purple) Steel keepers and steel plate target

The design consists of a set of three NdFeB PMs (Red) and three AlNiCo5 PMs (Green). Each single PM has a diameter of 5 mm and 20 mm length. AlNiCo5 PMs are coiled with 250 turns of 32 AWG insulated copper magnetic wire (Copper). Two 1008 steel discs of 20 mm diameter and 5 mm width, placed at both ends of the PMs, act as keepers for the EPM. Finally, a gap of 1 mm was left between the wheel and the steel 1008 target plate, with the intention to simulate a possible coating or dirt present in the target. The final wheel model is 30 mm in length and 20 mm diameter.

3.2   Simulations for the Off and On Initial States

FEM was employed to model the magnetic field density interactions of both types of PMs for the Off, On, and continuously varying configurations. Two sets of simulations analysis were performed: (1) for the initial Off and On states; and (2) in transient mode for the continuously varying state. For the initial configurations of Off and On states, results showed no difference for these two states.

For the Off state configuration, the initial magnetic field direction of both, NdFeB and AlNiCo5 PMs, was set in opposite directions. Simulation results, Fig. 3 (Left), show that the magnetic flux lines flow in a closed loop between the two types of PMs. The resulting effect is no adhesion force exerted to the target plate, thus, Off state.

Then, the On state configuration was explored. The initial magnetic field direction of both, NdFeB and AlNiCo5 PMs, was set aligned in the same direction. Results from the simulation, Fig. 3 (Right), show that the magnetic flux of both magnets flows from the PMs Poles through the steel keepers to the target. In this case maximum adhesion force is exerted to the target plate. It can be observed that, a small amount of the NdFeB PMs magnetic flux flows in a close loop through the AlNiCo5 PMs, reducing the possible maximum magnetic attracting force. This phenomenon is due to the difference of coercivity of the PMs.

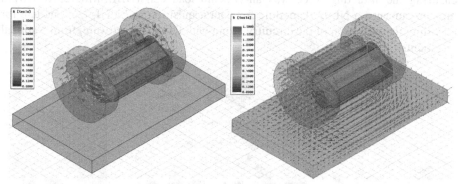

Fig. 3. (Left) Off state. Magnetic flux lines flowing in a close loop between NdFeB (red) and AlNiCo5 (green) PMs. (Right) On state. Magnetic flux lines flowing from PMs through steel keepers to the steel plate target. Simulations conducted using ANSYS Maxwell-3.

3.3   Simulations of the Continuous Varying State

For the continuously varying state, simulations were performed to analyse the transient behaviour. A series of increasing voltage pulses (10ms width and 5ms spacing) were applied through the windings. The applied voltage magnitude was set from 8V to 20V with 0.5V increment steps. Results, Fig. 4, show the input voltage increment steps (red) and their corresponding magnetic force output (blue), exerted to the target. It can be seen that the magnetic force output increments monotonically between 11V and 19V. Any voltage pulse below that range is meaningless since the magnetic field intensity generated by the coil is not strong enough to remagnetize the AlNiCo5 PMs. Any voltage pulse above it does not provide further final magnetic adhesion, due to the fact that AlNiCo5 PMs reached saturation. It can be observed, from the simulation results, that increasing magnetic force is not precisely linear in the working region, between minimum and maximum.

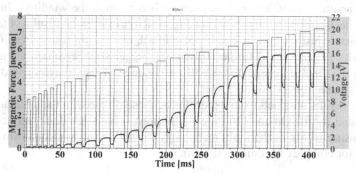

Fig. 4. Comparison of increasing magnetic force (Blue) vs corresponding voltage pulse (Red)

3.4   Experiments

The wheel design proposed in Section 3 was constructed, Fig. 5, and a set of experiments performed in order to validate the simulation results. The equipment used for measuring the attracting force was an S-beam load cell (DBBSMM 25KG) from Omni Instruments, a NI-9201 module for data acquisition and a NI-9263 module for controlling the activation of the circuitry along with LabView software, from National Instruments.

Fig. 5. (Top) Main EPM wheel components, (Bottom left) Basic wheel design before winding, (Bottom right) Final EPM wheel after winding

For the initial Off and On state cases, the configuration of the PMs was set as explained in Section 3.2. First, when the magnetic fields of both types of PMs were set in opposite directions no force was exerted to the target, thus initial Off state behaviour was validated. Then, the AlNiCo5 PMs were flipped, so their magnetic field was in same direction as the NdFeB PMs, generating an attractive magnetic force against the target, thus initial On state was validated.

Next, the EPM wheel in Off configuration was attached to the load cell and placed 1 mm from steel target. A set of 30V pulses, in order to reach the saturation region of the AlNiCo5 PMs, were applied through the windings. As result the EPM wheel made the transition from Off to On state. Results shown in Fig. 6 (Left).

Starting from the On state of the previous experiment, the voltage polarity was inverted and another set of 30V pulses were applied through the windings. In this case the EPM wheel turned Off, as shown in Fig. 6 (Right) From the results obtained it is observed that, it just requires a single pulse to achieve the transition from Off to On state, and vice versa. Even if further pulses, and bigger voltage values, are applied no increase in the final magnetic force will be obtained. This behaviour is due to AlNiCo5 PMs saturation.

From the results obtained from experiments, those obtained in the simulations have been validated for the On and Off cases. Therefore, it has been demonstrated the feasibility of using EPMs as the main technology for the adhesion mechanism in climbing robots, for complex ferromagnetic structures.

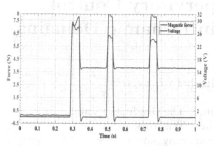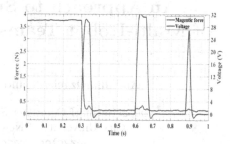

Fig. 6. (Left) Transition from Off to On state. Maximum magnetic force (Blue) vs input voltage (Red). (Right) Transition from On to Off state. Magnetic force (Blue) vs applied voltage (Red).

4   Conclusions

The work presented in this paper proposes a novel continuously varying magnetic adhesion mechanism to be implemented in mobile robots, used for inspecting complex ferromagnetic structures. The use of EPMs as the core technology for the adhesion mechanism was proven and validated through corresponding simulations and experiments. The use of this novel approach of EPM technology, as adhesion mechanism, enhances the mobile robot capability to overcome different types of obstacles and allows it to manoeuvre in complex environments, by just a fraction of the power need for EMs.

References

1. Elliott, M., Morris, W., Calle, A., Xiao, J.: City-Climbers at work. In: Proc. 2007 IEEE Int. Conf. Robot. Autom, pp. 2764–2765. IEEE (2007)
2. Santos, D., Heyneman, B., et al.: Gecko-inspired climbing behaviors on vertical and over-hanging surfaces. In: 2008 IEEE Int. Conf. Robot. Autom, pp. 1125–1131. IEEE (2008)
3. Wang, H., Yamamoto, A., Higuchi, T.: Electrostatic-motor-driven electroadhesive robot. In: 2012 IEEE/RSJ Int. Conf. Intell. Robot. Syst., pp. 914–919. IEEE (2012)
4. Xiao, J., Sadegh, A.: City-Climber: A New Generation Wall-climbing Robots. Climbing Walk. Robot. Towar. New Appl. Itech Education and Publishing, pp. 383–402 (2007)
5. Tavakoli, M., Viegas, C., et al.: Omni-Directional Magnetic Wheeled Climbing Robots for Inspection of Ferromagnetic Structures. Rob. Auton. Syst. **61**, 997–1007 (2013)
6. Zhang, Y., Dodd, T., et al.: Design and optimization of magnetic wheel for wall and ceiling climbing robot. In: 2010 IEEE Int. Conf. Mech. Autom, pp. 1393–1398. IEEE (2010)
7. Carrara, G., De Paulis, A., Tantussi, G.: SSR: A Mobile Robot on Ferromagnetic Surfaces. Autom. Constr. **1**, 47–53 (1992)
8. Grieco, J.C., Prieto, M., Armada, M., De Santos, P.: A six-legged climbing robot for high payloads. In: Proccedings 1998 IEEE Int. Conf. Control Appl, pp. 446–450. IEEE (1998)
9. Marchese, A.D., Asada, H., Rus, D.: Controlling the locomotion of a separated inner robot from an outer robot using EPM. In: 2012 IEEE Int. Conf. Robot. Autom, pp. 3763–3770. IEEE (2012)
10. Knaian, A.N.: Electropermanent Magnetic Connectors and Actuators: Devices and Their Application in Programmable Matter, p. 206

An Approach to Supervisory Control
of Multi-Robot Teams in Dynamic Domains

A. Tuna Özgelen[1] and Elizabeth I. Sklar[2,3]([envelope])

[1] Department of Computer Science, The Graduate Center,
City University of New York, New York, USA
aozgelen@gradcenter.cuny.edu
[2] Department of Computer Science, University of Liverpool, Liverpool, UK
[3] Department of Informatics, King's College London, London, UK
elizabeth.sklar@kcl.ac.uk

Abstract. This paper explores an approach to human/multi-robot team interaction where a human provides supervisory instruction to a group of robots by assigning tasks and the robot team coordinates to execute the tasks autonomously. A novel, human-centric graph-based model is presented which captures the complexity of task scheduling problems in a dynamic setting and takes into account the spatial distribution of the locations of the tasks and the robots that can complete them. The focus is on problem domains which involve inter-dependent and multi-robot tasks requiring tightly-coupled coordination, occurring in dynamic environments where additional tasks may arrive over time. A user study was conducted to assess the efficacy of this graph-based model. Key factors have been identified, derived from the model, which impact how the human supervisors make task-assignment decisions. The findings presented here illustrate how these key factors capture the complexity of the task-assignment situation and correlate to the mental workload as reported by human supervisors.

Keywords: Human-robot interaction · Multi-robot coordination · Task allocation

1 Introduction

This work investigates possible approaches to improve human performance in *Single-Operator Multi-Robot (SOMR)* control, in domains such as Urban Search and Rescue (USAR), where the human operator interacts with the team at a tactical level, while the lower level control decisions, such as trajectory planning, navigation and collision avoidance, are left to the autonomous platforms. In previous work, different interaction schemes have been developed that include automated planning agents to help a human operator [2]. These studies have shown that these agents may become a hindrance to the human operators *Situational Awareness (SA)* [3] if the proposed automated solutions are not understood by the operators or do not meet the operators' expectations [1].

© Springer International Publishing Switzerland 2015
C. Dixon and K. Tuyls (Eds.): TAROS 2015, LNAI 9287, pp. 198–203, 2015.
DOI: 10.1007/978-3-319-22416-9_24

In this work, we define a formal, graph-based model that captures the complexity of task scheduling problems *from the human operator's perspective*. Our conjecture is that such models are key to improving the interaction between autonomous robot teams and human operators.

The formal model defined here is extended from our earlier model derived for static environments [8]. This extension captures *dynamic* domains, where task allocation occurs simulateously with task execution and new tasks may arrive while allocation—and execution—are underway. Several key parameters are verified via a user study, which is described in Section 3.

2 Methodology

The application domains we consider are complex task environments, where tasks may require multiple robots to work in tight-coordination and/or tasks may depend on other tasks. Our experiment scenarios are inspired from RoboCup Rescue Simulation [6] where heterogeneous groups of agents attend to victims, fires and roadblocks in the aftermath of an earthquake in an urban environment. In our scenarios, a fixed number of robots and tasks are scattered around an office-like environment. The tasks can be either a *sensor-sweep* task, where a robot is expected to go to a specific location and send back sensor information (e.g., camera feed), a *fire-extinguishing* task, resulting from a fire in the environment or a *debris-removal* task, resulting from structural collapses that create debris. These latter two types of tasks may require multiple robots to execute, and they block any access to areas adjacent to those in which they appear. Sensor-sweep tasks can be executed by a single robot and do not block access.

Our problem domain falls into the *Multi-Robot (MR)* and *Instantaneous Assignment (IA)* categories within the taxonomy introduced by Gerkey and Matarić [4] and the *constrained tasks (CT)* and *dynamic allocation (DA)* categories within the extensions added by Landén et al. [7]. The CT dimension signifies the dependencies between tasks, which appear as precedence relationships in our experimental scenarios; for example, a sensor-sweep task that is located inside a room whose entrance is blocked by heavy debris is thus dependent on the completion of a debris-removal task before the sensor-sweep task location can be accessed and the task completed. The DA dimension corresponds to possible changes in the task-assignment problem due to arrival of new tasks while existing tasks are being allocated and/or executed.

We are studying ways in which humans interact with teams of robots faced with the range of *MR-CT-DA* scenarios mentioned above. In earlier work, we developed a graph-based data structure, called a *Task Assignment Graph (TAG)* [8], which represents spatial relationships between task locations and robots. Formally, in an environment containing m robots, $\mathcal{R} = \{r_1, ..., r_m\}$, and n tasks, $\mathcal{K} = \{k_1, ..., k_n\}$, we define $TAG = (V, E)$ as a set of vertices $V = \{v_1, ..., v_n\}$, where each vertex, v_i, represents a task, $k_i \in \mathcal{K}$, and a set of edges E. There exists an edge in E between any two task vertices v_i and v_j *iff* tasks k_i and k_j can be accessed from one to the other in the robots' physical environment. Each task vertex v_i contains a set of robots Acc_i that can access task k_i, and a domain Dom_i which consists of the sets of all possible assignments for that task. The cardinality of each set in Dom_i is defined by the number of robots, Req_i, required to execute task k_i. A vertex may be labelled *critical* if they are responsible for maintaining connectivity between components of the graph (see [8] for details). Essentially, a TAG is a hybrid graph structure that combines a connectivity graph and a constraint network. This level of abstraction allows us to focus on spatial relationships in our analysis, without paying further attention to other specifics of the mission domain or environment.

A TAG models an isolated assignment problem instance. In a dynamic environment, the addition of new tasks will lead to a sequence of TAGs. Here, we introduce a new model, which we call a *Mission Assignment Problem (MAP)*, to reflect the changes to the task-assignment problem space that occurs during a mission. A MAP is an ordered list, $MAP = \langle TAG_0, TAG_1, ... \rangle$, where TAG_i is added to the map at time t_i, and t_0 represents the time when the first task appears. Every entry in the MAP represents a *decision point* for a human who is assigning tasks to robots.

We are studying the impact of the MAP with respect to the human's *mental workload*. We surmise that a few small changes from any TAG_i to TAG_{i+1} will not be difficult to comprehend, whereas many significant changes will quickly overwhelm the human, particularly if these changes occur in rapid succession. Our hypothesis is that mental workload is directly affected by the number of solutions that can be produced for a scenario, which in turn is affected by: *(i.)* the ratio of tasks that require close coordination to the total number of tasks; *(ii.)* the dependencies between the tasks; *(iii.)* the spatial distribution of robot platforms across the environment; and *(iv.)* changes in the environment, such as new tasks that may prevent robots reaching previously assigned tasks.

In our earlier work [8], we identified two factors derived from a TAG which influence the human's mental workload. These are: the *Average Platform Requirement*, $APR = \sum_{i=1}^{n} req_i/n$; and the *Critical Task Ratio*, $CTR = |V_{critical}|/n$. We conducted a user study that involved static environments, and our results verified that both APR and CTR are significant factors with respect to the human's mental demand. In the work presented here, we consider dynamic environments. We identify two new metrics to account for conditions *(iii)* and *(iv)*, respectively. These are: the *Average Domain Density*, $ADD = \sum_{i=1}^{n} DD_i/n$, where $DD_i = |Dom_i|/\binom{req_i}{m}$; and the *Tag Disruption Ratio*, TDR, which is the ratio of the number of assignments removed to the number of assignments performed from one TAG to the next.

3   User Study

We ran a user study in which 30 participants were presented with several RoboCup-Rescue-like scenarios (described above) requiring real-time assignment of tasks to a team of 3 robots. The gender balance of participants was: 10 female (33%) and 20 male (67%). The average age of the participants was 31.5 years, and the average amount of computer experience was 14.9 years.

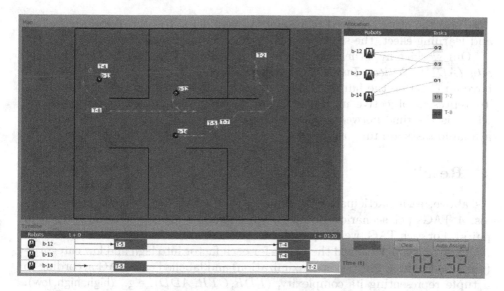

Fig. 1. TASC Interface. On the left, the map area shows the locations of the robots and tasks. Robots' immediate paths are displayed in green. On the right, the task assignment area is where users allocate robots to tasks. On the bottom, an expected timeline of plan execution is displayed.

The experiments were conducted with a robot simulation environment and the *Task Assignment Supervision and Control (TASC)* interface (Figure 1). This interface was integrated with the *HRTeam* [9] multi-robot framework, which is built on Player/Stage [5] and was developed in some of our earlier work. In brief, HRTeam facilitates communication between multiple robot controllers, the Stage simulator (or physical robots), and a user interface (e.g., TASC). The TASC interface communicates with the robot controller processes through a set of messages. The robots' locations, paths to their task locations and state information are updated in the interface based on the messages sent by the robots. All task assignments and removals are made via the TASC interface and are immediately updated on the robots. In this setup, the robots operate autonomously with respect to path planning, navigation and collision avoidance. As well, the robots rely fully on the TASC interface to dictate which tasks they will execute and the order in which they will execute them.

Each experiment started with a training session, followed by 4 experimental scenarios. Participants were instructed to distribute tasks to the robots in such a way that the execution of the plan would result in the fastest completion time of all the tasks. In addition, they were instructed to maintain a full assignment of tasks at all times, meaning that they should keep adding tasks to the robots' task queue, without waiting for the robots the complete their immediate tasks. Each experimental scenario contained 8 tasks, two of which were available initially. Every 45 seconds, two new tasks are introduced, one of which was a single-robot task and the other required two robots to complete. To control for order effect and learning effect, the scenarios were presented in randomized order.

Our working hypothesis is that, at the TAG-level, the *Average Domain Density (ADD)*, *Critical Task Ratio (CTR)* and *Tag Disruption Ratio (TDR)* all have an effect on the human's mental workload when assigning tasks to robots. To represent objective mental demand, we measured *Plan Completion Time*, which is the time between the arrival of the new TAG and the time when all available tasks are fully assigned.

4   Results

As above, each participant was exposed to 16 TAGs (4 experimental scenarios, 4 TAGs per scenario—because each scenario's 8 tasks were introduced in pairs). For each TAG, we computed the factors described earlier: TDR, CTR and ADD. We then partitioned the values for each factor into high and low categories (clustered using Expectation-Maximization) and labelled each TAG according to a tuple representing its complexity, $\langle TDR, CTR, ADD \rangle$, e.g., $\langle high, high, low \rangle$. Organised in this way, users were exposed to 0 or more instances of the 8 possible TAG complexities.

We analysed the *Plan Completion Time*, our mental workload metric, for each user, grouped according to TAG complexity. If a user was exposed to a particular TAG complexity more than once, then the average plan completion time was computed for that user. In order to evaluate the impact of each of the three factors on the users' plan completion times, we ran a 3-way repeated measures analysis of variance (ANOVA). The "repetition condition" for each user was considered to be exposure to different TAG complexity levels; thus each user may have experienced up to 8 different conditions[1].

All three factors (TDR, CTR and ADD) were found to have a significant effect on the Plan Completion Time, as shown in Figure 2. For TDR, $F(1, 21) = 12.02$ and $p = 0.0023$; for CTR, $F(1, 25) = 15.59$ and $p = 0.001$; for ADD $F(1, 26) = 8.39$ and $p = 0.0075$. There was no significant interaction found among variables. These intuitive results show clearly that the selected model parameters have significant effect on human subjects' task-assignment time, which was also confirmed during post-experiment inverviews with participants.

5   Summary

In this work, we presented the MAP model for capturing human cognitive workload for dynamic task allocation environment and validated three key factors derived from the MAP, namely: *TDR*, *CTR* and *ADD*. Planned future work includes utilizing the MAP model features and the validated factors for steering an automated decision support agent, in order to improve the interaction between the human operator and the agent.

---

[1] In reality only 7, because there were no instances of $\langle high, high, high \rangle$ TAGS here.

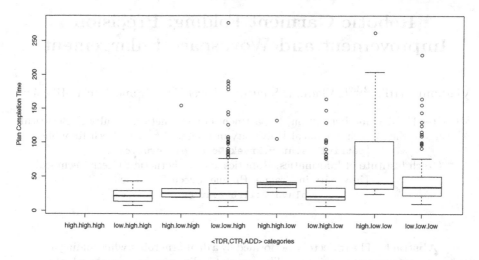

Fig. 2. Plan Completion Time vs. Scenario categories based on TDR, CTR and ADD

References

1. Clare, A., Cummings, M.L.: Task-based interfaces for decentralized multiple unmanned vehicle control. In: Proceedings of AUVSI (2011)
2. Cummings, M., Bruni, S.: Human-automated planner collaboration in complex resource allocation decision support systems. Intelligent Decision Technologies **4**(2) (2010)
3. Endsley, M.R.: Design and evaluation for situation awareness enhancement. In: Proceedings of the Human Factors Society 32nd Annual Meeting (1988)
4. Gerkey, B.P., Matarić, M.J.: A Formal Analysis and Taxonomy of Task Allocation in Multi-Robot Systems. The International Journal of Robotics Research **23**(9) (2004)
5. Gerkey, B.P., Vaughan, R.T., Howard, A.: The player/stage project: Tools for multi-robot and distributed sensor systems. In: Proceedings of the 11th International Conference on Advanced Robotics (2003)
6. Kitano, H., Tadokoro, S., Noda, I., Matsubara, H., Takahashi, T., Shinjou, A., Shimada, S.: RoboCup rescue: search and rescue in large-scale disasters as a domain for autonomous agents research. In: IEEE International Conference on Systems, Man, and Cybernetics (1999)
7. Landén, D., Heintz, F., Doherty, P.: Complex task allocation in mixed-initiative delegation: A UAV case study (Early Innovation). In: Desai, N., Liu, A., Winikoff, M. (eds.) PRIMA 2010. LNCS, vol. 7057, pp. 288–303. Springer, Heidelberg (2012)
8. Özgelen, A.T., Sklar, E.I.: Modeling and analysis of task complexity in human-robot teams (Late Breaking Report). In: Proceedings of the 9th ACM/IEEE International Conference on Human Robot Interaction (HRI) (2014)
9. Sklar, E., Parsons, S., Ozgelen, A.T., Schneider, E., Costantino, M., Epstein, S.L.: Hrteam: A framework to support research on human/multi-robot interaction. In: Proc. of Autonomous Agents and Multiagent Systems (AAMAS) (2013)

Robotic Garment Folding: Precision Improvement and Workspace Enlargement

Vladimír Petrík[1,2]([✉]), Vladimír Smutný[1], Pavel Krsek[1], and Václav Hlaváč[2]

[1] Center for Machine Perception, Department of Cybernetics, Faculty of Electrical Engineering, Czech Technical University in Prague, Prague, Czech Republic
{petrivl3,smutny,krsek}@cmp.felk.cvut.cz
[2] Czech Institute of Informatics, Robotics, and Cybernetics, Czech Technical University in Prague, Prague, Czech Republic
hlavac@ciirc.cvut.cz

Abstract. The trajectory performed by a dual-arm robot while folding a piece of garment was studied. The garment folding was improved by adopting here proposed novel circular folding trajectory, which takes the flexibility of the garment into account. The benefit lies in an increased folding precision. In addition, several relaxations of the folding trajectory were introduced, thus enlarging the working space of the dual-arm robot.

The new folding trajectory was experimentally verified and compared to the state-of-the-art methods. The advocated approach assumes that the folding trajectory and the robot arms constitute a closed kinematic chain. Closed loop planning techniques introduced recently enable planning of the folding task without solving the inverse kinematics in each planning step. This approach has favourable properties because the model used is extensible. For instance, it is possible to take into account the force/tension applied by the robot grippers to the held garment.

Keywords: Gravity based folding · Garment folding · Closed kinematic chain planning

1 Introduction

Humans manipulate soft materials intuitively while getting dressed, lacing their shoes, bending a piece of paper or folding a garment. The garment folding is studied in this contribution. The mentioned manipulation tasks are difficult for robots because of the high degree of freedom allowed by the soft materials. This work improved and experimentally verified the garment folding.

The robotic garment manipulation can be divided into two consecutive operations: unfolding and folding. The unfolding precedes the folding; it starts from the garment in a random configuration and it finishes in the configuration in which the garment is spread flat on the table.

Both operations were studied, starting from the simplest case: towel folding [7]. The unfolding was studied by several researchers using either data-driven methods [2] or model-driven methods [3,4] for the garment state recognition. The

© Springer International Publishing Switzerland 2015
C. Dixon and K. Tuyls (Eds.): TAROS 2015, LNAI 9287, pp. 204–215, 2015.
DOI: 10.1007/978-3-319-22416-9_25

data-driven method uses depth data in order to estimate the grasp position as well as the following action. The model-driven method operates on the 3D point cloud, fitting the precomputed mesh models of the garment to the observation. All the mentioned approaches manipulate the garment in the air and finish with the unfolded garment grasped in both arms. Additional operation needs to be performed in order to put the garment on the table, which might lead to it being wrinkled. It is possible to flatten a wrinkled garment by stroking it using a robot arm/gripper [13].

The folding operation assumes that the garment lies flat on the table with unknown garment pose. Miller et al. in [8,9] demonstrated that the folding can be split into the individual folds while the garment pose is estimated for each fold separately. They used a parametrized shape model [9] for representing the garment. The garment features were used while computing the fold. The fold can be completed according to the gravity based folding approach [8], which divides the gripper trajectory into several straight line segments as shown in Fig. 2. The used polygonal model and features sensed/computed from it were extended. A significant speed-up by a factor of several tens was achieved [11,12]. The novel solution for the garment-table segmentation was also proposed in [12] in order to make garment folding possible on a regular table.

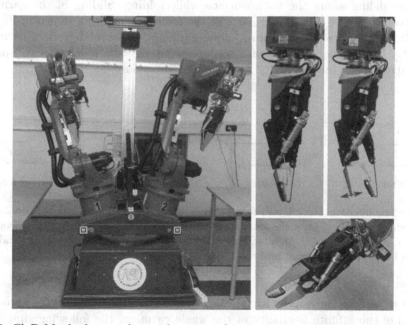

Fig. 1. CloPeMa dual-arm robot with grippers for garment manipulation. The grippers are compliant in the direction shown in the top right figure.

We propose an alternative strategy to the single gravity based fold. The robot trajectory for the fold execution is reformulated as the closed loop planning task, which allows taking the garment flexibility into account. Several constraints relaxations, based on the garment flexibility, are described. Designed modifications were experimentally verified on the CloPeMa robot.

The CloPeMa robot, shown in Fig. 1, consists of two industrial welding robotic arms Yaskawa/Motoman MA1400 mounted on the torso. Its grippers were specially designed for the garment manipulation [6]. The grippers have variable stiffness which facilitates the grasping of the garment from a non-deformable table. The grippers have two asymmetrical fingers. A thin finger is used to slide under the garment, while the second finger incorporates various sensors.

This paper is structured as follows. Sec. 2 proposes the novel strategy for the single gravity based fold. Sec. 3 deals with the closed loop planning for the purpose of folding. Sec. 4 describes our experimental results. Sec. 5 concludes the paper and suggests future work.

2   Improvement of the Gravity Based Folding

The gravity based folding procedure [8] was built on several assumptions, of which the following are going to be examined in detail: *The garment has an infinite flexibility. Friction between the garment and the table surface is infinite.* These assumptions are not satisfied in the real environment, which leads to the garment sliding along the table surface while folding. Sliding of the garment leads to incorrectly overlapped parts of the folded garment as shown in Fig. 2. The sliding is worse for multiply folded garments because they become more rigid. By taking the flexibility into account, we designed a new trajectory as an alternative to the linear gravity based folding procedure [8].

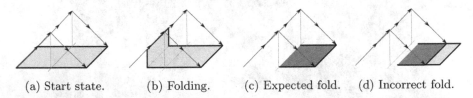

(a) Start state.   (b) Folding.   (c) Expected fold.   (d) Incorrect fold.

Fig. 2. The linear gravity based folding and the garment sliding visualization. One fold is shown, starting from the start state and ending in the expected or the incorrect position. The success of the folding depends mostly on the garment flexibility and the friction between the garment and the table surface.

Let us assume the infinite friction between the table surface and the garment. Instead of the infinite flexibility of the whole garment, the following simplified model is considered. It is assumed that the material itself is rigid. The flexibility is expected in the folding line only. A circular folding trajectory along the folding straight line is the natural consequence of the assumption. In a frictionless case and a non-circular trajectory case, the material slides on the surface. The sliding motion is caused by the varying distance between the gripper and the folding

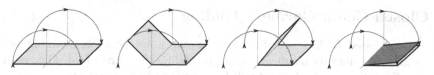

Fig. 3. The circular gravity based folding which preserves the constant distance between the grippers and the folding line

line, which results in pulling/pushing the garment along the table surface. The circular trajectory for one fold is shown in Fig. 3.

Let us come back to the initial assumption, i.e. consider the infinitely flexible material and the infinite friction as in [8]. For the linear trajectory proposed in [8], the only reaction to the gravity force is exhibited by the grippers, which results in folding as shown in Fig. 2b. If the circular trajectory is considered, the garment bends due to the gravity force and additional reaction force caused by the friction. It results in applying the force in the direction of garment sliding. Due to the infinite friction the garment will not slide. When the friction force is smaller than the horizontal component of the gripper force, the sliding occurs resulting in the incorrect fold.

Both the infinitely flexible and the rigid materials are extreme models of the real world materials. The corresponding folding trajectories form boundaries of valid folding trajectories. The correct trajectory depends on the material properties and lies in between these limits as shown in Fig. 4d. In the finite friction case, the space of valid trajectories shrinks depending on the garment rigidity as well. In the original gravity based folding, the folding trajectory corresponds to the lower bound. We observed that the upper bound trajectory provides better folding results for real life garments. In the experimental section, both trajectories are tested on different garment types and compared in terms of the folding precision.

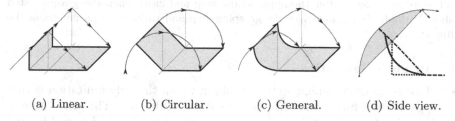

(a) Linear.   (b) Circular.   (c) General.   (d) Side view.

Fig. 4. Possible folding trajectories bounded by linear and circular trajectories. Folding trajectories state space is visualized in gray in Fig. 4d, in which the garment position for the circular and linear folding is shown dashed and dotted, respectively. The general trajectory represents one of the trajectories from this state space.

3 Closed Chain Garment Folding

The gravity based folding specifies the grasping point position unambiguously but the gripper orientation can vary while providing the correct fold. The possible gripper orientations while folding will be examined in this section.

3.1 Grasping and Folding

First, the folded garment needs to be grasped from the table surface. CloPeMa robot grippers were designed to slide one finger under the garment on the flat surface. The finger is compliant to reflect a small uncertainty in the table and the robot relative position. The grasping is shown in Fig. 5a. The whole trajectory including the grasping and the folding is shown in Fig. 5b. The grasping and folding trajectory defines a planning task. The required gripper orientation while grasping the garment constrains the space of achievable folding trajectories. A possible extension of the robot working space can be gained by relaxing the requirement of fixed gripper orientation. Let us introduce several relaxations.

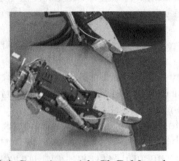

(a) Grasping with CloPeMa robot.

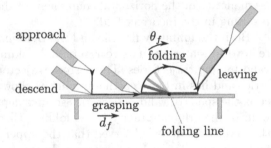

(b) Individual fold steps.

Fig. 5. Grasping (a) and splitting the garment folding into steps. The diagram (b) reads from its left side in time. The shown grippers (a) are in the middle of the grasping step (b). The grippers open after the approaching step and close when the grasping step finishes. Variable d_f parametrizes the grasping step and variable θ_f parametrizes the folding step.

3.2 Azimuth Relaxation

We will focus on the grasping step, Fig. 5b, in which the only limitation is that the gripper's lower finger must be aligned with the table top. The orientation of the gripper in the horizontal plane can be arbitrary within the interval limited by the garment contour as shown in Fig. 6a. The angular offset δ_α reflects the angular dimension of the gripper finger tip. The value is constant and for the grippers mounted on the CloPeMa robot it is set to $30°$. This relaxation, named as the azimuth relaxation, is considered for the grasping phase only. During the fold, the relative orientation of the gripper and the held part of the garment has to be fixed not to crease the garment. Azimuth relaxation gives an additional degree of freedom per arm.

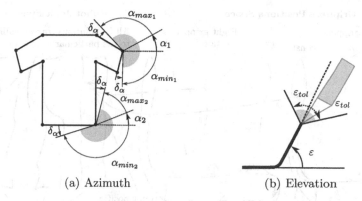

(a) Azimuth   (b) Elevation

Fig. 6. Gripper constraints relaxation. The left drawing shows the azimuth relaxation limited by the garment contour. The right drawing shows the elevation relaxation limited by constant angles which depend on the garment flexibility.

3.3 Elevation Relaxation

The orientation of the gripper can be relaxed along the fold trajectory too. Instead of considering the fixed elevation angle (Fig. 6b), it could be relaxed slightly even for materials with low flexibility like jeans. The relaxation concerns the folding step only. For the grasping step, the elevation is determined by the allowed relative position of the table top and the gripper. A planning algorithm is used to find a feasible trajectory under the elevation relaxation.

3.4 Fold Planning by Solving the Inverse Kinematics

Let us start with the current fold planning procedure without relaxations. Trajectories, which grippers should follow, are parametrized in Cartesian space and then discretized using fixed-time steps. For each discretization step, the inverse kinematics (IK) is solved returning several robot configurations. The configuration of the robot should not change along the folding trajectory. The trajectory has to be collision free, too. If the set of solutions satisfying the above conditions is not empty, one of the solutions is selected. Otherwise, the fold is not accomplishable by the given robot.

The extended formulation, which deals with the relaxation, requires solving a planning task in a space of relaxed grippers positions. The IK is computed in each step of the planner to obtain the robot position, which is checked for feasibility. The IK needs to be solved many times in this formulation. A computationally efficient IK is thus required. For 6 DOF arms, the closed form solver is often available but for redundant robots the IK has infinitely many solutions. CloPeMa robot is redundant with 13 degrees of freedom (DOF).

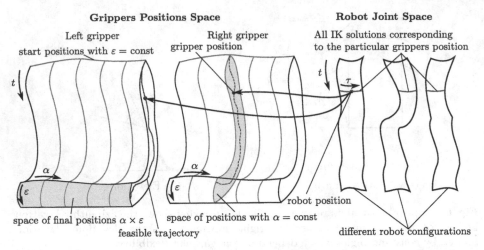

Fig. 7. The planning task for 13 DOF redundant robot with constraints relaxation. The leftmost two columns shows the left and right gripper positions space. The robot joint space is depicted on the right. The motion starts set (shown in green) has a fixed elevation ε but a relaxed azimuth α. Both azimuth and elevation are relaxed in the final position (light orange region). The time evolves roughly from top to bottom. A fixed azimuth is selected at the beginning of the motion. The manifold of the fixed azimuth is shown in light pink. Feasible trajectories (example shown in blue) never leave the manifold. The particular grippers position corresponds to a set of 1D curves (red) in the joint space as the grippers position has 12 DOF and robot has 13 joints. These curves can be parametrized, e.g. by the torso angle τ.

3.5   Fold Planning in Joint Space

In order to avoid solving the inverse kinematics, the planner operates in joint coordinates. The randomized planning techniques like rapidly exploring random trees [5] are usually used but they require a valid state sampling algorithm. Unfortunately, the folding task forms a family of disjoint manifolds in the space of grippers positions. They are related to the manifolds in the joint space via the robot kinematics (Fig. 7). The probability to sample from the manifold approaches zero when a continuous state space is considered. The sampling on the manifold for the purpose of planning was studied in [10] where the closed loop kinematics planning was considered as well. Reformulating the single fold to the closed loop kinematics form enables the use of the recently developed planning techniques like the planning by rapidly exploring manifolds [10] or the iterative relaxation of constraints [1].

The single fold together with grasping expressed in the kinematic model for the CloPeMa robot and circular gravity based folding is visualized in Fig. 8. Two closed loops are formed in frames structure while all relaxations described before are considered. Such a model is used for sampling the valid states as well as for planning using the CuikSuite software [10]. Different joint limits are used for each folding step (Fig. 5b), specifying kinematic restrictions of the manipulator and garment while sampling and planning. These limits form the complete fold task starting from garment grasping, through circular fold and ending in the release state where garment is not in the grippers. Relaxations are easily adjustable via joint limits to reflect different garment flexibility if necessary. The model is easily extensible if another relaxation is designed in the form of an additional joint. Note that the linear gravity based folding can be expressed as a closed loop kinematic planning task as well. The method does not require inverse kinematics which makes it more general.

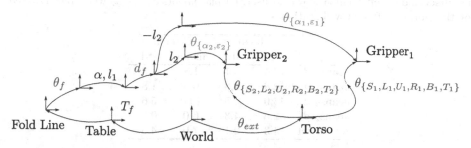

Fig. 8. The closed kinematic chain of the CloPeMa robot. Blue variables represent the position and parameters of the fold. They define a planning task. Red variables are individual joints of the robot. Orange color stands for the azimuth and elevation relaxations. The folding trajectory is parametrized by d_f and θ_f, respectively (Fig. 5b).

4   Experiments

4.1   Minimizing Folding Misalignment

The experiment measuring the displacement of the garment from the expected position compared the linear and the circular gravity based folding procedures. The experiment consists of the following steps: First, the garment is placed manually into the fixed position on the table. The fold is performed using one of the mentioned methods. The expected behavior is that the garment does not slide on the table and thus the non grasped edge of the garment will be in the same position after the fold. Due to the imprecise manipulation and the softness

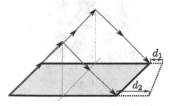

Fig. 9. The measurement of the garment displacement after incorrect fold. Dotted line represents the garment position after the fold is performed.

of the material, this is mostly not true. The displacement of the non grasped edge with respect to the expected position is measured as shown in Fig. 9. The displacements are zero for both d_1 and d_2 for the precise fold. The results of the experiments for different garments (Fig. 10) are shown in Table 1. Based on these results, it is clear that the circular folding outperformed the linear one in precision.

Table 1. Displacement (in millimeters) of the garment after the fold estimated from 5 folds per garment and per method. For jeans measurements, d_1 represents the leg opening offset and d_2 represents the waist offset. Displacements d_1 and d_2 were concatenated for the purpose of the towel statistics.

	Linear		Circular	
	mean	std	mean	std
Jeans$_1$ d_1	16.4	5.1	1.8	1.8
Jeans$_1$ d_2	32.8	4.8	0.8	1.8
Jeans$_2$ d_1	43.0	4.5	19.0	5.6
Jeans$_2$ d_2	32.6	4.3	6.0	0.7
Towel	14.3	2.1	7.0	4.5

4.2 Workspace Enlargement

The next experiment deals with the workspace volume analysis for the purpose of the folding. The relaxations together with all active joints being used for the planning should enlarge the working space of the robot. The towel folding scenario with varying towel position is simulated to examine the working space of the CloPeMa robot.

The multidimensional workspace hyper volume is difficult to evaluate numerically. For demonstration purposes we have chosen to sample the working space along chosen axis while keeping other parameters constant. The results and the scenario are shown in Fig. 11. Three different cases were investigated and in all cases the elevation relaxation was enabled. The first case shows the working space for fixed torso axis and for relaxed azimuth constraint with limits visualized in

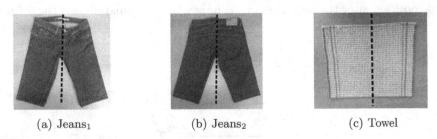

(a) Jeans$_1$ (b) Jeans$_2$ (c) Towel

Fig. 10. The garments used for the experiment shown in the start state. The folding line is shown as dashed. The first and the second jeans differ in facing up and down only.

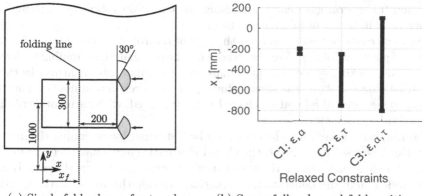

(a) Single fold scheme for towel   (b) Succesfully planned fold positions

Fig. 11. Workspace analysis experiment with the torso axis located in the origin. The fold was planned for varying folding line position x_f. The range of x_f was investigated under different relaxations. In the first case, the elevation and the azimuth were relaxed. In the second case, only the elevation was relaxed but the torso motion was allowed. The last case considered the azimuth and elevation relaxations, as well as the torso motion.

the left drawing. In the second case the torso axis was relaxed but the azimuth angle was fixed for both arms. Note that working space for the second case varies depending on the selected azimuth angles. From all the tested angles, we selected one producing the largest working space. The third case visualizes the possible positions when all the relaxations were allowed (i.e. elevation, azimuth, torso).

Based on these results, it is clear that the space of possible towel locations is significantly larger when the relaxations are allowed. Asymmetrical results are caused by the fact that grasping and releasing positions of the gripper are not symmetrical. Reachable distance of the robot is thus larger in the direction of folding.

The method generality and workspace enlargement is at the cost of longer computational time. The CuikSuite software valid states sampling takes several minutes compared to several seconds in the closed form inverse kinematics case. The computational cost can be overcome by another sampling technique or using the combination of planning approaches we mentioned. We will investigate these possibilities in the future.

5   Conclusions

This paper examined the planning of a single fold within the garment folding procedure performed by two robotic arms. It has been demonstrated that the originally proposed gravity based folding can be improved by the use of a circular trajectory, which is more suitable for real life garments. Experiments have been performed for several garment types using the CloPeMa robot, quantifying the improvement in terms of folding precision.

The next contribution concerned the size of the robot working space for the purpose of garment folding. We observed with the original linear folding trajectories that the working space is rather small even for bulky robotic arms like those of the CloPeMa robot. It was shown that constraints relaxation may extend the working space. Several relaxations have been introduced, with garment flexibility being taken into account.

Planning with relaxations has been studied, starting by extending the inverse kinematics based solution. Next, the closed kinematic chains have been proposed for the planning under relaxations and used to perform our experiments. It can be concluded that the constraints relaxation extends the robot working space significantly.

The results may be important for the future of robot garment folding because they can lead to smaller sized robots for the same working volume and the size of the manipulated garment.

In future work, the garment flexibility can be estimated in the course of the folding. The force/torque sensor mounted between the gripper and the robot flange (as in CloPeMa robot) or a fully force compliant robot as KUKA LBR can be used for the estimation. Based on such estimation, the relaxations limits can be updated online to provide the largest working space as well as a more precise fold. In such an extension, the folding trajectory will be replanned to reflect the properties of the previously unknown garment.

Acknowledgments. Thanks to Libor Špaček for discussions, proof reading and text corrections. This work was supported by the Technology Agency of the Czech Republic under Project TE01020197 Center Applied Cybernetics, the Grant Agency of the Czech Technical University in Prague, grant No. SGS15/203/OHK3/3T/13 and the European Commission under the grant agreement FP7-ICT-288553.

References

1. Bayazit, O., Xie, D., Amato, N.: Iterative relaxation of constraints: a framework for improving automated motion planning. In: 2005 IEEE/RSJ International Conference on Intelligent Robots and Systems (IROS 2005), pp. 3433–3440, August 2005
2. Doumanoglou, A., Kargakos, A., Malassiotis, T.K.K.S.: Autonomous active recognition and unfolding of clothes using random decision forests and probabilistic planning. In: Proc. IEEE Int. Conf. on Robotics and Automation (ICRA), pp. 987–993 (2014)
3. Kita, Y., Kita, N.: A model-driven method of estimating the state of clothes for manipulating it. In: Proc. IEEE Workshop on Applications of Computer Vision (WACV), pp. 63–69 (2002)
4. Kita, Y., Ueshiba, T., Neo, E.S., Kita, N.: Clothes state recognition using 3D observed data. In: Proc. IEEE Int. Conf. on Robotics and Automation (ICRA), pp. 1220–1225 (2009)

5. Lavalle, S.M.: Rapidly-exploring random trees: A new tool for path planning. Tech. rep. (1998)
6. Le, T.H.L., Jilich, M., Landini, A., Zoppi, M., Zlatanov, D., Molfino, R.: On the development of a specialized flexible gripper for garment handling. Journal of Automation and Control Engineering 1(2), 255–259 (2013)
7. Maitin-Shepard, J., Cusumano-Towner, M., Lei, J., Abbeel, P.: Cloth grasp point detection based on multiple-view geometric cues with application to robotic towel folding. In: Proc. IEEE Int. Conf. on Robotics and Automation (ICRA), pp. 2308–2315 (2010)
8. Miller, S., van den Berg, J., Fritz, M., Darrell, T., Goldberg, K., Abbeel, P.: A geometric approach to robotic laundry folding. International Journal of Robotics Research (IJRR) 31(2), 249–267 (2012)
9. Miller, S., Fritz, M., Darrell, T., Abbeel, P.: Parametrized shape models for clothing. In: Proc. IEEE Int. Conf. on Robotics and Automation (ICRA), pp. 4861–4868 (2011)
10. Porta, J., Ros, L., Bohigas, O., Manubens, M., Rosales, C., Jaillet, L.: The CUIK Suite: Analyzing the Motion Closed-Chain Multibody Systems. IEEE Robotics & Automation Magazine 21(3), 105–114 (2014). http://dx.doi.org/10.1109/mra.2013.2287462
11. Stria, J., Průša, D., Hlaváč, V., Wagner, L., Petrík, V., Krsek, P., Smutný, V.: Garment perception and its folding using a dual-arm robot. In: 2014 IEEE/RSJ International Conference on Intelligent Robots and Systems (IROS 2014), pp. 61–67, September 2014
12. Stria, J., Průša, D., Hlaváč, V.: Polygonal models for clothing. In: Mistry, M., Leonardis, A., Witkowski, M., Melhuish, C. (eds.) TAROS 2014. LNCS, vol. 8717, pp. 173–184. Springer, Heidelberg (2014)
13. Sun, L., Aragon-Camarasa, G., Cockshott, P., Rogers, S., Paul Siebert, J.: A heuristic-based approach for flattening wrinkled clothes. In: Natraj, A., Cameron, S., Melhuish, C., Witkowski, M. (eds.) TAROS 2013. LNCS, vol. 8069, pp. 148–160. Springer, Heidelberg (2014)

Computing Time-Optimal Clearing Strategies for Pursuit-Evasion Problems with Linear Programming

Hongyang Qu[✉], Andreas Kolling, and Sandor M. Veres

Department of Automatic Control and Systems Engineering,
University of Sheffield, Sheffield, UK
{h.qu,a.kolling,s.veres}@shef.ac.uk

Abstract. This paper addresses and solves the problem of finding optimal clearing strategies for a team of robots in an environment given as a graph. The graph-clear model is used in which sweeping of locations, and their recontamination by intruders, is modelled over a surveillance graph. Optimization of strategies is carried out for shortest total travel distance and time taken by the robot team and under constraints of clearing costs of locations. The physical constraints of access and timely movements by the robots are also accounted for, as well as the ability of the robots to prevent recontamination of already cleared areas. The main result of the paper is that this complex problem can be reduced to a computable LP problem. To further reduce complexity, an algorithm is presented for the case when graph clear strategies are a priori available by using other methods, for instance by model checking.

1 Introduction

Search and pursuit-evasion problems appear in a variety of formulations in the literature. A survey of methods can be found in [2] where taxonomies of approaches and assumptions based on attributes of searchers, targets and the environments are reviewed. A large class of problems assumes that there is a graph-based description, an "abstraction", of the environment which models it into locations, represented by vertices, and passages between locations, represented by the edges of a graph. Once such an abstraction is available one can apply the techniques developed in graph-searching for robotic search. Probably the first related analysis of graph searching has been published in [12]. The searchers' objective was to catch evaders along the edges the graph with the minimum number of searchers. A large number of variants of graph searching problems appeared which differed in their assumption on the mobility and cost models of a team of search robots clearing edges and vertices of a graph. A review of these is found in [3]. Many of the methods and results developed in the context of graph-searching were primarily concerned with minimizing the number of searchers rather than the distance or time travelled. In this paper we present

This Work Was Supported by the EPSRC Project EP/J011894/2.

C. Dixon and K. Tuyls (Eds.): TAROS 2015, LNAI 9287, pp. 216–228, 2015.
DOI: 10.1007/978-3-319-22416-9_26

a solution for minimizing time, in addition to the number of searchers, using an linear programming formulation for a graph-based pursuit-evasion problem.

The recent framework of interest in this paper is the Graph-Clear(GC) model [5,7,10] for pursuit-evasion problems with teams of robots. GC is particular useful for search problems where multiple robots cooperate to detect intruders in complex environments with limited sensing capabilities. Following the GC approach, in this paper we consider the problem of searching a graph for an unknown number of omniscient and smart targets moving with unbounded speed. Such targets represent a worst-case scenario and are represented by the formal concept of contamination [12], defined in Section 3. The searchers can execute actions that clear contamination or block it from spreading through parts of the environments. These actions have an associated cost that reflects the number of robots needed to prevent targets from passing through undetected.

Linear programming (LP) techniques have previously been applied to robot pursuit-evasion problems in [14] and [13]. In [14] the authors considered polygonal environments partitioned into convex regions. From a region a searcher is able to detect targets in all other visible regions. The resulting visibility relations between regions are considered in the LP formulation. Searchers can move between adjacent regions at each time step but, to reduce the complexity, multiple pursuers cannot occupy the same region. A receding horizon approach was applied in order to solve the LP formulation, albeit without guarantees for optimality. Our work presented here is a continuation of [13] which considered the computation of strategies for Graph-Clear with the minimal number of searchers using an LP formulation. In this paper we focus on computing time optimal strategies. In contrast to [14] target detection in our scenario may require multiple searchers and our environmental model is a graph with vertices that do not necessarily represent convex regions and can exhibit more complex neighborhood relations. In general, the problem of computing time optimal strategies is shown to be harder than minimizing the number of searchers [1], e.g., it is already NP-hard on star-shaped trees.

In the next section a formal description of surveillance graphs and of the GC problem is provided. Section 3 develops the LP model to obtain minimal total travel time of the robots to clear a graph in a physically feasible way. Section 4 deals with optimisation heuristics to reduce the computational complexity of the optimal strategy. Section 5 applies the LP system to find the shortest execution plan for a given clear strategy. Section 6 presents a computational example and the last section concludes the paper.

2   Pursuit-Evasion Problem

For our purposes we adopt the model called Graph-Clear introduced and formalized in [10]. Therein the environment is given by a surveillance graph which is a weighted graph $G = (V, E, w)$ with an undirected graph with vertex set V, edge set E, and $w : V \cup E \to \mathbb{N}^+$ as a weight function. To model contamination and how it is spreading the vertices and edges have an associated state with vertices either being *clear* or *contaminated* and edges either being *clear, contaminated,* or *blocked*. These are abbreviated as $\mathcal{R}, \mathcal{C}$, and $\mathcal{B}$ for clear, contaminated, and blocked, respectively. The state of the surveillance graph with n vertices and m edges is then given by $\nu \in \mathcal{V}(G) = \{\mathcal{R}, \mathcal{C}\}^n \times \{\mathcal{R}, \mathcal{C}, \mathcal{B}\}^m$. As a shorthand, we write $\nu(v_i)$ and $\nu(e_j)$ for the state of a particular vertex or edge. Contamination spreads on recontamination paths. These are paths of vertices and edges on which no edge is blocked. Finally, searchers can execute actions which are either a sweep on a vertex or a block on an edge. The executed actions on G can be represented by $a = \{a_1, \ldots, a_{n+m}\} \in \{0,1\}^{n+m} = \mathcal{A}(G)$ where a 1 for an associated vertex indicates a sweep and a 1 for an associated edge indicates a block. The cost of an action a is given by $c(a) = \sum_{i=1}^n a_i w(v_i) + \sum_{j=1}^m a_{n+j} w(e_j)$, representing the number of robots needed to execute all sweeps and blocks for the action. The spread of contamination and the clearing of actions can now be formalized via a transition function ζ, defined in [10] as follows:

Definition 1 (Transition function). *Let G, $\mathcal{V}(G)$ and $\mathcal{A}(G)$ be defined as above. The* transition function ζ *maps a state and an action into a new state:*

$$\zeta : \mathcal{V}(G) \times \mathcal{A}(G) \to \mathcal{V}(G).$$

Given $a \in \mathcal{A}(G)$ and $\nu \in \mathcal{V}(G)$, the new state ν' is defined as follows:

1. *if $a_{n+j} = 1$, $1 \le j \le m$, then $\nu'(e_j) = \mathcal{B}$*
2. *if $a_i = 1$, $1 \le i \le n$, then $\nu'(v_i) = \mathcal{R}$*
3. *if $\nu_{n+j} = \mathcal{B}$, $a_{n+j} = 0$, $1 \le j \le m$, and no recontamination path between e_j and $x \in V \cup E$ with $\nu(x) = \mathcal{C}$ exists, then $\nu'_{n+j} = \mathcal{R}$*
4. *if there exists a recontamination path between $x \in V \cup E$ and $y \in V \cup E$ with $\nu(y) = \mathcal{C}$, then $\nu'(x) = \mathcal{C}$*
5. *$\nu'_i = \nu_i$ otherwise.*

In colloquial terms the above describes the following rules stated in [10]:

1. edges where a block action is applied become blocked;
2. vertices where a sweep action is applied become clear;
3. blocked edges where a block action is not applied anymore become clear if there is no recontamination path involving them;
4. vertices or edges for which a recontamination path towards a contaminated vertex or edge exists become contaminated;
5. vertices or edges keep their previous state if none of the former cases apply.

A strategy to clear an initially fully contaminated surveillance graph G is a sequence of actions $\mathcal{S} = \{a_1, a_2, \ldots, a_k\}$. A strategy that is a solution to the Graph-Clear problem has in addition a minimal cost, i.e. $ag(\mathcal{S}) = \max_{i=1\ldots k} c(a_i)$ is minimal. In [10] it has been shown that solving the Graph-Clear problem on graphs is NP-hard and a polynomial time algorithms for trees was presented. The algorithms has been applied to robotic search in [6,8–11]. In [6] a method to extract instances of the Graph-Clear problem from robot maps was presented, validating the graph-based model for practical use.

The above definitions are best illustrated with the simple example shown in Fig. 1. In this example vertices are associated with rooms, and edges with

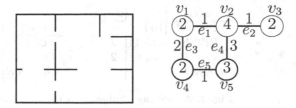

Fig. 1. A simple example environment and a possible surveillance graph that can model the search for a target. Numbers on vertices are clearing costs and numbers on edges are blocking costs.

connections between rooms. Edges between vertices are blocked by placing a robot in the connection between rooms. All contaminated parts can hide an intruders while cleared parts are guaranteed to be free of undetected intruders. A room is cleared by using the specified number of robots to sweep through it.

Much of the prior work in graph-searching by [3] focused on actions that are executed by a single searcher, while the Graph-Clear model explicitly considers actions that require multiple searchers. In this paper we consider connected searching (sometimes also named as contiguous search) to be a search in which the cleared edges and vertices always form a connected subgraph. In Graph-Clear, it is obligatory to block every connected edge when sweeping a node.

Fig. 2 presents a surveillance strategy to solve the Graph-Clear problem associated with the graph shown in Fig. 1. The first column displays the status of the graph in the form of "$\nu(v_1) \cdots \nu(v_5) \, \nu(e_1) \cdots \nu(e_5)$", the second the applied action of the form "$a_{v_1} \cdots a_{v_5} \, a_{e_1} \cdots a_{e_5}$", and the third the cost. In the third row an action sweeps two vertices at the same time. A final action removing all blocks is executed in the end (with 0 cost). The cost of this strategy is 12, i.e., the maximum value read in the third column, and as such not optimal.

$\nu(G)$	a	$c(a)$
$CCCCC\ CCCCC$	$10000\ 10100$	5
$RCCCC\ BCBCC$	$00010\ 10101$	6
$RCCRC\ BCBCB$	$01100\ 11011$	12
$RRRRC\ BBRBB$	$00001\ 00011$	7
$RRRRR\ RRRBB$	$00000\ 00000$	0
$RRRRR\ RRRRR$		

Fig. 2. A Graph-Clear strategy

3   Strategy with Global Minimal Execution Time

The previous section introduces an optimal clearing strategy that has minimal cost. When such a strategy is implemented in practice, we have to decompose

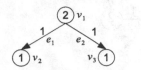

$\nu(G)$	a	$c(a)$
$\mathcal{CCC}$ $\mathcal{CC}$	100 11	4
$\mathcal{RCC}$ $\mathcal{BB}$	010 11	3
$\mathcal{RRC}$ $\mathcal{BB}$	001 01	2
$\mathcal{RRR}$ $\mathcal{RR}$		

$\nu(G)$	a	$c(a)$
$\mathcal{CCC}$ $\mathcal{CC}$	100 11	4
$\mathcal{RCC}$ $\mathcal{BB}$	011 11	4
$\mathcal{RRR}$ $\mathcal{RR}$		

Fig. 3. An example   **Fig. 4.** A clearing strategy   **Fig. 5.** A shortest strategy

it into execution paths for each individual robot. In general, there are many ways to decompose a strategy. In this paper, we are interested in the executions that have the shortest execution time. For example, Fig. 4 shows an optimal strategy for the model in Fig. 3. Fig. 6 illustrates an execution of the strategy in Fig. 4 with 4 robots. In this figure, two robots are marked with solid discs and the other two are marked by circles. Clearly, this graph does not represent the shortest solution because step 2 can be skipped by asking the two robots marked with solid discs in v_1 in Fig. 6(b) to move with the other two at the same time. This would make the system evolve from step 1 in Fig. 6(b) to step 3 directly in Fig. 6(d). Indeed, by skipping step 2, we change this strategy into a new strategy demonstrated in Fig. 5.

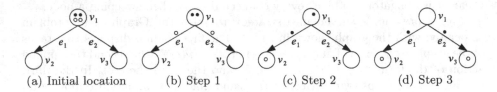

(a) Initial location   (b) Step 1   (c) Step 2   (d) Step 3

Fig. 6. An execution of a clearing strategy

In this section, we propose a linear programming system to compute a clearing strategy that has the minimal cost and its execution plan for each robot with minimal execution time. We first present the general constraints that can be applied in a broader context, and second the specific constraints for computing the shortest execution path. We assume that all robots move at the same speed, and therefore, execution time in one step can be measured by the maximum travel distance during the step. We also assume that the distance between a vertex and an adjacent edge is constant.

3.1   General Constraints

As stated in the previous section, a graph with n vertices and m edges can be cleared in n steps, each of which clears one vertex. Suppose that at least k robots are needed to clear a graph[1]. In each step, a robot can be located in one of the

---

[1] The minimal number of robots can be calculated using the algorithm in [10] or [13].

$n+m$ places. Let $l = n+m$ be the number of possible locations. We also assume that initially the robot has been placed into a vertex or edge. Therefore, we need $l \cdot (n+1)$ binary LP variables $X_1, \ldots, X_{l \cdot (n+1)}$ to encode all possible locations of each robot in every step and the initial locations. We assume that the edges are numbered after vertices, i.e., from $n+1$ to $n+m$, and that the robots are numbered from 0 to $k-1$. The initial location of robot j $(0 \le j \le k-1)$ is constrained by the following equation:

$$X_{j \cdot l+1} + \cdots + X_{(j+1) \cdot l} = 1. \tag{1}$$

As each variable is binary, the above equation captures the fact that the robot locates exactly in one place.

The constraint for the location of robot j at the i-th step $(1 \le i \le n)$ is formulated as follows.

$$X_{i \cdot k \cdot l+j \cdot l+1} + \cdots + X_{i \cdot k \cdot l+(j+1) \cdot l} = 1. \tag{2}$$

The expression $i \cdot k \cdot l$ above indicates the total number of LP variables from step 0 to step $i-1$, where the initial location is seen as step 0. We call the variables appearing in Equations (1) and (2) *location variables*. As n steps are required to clear the graph, there are

$$\Delta_1 = (n+1) \cdot k \cdot l \tag{3}$$

location variables in the LP system for k robots.

The following constraint models the travel distance of a robot moving in a step. For a robot j and two locations p and q $(1 \le p, q \le l)$, we generate a binary LP variable $X_{f(p,q)}$ to represent the possibility that the robot moves from p to q at the i-th step $(1 \le i \le n)$.

$$2 \cdot X_{f(p,q)} - X_{(i-1) \cdot k \cdot l+j \cdot l+p} - X_{i \cdot k \cdot l+j \cdot l+q} \le 0, \tag{4}$$

where

$$f(p,q) = \Delta_1 + (i-1) \cdot k \cdot l^2 + j \cdot l^2 + (p-1) \cdot l + q. \tag{5}$$

In this constraint, $X_{(i-1) \cdot k \cdot l+j \cdot l+p}$ encodes the location of the robot j at the $(i-1)$-th step and $X_{i \cdot k \cdot l+j \cdot l+q}$ its location at the i-th step. When both variables are set to one, which means that indeed the robot moves from p to q at the i-th step, $X_{f(p,q)}$, a binary *move variable*, can be set to one without violation of Equation (4). In other cases, i.e., at least one location variable is zero, $X_{f(p,q)}$ has to be set to zero, indicating that this is not an actual move. For each robot and each step, there are l^2 possible moves, and hence, we need l^2 move variables. The following constraint requires that exactly one of the l^2 move variables is set to one representing the actual move.

$$\sum_{1 \le p,q \le l} X_{f(p,q)} = 1. \tag{6}$$

The move variables constitute the majority of total variables used in the LP system: there are

$$\Delta_2 = n \cdot k \cdot l^2 \tag{7}$$

move variables.

Now we introduce *distance variables* of integer type, one for each step. Let D_i be the distance variable for the i-th step $(1 \le i \le n)$. As each robot has one move variable set to one, the following constraint states that D_i is no less than the maximum distance a robot can move in one step:

$$\bigwedge_{0 \leq j < k} \{ (\sum_{1 \leq p,q \leq l} d_{p,q} \cdot X_{f(p,q)}) - D_i \leq 0 \}, \tag{8}$$

where $d_{p,q}$ is the minimum travel distance between p and q. Clearly, there are

$$\Delta_3 = n \tag{9}$$

distance variables. The objective function is to minimise the sum of the distance variables, i.e.,

$$\sum_{1 \leq i \leq n} D_i. \tag{10}$$

3.2 Constraints for Graph-Clear Strategies

In this subsection, we model the constraints required by Graph-Clear strategies. The first constraint is to block an edge. Let c_{e_r} be the cost of edge e_r $(1 \leq r \leq m)$. At the i-th step, edge e_r has a binary *edge blocking* variable $Y_{i \cdot m + r}$ to represent whether it is being blocked at this step. Equation (11) enforces that the edge blocking variable cannot be set to one if the number of robots staying in the edge is fewer than the cost of blocking the edge.

$$c_{e_r} \cdot Y_{i \cdot m + r} - \sum_{0 \leq j < k} X_{i \cdot k \cdot l + j \cdot l + n + r} \leq 0. \tag{11}$$

There are in total

$$\Delta_4 = n \cdot m \tag{12}$$

of edge blocking variables. Equation (13) guarantees that the edge blocking variable is set to one as far as the number of robots in the edge reaches the required number for blocking the edge.

$$\sum_{j=0}^{k-1} X_{i \cdot k \cdot l + j \cdot l + n + r} + (c_{e_r} - 1 - k) \cdot Y_{i \cdot m + r} \leq c_{e_r} - 1. \tag{13}$$

The correctness of Equation (13) can be proved by contradiction. Suppose the number of robots in the edge is large enough to block the edge and the edge blocking variable is set to zero. Then, Equation (13) is transformed into

$$\sum_{j=0}^{k-1} X_{i \cdot k \cdot l + j \cdot l + n + r} \leq c_{e_r} - 1,$$

which is clearly unsatisfiable. Therefore, $Y_{i \cdot m + r}$ has to be set to 1 and Equation (13) becomes

$$\sum_{j=0}^{k-1} X_{i \cdot k \cdot l + j \cdot l + n + r} \leq k,$$

which is tautology.

Clearing a vertex at the i-th step can be modelled in a similar way. Let c_{v_r} be the cost of clearing vertex v_r $(1 \leq r \leq n)$, and $Z_{i \cdot n + r}$ the binary *vertex clearing* variable. Equations (14) and (15) formalise the constraint. In addition, all adjacent edges to v_r have to be blocked at the same time. For each adjacent edge e_s, Equation (16) requests e_s to be blocked.

$$c_{v_r} \cdot Z_{i \cdot n + r} - \sum_{0 \leq j < k} X_{i \cdot k \cdot l + j \cdot l + r} \leq 0. \tag{14}$$

$$\sum_{j=0}^{k-1} X_{i \cdot k \cdot l + j \cdot l + r} + (c_{v_r} - 1 - k) \cdot Z_{i \cdot n + r} \leq c_{v_r} - 1. \tag{15}$$

$$Z_{i \cdot n + r} - Y_{i \cdot m + s} \leq 0. \tag{16}$$

There are in total

$$\Delta_5 = n^2 \tag{17}$$

of vertex clearing variables.

A Graph-Clear strategy clears one vertex at each step, which can be expressed as follows.

$$\sum_{r=1}^{n} Z_{i \cdot n + r} \geq 1. \tag{18}$$

The above equation states that at the i-th step, there exists a node r ($1 \leq r \leq n$) that is being cleared. When a Graph-Clear strategy is executed completely, we would expect that all vertices have been cleared, which is expressed by the following equation on all vertices: for each node r, there exists a step i ($1 \leq i \leq n$) at which r is cleared.

$$\sum_{i=1}^{n} Z_{i \cdot n + r} \geq 1. \tag{19}$$

The strategy also requires that the node being cleared at the i-th step ($i > 1$) should be adjacent to a node that has been cleared at a previous step. Let $V_r = \{r_1, \cdots, r_t\}$ be the set of indices of adjacent nodes for node v_r. We have

$$Z_{i \cdot n + r} - \sum_{j=1}^{i-1} \sum_{p \in V_r} Z_{j \cdot n + p} \leq 0. \tag{20}$$

The correctness of Equation (20) is obvious: $Z_{i \cdot n + r}$ cannot be set to 1 if none of $Z_{j \cdot n + p}$ variables is 1, representing no adjacent nodes in V_r has been cleared before. Similarly, an edge e_r cannot be blocked at the i-th step until one of its adjacent vertices has been swept before or being swept at the same step. Let $E_r = \{r_1, r_2\}$ be the adjacent vertices of e_r. We have

$$Y_{i \cdot m + r} - \sum_{j=1}^{i} \sum_{p \in E_r} Z_{j \cdot n + p} \leq 0. \tag{21}$$

To prevent recontamination, an edge e_r that is blocked for sweeping a vertex has to be continuously blocked until both adjacent vertices have been swept. Let v_p be an adjacent vertex of e_r. This constraint is characterised by the equation:

$$Y_{(i-1) \cdot m + r} - \sum_{j=1}^{i} Z_{j \cdot n + p} - Y_{i \cdot m + r} \leq 0. \tag{22}$$

The correctness of Equation (22) is straightforward: at the i-th step, $Y_{i \cdot m + r}$ has to be set to 1, meaning that e_r is being blocked at this step, if it is blocked at the previous step ($Y_{(i-1) \cdot m + r} = 1$) and vertex v_p has not been swept yet (all Z variables are zero). If one of the Z variable is 1, then there is no constraint on $Y_{i \cdot m + r}$ enforced by v_p.

Complexity. The total number of LP variables is
$$\Delta = \Delta_1 + \Delta_2 + \Delta_3 + \Delta_4 + \Delta_5, \tag{23}$$
where Δ_2 is in the dominant position, which means that the number of LP variables is proportional to $\mathcal{O}(k \times n \times (n + m)^2)$. This suggests that the LP program for a large surveillance graph can easily be beyond the capacity of the state-of-the-art LP solvers.

4   Optimisation Heuristics

Equation (1) allows the robots to be scattered all over the graph when they start to clear the graph. A reasonable assumption in practice is to place them in a vertex initially. Therefore Equation (24) can be modified to exclude edges.
$$X_{j \cdot l + 1} + \cdots + X_{j \cdot l + n} = 1. \tag{24}$$
Similarly, we can assume all robots initially gather in an edge if requested.

After taking Equation (24), we could make another assumption that all robots start from the same location. This can be modelled by the following constraints: for all vertex $1 \leq i \leq n$ and all robots $0 < j < k$, we have
$$X_i - X_{j \cdot l + i} = 0. \tag{25}$$
The equations in this category set all variables corresponding to the same initial location to the same value by letting all robots from 1 to $k - 1$ stay in the same location of robot 0.

To speed up the search for an optimal clear strategy with shortest execution time, it is essential to get a tight upper-bound for each distance variables. If all robots are initially positioned in the same vertex, then it is reasonable to assume they would sweep the initial vertex first. Therefore, the shortest time to execute the first step is the same as the time for some robots moving to the adjacent edges to block. Constraints specified in Equations (6) and (8) can be simplified by removing impossible moves between vertices and edges. For example, suppose that robots initially gather at vertex v_3 in Fig. 1. To sweep v_3, there is no need to allow robots to move to vertex v_2 and beyond. Hence, the move variables for those unrealistic moves can be removed from Equations (6) and (8).

For other steps, it is also very helpful if we can estimate the maximum distance that a robot needs to move. For the example in Fig. 1, we can get an optimal execution even if we do not allow a robot to move across three edges if it starts from an edge, or across three vertices if it starts from a vertex. For example, a robot cannot move from e_2 to e_3 or from v_2 to v_4 in one step.

5   Executing a Predefined Strategy with Minimal Execution Time

The LP system defined in Section 3 is still very hard to solve even if we apply the heuristics in Section 4. In this section, we apply the LP system to find a shortest execution plan for a given clearing strategy. The strategy specifies the vertex to sweep and edges to block at each step, and thus, greatly reduces the search space. To achieve this, the constraints listed in Section 3.2 are not needed any more. Instead, we add the following constraints.

As in the previous section, we assume that initially the robots stay in the vertex that is swept at the first step. For each edge e_r being blocked at the i-th step, we generate the following inequality.

$$\sum_{j=1}^{n} X_{i \cdot n \cdot l + j \cdot l + n + r} \geq c_{e_r}. \tag{26}$$

Similarly, for vertex v_r being swept at the i-th step, we have

$$\sum_{j=1}^{n} X_{i \cdot n \cdot l + j \cdot l + r} \geq c_{v_r}. \tag{27}$$

The intuition in these equations is that the number of robots in e_r (v_r) should exceed the cost of blocking e_r (sweeping v_r).

Note that the LP system does not impose any constraints on edges that are in the clear, i.e., $\mathcal{R}$, state. A clear edge may still be blocked during the execution when necessary for saving execution time.

Complexity. The extra constraints in Equations (26) and (27) do not introduce new LP variables. Thus, the total number of LP variables for computing the execution of a predefined strategy is

$$\Delta = \Delta_1 + \Delta_2 + \Delta_3. \tag{28}$$

This suggests that the complexity for finding an optimal execution plan for a given strategy is in the same level of that for computing a global optimal execution plan. In practice, however, this kind of LP programs can be solved much faster than that in Section 3 due to fewer LP variables and more constraints.

6   Experiment

The LP system described in this paper can be automatically generated by providing the adjacency matrix of the graph, and the distance matrix which records the minimal travel time/distance between any two locations in the graph. The number of robots, i.e., the cost of clearing a graph, also needs to be specified. In [13], we proposed to apply model checking techniques to compute the minimal cost and generating a corresponding strategy. Fig. 7 shows such a strategy.

Our prototype implementation takes this example strategy and generates the LP system described in Section 5, which is then solved by the LP solver Gurobi Optimizer [4]. We assume that the travel distance between a vertex and an adjacent edge is one and takes one unit of time to move. In a computer with dual Intel Xeon E5-2643 v2 processors and 384GB RAM, it took 112 seconds to find an optimal execution with the execution time 9 units of time, which is the minimal execution time for this strategy. The detail of the execution is illustrated in Fig. 7, where we list the location of each robot at every step. Note that the location at step 0 is the initial location. The last column gives the number of time units needed for every step.

$\nu(G)$	a	$c(a)$
$CCCCC\ CCCCC$	00001 00011	7
$CCCCR\ CCCBB$	00010 00111	8
$CCCRR\ CCBBB$	10000 10110	8
$RCCRR\ BCBBR$	01000 11010	9
$RRCRR\ BBRBR$	00100 01000	3
$RRRRR\ RBRRR$		

Fig. 7. A Graph-Clear Strategy

Step	Robot									Time
	1	2	3	4	5	6	7	8	9	
0	v_5	v_5	v_5	v_5	v_5	v_5	v_5	v_5	v_5	
1	v_5	e_5	v_5	e_5	e_4	e_4	e_4	v_5	e_5	1
2	v_4	e_3	e_4	e_3	e_4	e_1	e_4	e_5	v_4	2
3	e_3	v_1	e_4	v_1	e_4	e_1	e_4	v_5	e_3	1
4	v_2	v_2	e_4	e_1	e_4	e_2	v_2	e_4	v_2	3
5	e_2	v_2	e_4	e_1	e_4	v_3	e_2	e_4	v_3	2

Fig. 8. An execution

However, if we use the LP system defined in Section 3 and the heuristics in Section 4, then the solver managed to find an execution plan with 8 time units. The locations at each step is listed in Fig. 9 and the corresponding strategy is shown in Fig. 10. We can see that vertices v_1 and v_4 can be cleared in one step, which makes the fifth step redundant. However, the time for solving the LP system was increased to 60143 seconds.

Step	Robot									Time
	1	2	3	4	5	6	7	8	9	
0	v_3	v_3	v_3	v_3	v_3	v_3	v_3	v_3	v_3	
1	e_2	v_3	e_2	v_3	e_2	e_2	e_2	e_2	e_2	1
2	e_4	e_2	e_1	v_2	v_2	e_4	e_4	v_2	v_2	2
3	e_5	e_4	e_4	v_5	e_4	v_5	v_5	e_1	v_5	2
4	v_4	v_4	v_1	e_1	v_1	v_5	e_5	e_3	e_3	3

Fig. 9. An optimal execution

$\nu(G)$	a	$c(a)$
$CCCCC\ CCCCC$	00100 01000	3
$CCRCC\ CBCCC$	01000 11010	9
$CRRCC\ BBCBC$	00001 10011	8
$CRRCR\ BRCBB$	10010 10101	8
$RRRRR\ BRBRB$		

Fig. 10. A Graph-Clear Strategy

Discussion. During at any point of the process of solving the LP system, Gurobi Optimizer maintains an upper bound and a lower bound (also called *best bound*) for the value of the objective function. The upper bound is the value obtained by the temporary best solution, called *incumbent solution*, up to that point. The solving process terminates when the upper bound matches the lower bound. However, this process can be interrupted at any point and the incumbent solution is returned. The incumbent solution may not be an optimal solution, but in this example, the incumbent solution found at 110 seconds represents an execution plan with 8 time units, which is the upper bound. It means that this solution is an optimal solution, although the best bound at that point is lower than the upper bound. Therefore, the solving process can be stopped at any point after the upper bound reaches 8. The solving process for a given strategy can be stopped in the same way. When the global minimal value of the objective function is not known, we can use the following two steps to acquire an acceptable solution. First, we compute a strategy with a minimal number of pursuers and obtain the minimal value for the objective function of this strategy. Second, we run the LP solver to solve the LP system for the global optimal solution and terminate the solving process at any point when a better strategy than the known strategy is found.

7   Conclusion

This paper solves the optimal design of movements by robots in an enviroment modelled by a surveillance graph under the constraints of robot movements and resources. A method to find a shortest execution plan for a given clear strategy is also presented. It is assumed that the moving distance between a vertex and an adjacent edge is constant. As this assumption may not hold in practice when a vertex represents a fairly large area, further work can be carried out to improve the modelling approach taken in this paper.

References

1. Borie, R., Tovey, C., Koenig, S.: Algorithms and complexity results for pursuit-evasion problems. In: Proceedings of the International Joint Conference on Artificial Intelligence, pp. 59–66 (2009)
2. Chung, T.H., Hollinger, G.A., Isler, V.: Search and pursuit-evasion in mobile robotics. Autonomous Robots **31**(4), 299–316 (2011)
3. Fomin, F.V., Thilikos, D.M.: An annotated bibliography on guaranteed graph searching. Theoretical Computer Science **399**(3), 236–245 (2008)
4. Gurobi Optimization, Inc., Gurobi optimizer reference manual (2014)
5. Kolling, A., Carpin, S.: The graph-clear problem: definition, theoretical properties and its connections to multirobot aided surveillance. In: Proc. of IEEE/RSJ Intl. Conf. on Intelligent Robots and Systems, pp. 1003–1008 (2007)
6. Kolling, A., Carpin, S.: Extracting surveillance graphs from robot maps. In: Proceedings of IROS 2008, pp. 2323–2328 (2008)
7. Kolling, A., Carpin, S.: Multi-robot surveillance: an improved algorithm for the graph-clear problem. In: Proc. IEEE Int. Conf. on Robotics and Automation, pp. 2360–2365 (2008)
8. Kolling, A., Carpin, S.: Surveillance strategies for target detection with sweep lines. In: Proceedings of IROS 2009, pp. 5821–5827 (2009)
9. Kolling, A., Carpin, S.: Multi-robot pursuit-evasion without maps. In: Proceedings of ICRA 2010, pp. 3045–3051 (2010)
10. Kolling, A., Carpin, S.: Pursuit-evasion on trees by robot teams. IEEE T. Robot. **26**(1), 32–47 (2010)

11. Kolling, A., Kleiner, A.: Multi-uav motion planning for guaranteed search. In: Proceedings of AAMAS 2013, pp. 79–86 (2013)

12. Parsons, T.D.: Pursuit-evasion in a graph. In: Alavi, Y., Lick, D.R. (eds.) Theory and Applications of Graphs. LNCS, vol. 642, pp. 426–441. Springer, Heidelberg (1992)

13. Qu, H., Kolling, A., Veres, S.M.: Formulating robot pursuit-evasion strategies by model checking. In: Proceedings of IFAC 2014, pp. 3048–3055 (2014)

14. Thunberg, J., Ögren, P.: A mixed integer linear programming approach to pursuit evasion problems with optional connectivity constraints. Auton. Robots **31**(4), 333–343 (2011)

The Benefits of Explicit Ontological Knowledge-Bases for Robotic Systems

Zeyn Saigol[1](✉), Minlue Wang[2], Bram Ridder[3], and David M. Lane[1]

[1] Ocean Systems Lab, Heriot-Watt University, Edinburgh, UK
zs@zeynsaigol.com, D.M.Lane@hw.ac.uk
[2] Intelligent Robotics Lab, University of Birmingham, Birmingham, UK
[3] Department of Informatics, King's College London, London, UK

Abstract. With the increasing abilities of robots comes a corresponding increase in the complexity of creating the software enabling these abilities. We present a case study of a sophisticated robotic system which uses an ontology as the central data store for all information processing. We show how this central, structured and easily human-understandable knowledge-base makes for a system that is easier to develop, understand, and modify.

1 Introduction

Robotic systems are becoming more capable, but this makes the software engineering involved harder. Systems must rely on multiple components each with a different function, often written by different people with different technical specialities, and frequently wrapping "third-party" modules developed completely independently. This is a problem that can be tackled from many different directions: clear architecture and documentation (for both the system being developed, and any third-party libraries it relies on); good communication and a clear reporting structure within the development team; following best-practice for software development; and defining clear interfaces between the different components of the system. We address this last problem. According to Brooks [2]

The most pernicious and subtle bugs are system bugs arising from mismatched assumptions made by the authors of various components.

which illustrates the importance of the issue.

Our solution is to use a central ontological knowledge-base, shared between all components in the system. A novelty of our approach is that we attempt to store all the high-level knowledge of the robot in the knowledge-base, including observed objects, the robot's plans, and execution data such as inspection waypoints.

We consider the scenario of a mobile robot trying to find an item of interest in an unmapped area, which has general application to domains such as searching for survivors following a disaster, finding suspected bombs in sensitive locations, or performing autonomous science surveys in the oceans or on

© Springer International Publishing Switzerland 2015
C. Dixon and K. Tuyls (Eds.): TAROS 2015, LNAI 9287, pp. 229–235, 2015.
DOI: 10.1007/978-3-319-22416-9_27

extraterrestrial planets. The complete robotic system is presented along with experimental results in [12]; the aim of this paper is to explore in more detail how the components interact using the knowledge-base, and to highlight how this has benefited the development process.

2   Ontologies Background

We consider an ontological knowledge-base to be (i) a definition of the classes of things that may exist, properties they may have and potential relationships between them, and (ii) a store of instances of these classes and relations. While our system uses the W3C standard format for ontologies, OWL [9], we expect the benefits described in this paper would accrue to users of other systems for storing structured, semantically-tagged knowledge.

Ontologies are strongly associated with the semantic web [1], as they provide an ideal common language for the exchange of data between disparate web-enabled systems. However, they were used before that by the AI community to solve exactly the kind of issues outlined in the introduction (see for example [3]). The advantages of ontologies include the re-use of domain engineering outputs, readily available editing, consistency checking and reasoning tools, and an easily human-interpretable format for storing and amalgamating knowledge.

Recently they have gained in popularity in robotics, as shown by the IEEE working group developing a common ontology for robotics and autonomous systems [10,11]. Ontologies have been used to represent domain and common-sense knowledge for household robots, in the KnowRob [13] and ORO [5] systems. They have been used as a data store for underwater robots [7,8], been integrated with probabilistic reasoning systems [4], and as a bridge between perceptual data and semantic concepts for service robots [6]. We have adopted KnowRob for our ontology implementation.

3   System and Ontology

Our test setup is an indoor robot exploring an unknown area, but with knowledge of all of the object types it may encounter. We make the simplifying assumption that all faces of all objects are rectangular and of a single colour, but note that this framework extends to any object class which can be decomposed into recognisable sub-components. Our perception system can detect the colour, size and pose of rectangular surfaces, but observing one face of an object may not uniquely determine the object's class. Further observations will restrict the space of possible classes, and the set of possible worlds the robot might be in is maintained in the knowledge-base. Based on these possible worlds, the system creates a plan to explore the area so as to find a target object (of a known class) as quickly as possible.

The system is described and evaluated in [12], but here we give an overview of its main software components:

knowledge-base the central ontological store containing the robot's knowledge of the world.

perception extracts rectangular faces from RGB-D point clouds.

scene-generator builds up a set of scenes, each containing a concrete instantiation of shapes which is consistent with the faces observed so far.

waypoint-generator creates view-poses to observe locations likely to contain the target, using the scenes data in the knowledge-base.

contingent-planner uses the scenes representation and waypoints to define a plan to efficiently find a target, taking into account candidate objects that may restrict the robots path, and the different possible outcome from observation actions.

observation-action-planner accounts for noisy sensor observations by monitoring, at execution time, the improvement in value of the future plan if additional observation actions are performed.

scenes-to-belief-state converts the scenes into a probabilistic map of target location, for use by the observation-action-planner.

executor executes a mission by invoking other components in turn, generating waypoints, a contingent plan, executing the plan, and then repeating these steps until the target is found.

goal-evaluator determines if the knowledge-base has converged on a single, fixed location for the target.

Other functionality (including moving the robot and building an obstacle-avoidance map using SLAM) is handled by standard ROS components.

The ontology schema backing the system is shown in Figure 1, illustrating that we have used base classes from KnowRob where possible. The environment-model parts of the ontology are shown on the left of Figure 1, and include object templates (**Shape**) and observed objects (**ObjectInScene**). Multiple candidate **Scenes** are created by applying collision constraints with other objects. Note that we do not simply attach the pose corresponding to an **ObservedFace** to a **Shape**, as there may be several instances of a **Shape**, both within and across **Scenes**.

A significant aspect of our ontology is that it also stores mission-related information, as shown on the right of Figure 1. The **AreaOfInterest** defines our search area, **Pose2D** instances are waypoints, and the branching **ContingentPlan** is stored in the ontology. This allows, for example, the waypoint-generator to store waypoints in the ontology, the contingent-planner to access them and store its output plan into the ontology, and finally the execution system to read and execute this plan.

The core knowledge-base architecture is shown in Figure 2, and consists of a custom C++ node which starts an embedded Prolog engine to run KnowRob. Some manipulation of scenes is a good fit for logical operations and runs in Prolog, but other components access the knowledge-base via a C++ interface. This interface uses SWI-Prolog's foreign language interface for fast access to the ontology (and is a more performant replacement for KnowRob's `json_prolog` module), and has a ROS service wrapper. A stub class calls either the ROS

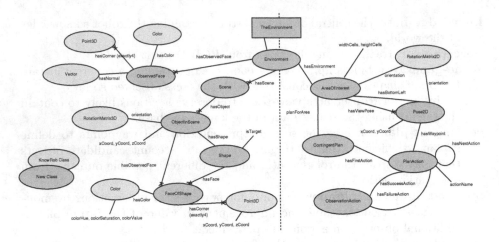

Fig. 1. Ontology layout, showing environment-modelling classes to the left of the dashed line, and mission-related classes to the right

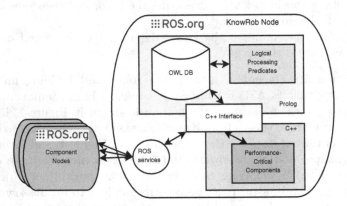

Fig. 2. System architecture

services or the native interface directly, depending on where the code is running, meaning performance-critical components can be moved inside the knowledge-base node by simply editing launch files.

4   Evaluation

The evaluation of architectural aspects of robotic systems is extremely difficult. We cannot demonstrate that using an ontology allows our system to do things that would otherwise be impossible, as we do not claim this to be true. Instead we hope to convince the reader that our system is more robust, simpler to build, and easier to modify and maintain due to the use of a central ontological knowledge-base.

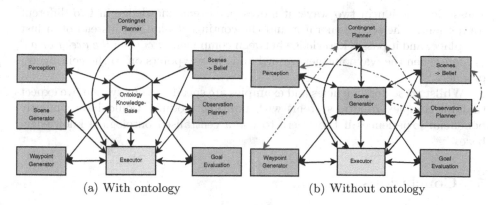

(a) With ontology   (b) Without ontology

Fig. 3. Inter-component dependencies for the complete system. (a) the system as it is implemented, (b) a possible dependency set resulting from removing the ontology. The dash-dotted red arrows indicate new dependencies, and the dotted blue arrow indicates a dependency that is no longer needed.

Figure 3(a) shows the inter-component dependencies of our system, and Figure 3(b) the expected dependencies a system without a central data store would have. For the non-ontology system we have assumed both the scene-generator and the contingent-planner will have their own persistent storage, whereas the waypoint-generator will pass data directly to the contingent-planner, which seems a reasonable design choice.

It turns out that for our implementation there is only a small reduction in the total number of inter-component connections compared to the non-ontology version (partly as there is one component fewer, namely the knowledge-base itself). However, the *complexity* of the interactions is much higher; instead of two fixed interactions (with the knowledge-base and the executor), components will have a varying number of interactions with unpredictable other components. Further, adding a new component may require changes to several other components, depending on which parts of the robot's knowledge the new component needs to make use of.

We present two examples of this: first, when the observation-action-planner was integrated, we only had to interface it to the knowledge-base, despite the fact it relies on knowing the plan produced by the contingent-planner, and the probability distribution over target locations produced by the scenes-to-belief-state which itself relies on the scenes representation. The second example is that we realised we needed to record the execution time of each processing component for results in [12]. These could only be produced by the components themselves, and were needed by the executor – but instead of needing new interfaces between the executor and each component, we could simply store the execution times in the knowledge-base.

Finally we note that having a central knowledge-base enables the executor to call each component directly. The non-ontology version in Figure 3(b) has a less clear design in two ways: it stores persistent knowledge in two different components (the scene-generator and the contingent-planner) instead of in just one place, and it passes knowledge between components, *cutting the executor out of loop*, when the waypoint-generator hands its waypoints on to the contingent-planner.

Whilst these dependencies and examples are specific to our system, we expect that all complex robotic systems with multiple modules would have similarly beneficial experiences if they were to use a centralised ontological knowledge-base.

5   Conclusions

We have outlined a robotic exploration system, and demonstrated how a centralised, semantic knowledge-base helps reduce the complexity of the interactions between components. With our ontological knowledge-base, a new developer does not need to guess at the interactions of a component: they are all invoked by the executor, perform some processing, and store any results into the knowledge-base.

Future work will include co-operation with a second robot, and we expect that having a well-defined ontology which can be easily shared (using ROS) with this second robot will make our task much easier.

References

1. Berners-Lee, T., Hendler, J., Lassila, O.: The semantic web. Scientific American (2001)
2. Brooks, F.P.: The Mythical Man-month: Essays on Software Engineering. Addison Wesley (1995)
3. Gruber, T.R.: Toward principles for the design of ontologies used for knowledge sharing. International Journal of Human Computer Studies 43(5–6), 907–907 (1995)
4. Hanheide, M., Gretton, C., Dearden, R., Hawes, N., Wyatt, J., Pronobis, A., Aydemir, A., Gbelbecker, M., Zender, H.: Exploiting probabilistic knowledge under uncertain sensing for efficient robot behaviour. In: IJCAI 2011 (2011)
5. Lemaignan, S., Ros, R., Mosenlechner, L., Alami, R., Beetz, M.: Oro, a knowledge management platform for cognitive architectures in robotics. In: IROS 2010 (2010)
6. Lim, G.H., Suh, I.H., Suh, H.: Ontology-based unified robot knowledge for service robots in indoor environments. IEEE Transactions on Systems, Man and Cybernetics, Part A: Systems and Humans 41(3), 492–509 (2011)
7. Maurelli, F., Saigol, Z., Lane, D.M.: Cognitive knowledge representation under uncertainty for autonomous underwater vehicles. In: ICRA 2014 Workshop on Persistent Autonomy for Underwater Robotics (2014)
8. Miguelanez, E., Patron, P., Brown, K.E., Petillot, Y.R., Lane, D.M.: Semantic knowledge-based framework to improve the situation awareness of autonomous underwater vehicles. IEEE Transactions on Knowledge and Data Engineering 23(5) (2011)

9. Motik, B., Patel-Schneider, P.F., Parsia, B., Bock, C., Fokoue, A., Haase, P., Hoekstra, R., Horrocks, I., Ruttenberg, A., Sattler, U., Smith, M.: OWL 2 web ontology language: Structural specification and functional syntax. Tech. rep., W3C (2009)
10. Paull, L., Severac, G., Raffo, G., Angel, J., Boley, H., Durst, P., Gray, W., Habib, M., Nguyen, B., Ragavan, S., Sajad Saeedi, G., Sanz, R., Seto, M., Stefanovski, A., Trentini, M., Li, H.: Towards an ontology for autonomous robots. In: IROS 2012 (2012)
11. Prestes, E., Carbonera, J.L., Fiorini, S.R., Jorge, V.A.M., Abel, M., Madhavan, R., Locoro, A., Goncalves, P., Barreto, M.E., Habib, M., Chibani, A., Gerard, S., Amirat, Y., Schlenoff, C.: Towards a core ontology for robotics and automation. Robotics and Autonomous Systems **61**(11), 1193–1204 (2013)
12. Ridder, B., Saigol, Z., Wang, M., Dearden, R., Fox, M., Hawes, N., Kumar, A., Lane, D.M., Long, D.: Efficient search for known objects in large, unknown environments using autonomous indoor robots (submitted 2015)
13. Tenorth, M., Beetz, M.: Knowrob: A knowledge processing infrastructure for cognition-enabled robots. International Journal of Robotics Research **32**(5) (2013)

Task-Based Variation of Active Compliance of Arm/Hand Robots in Physical Human Robot Interactions

Iason Sarantopoulos, Dimitrios Papageorgiou, and Zoe Doulgeri[✉]

Department of Electrical and Computer Engineering,
Aristotle University of Thessaloniki, 54124 Thessaloniki, Greece
iasons@auth.gr, {dimpapag,doulgeri}@eng.auth.gr

Abstract. In this work a control strategy combining admittance control with haptic cues is proposed for varying robot arm-hand target impedances in tasks involving physical Human Robot Interaction (pHRI). External force and robot state measurements are employed to initiate task phase transitions which further involve switching to appropriate target impedances. Three tasks typical for domestic robot assistance are demonstrated experimentally showing safe and natural interactions.

1 Introduction

Most of the existing arm/hand systems are usually position controlled systems that can be accurate but rigid and hence, potentially unsuitable for contact tasks with the environment and/or pHRI. Contact initiation and maintenance with the environment in the presence of uncertainties requires compliant motions to ensure environmental and robot safety. Moreover, it is important for the robot to present the required compliance in order to accomplish a task which involves human-robot interaction forces [1][2][3]. Compliance is either built within the structure of the robot via flexible joints, links and soft covers or achieved actively by the control action. Impedance control is typically employed to combine accuracy in free motion and a stable equilibrium in the case of interaction. This characteristic is clearly beneficial for avoiding transitions between position and force control at contact state changes which may induce instabilities in the presence of transition delays. However, the critical issue is the specification of the appropriate desired impedance for a given task which may need to be adapted to the environment variation or the task phase requirements. For example, when handing over an object to a human, a stiff grasp is required during the object's transfer and a compliant release triggered by the human hand-robot interaction forces, is needed for a natural result. Tactile and force feedback plays an important role in signaling transitions from one task phase to another. This fact has

This research is co-financed by the EU-ESF and Greek national funds through the operational program "Education and Lifelong Learning" of the National Strategic Reference Framework (NSRF) - Research Funding Program ARISTEIA I under Grant PIROS/506.

C. Dixon and K. Tuyls (Eds.): TAROS 2015, LNAI 9287, pp. 236–245, 2015.
DOI: 10.1007/978-3-319-22416-9_28

been early recognised in robotics; an example of its utilization in grasping can be found in [4]. The major challenges in resolving the general problem of task based impedance variation lie on a) the detection of task phase transitions, b) the utilization of appropriate target impedance at each task phase and c) the continuity of the state variables and the stability of the overall system.

This paper is a first attempt to address the general problem by utilizing preset target impedance values for each task phase and simple thresholding on selected signals for task phase transition detection. We exploit the concept of task phase transitions with haptic cues for tasks involving physical human robot interaction. Our system is composed by an arm and a fingered hand equipped with external force sensing that operate under an admittance control scheme. External force sensing allows us to develop a glossary consisted of natural haptic cues to perform automatically a variety of tasks involving pHRI phases. Three different tasks, typical in human centric environments, are presented focusing on force sensing; the handing over and handing in of an object from/to the robotic hand to/from the human hand and holding hands with support against gravity. Utilizing external force and robot state as initiators for a task phase transition we change arm and hand motion planning and compliance. The rest of the paper is organized as follows. Section 2 briefly describes robot control method preliminaries. Section 3 is denoted to the description of the proposed control strategy. Section 4 describes the experimental set up and results, while conclusions are drawn in Section 5.

2   Robot Control Method Preliminaries

The proposed control strategy, discussed in the following section, employs admittance control, which is a position based impedance control implementation. In impedance control, the robotic system is controlled so that it acts in closed-loop as an equivalent mass-damping-spring system having the contact force as input. There are two impedance control implementations, the torqued based and the position based implementation. In the torque based implementations, the controller is a mechanical impedance (i.e a physical system that accepts motion inputs and yields force outputs) and the controlled plant is treated as a mechanical admittance (i.e. a physical system that accepts force inputs and yields motion outputs) [5][6]. The opposite is true in position-based implementations known as admittance control where the plant is position controlled. Admittance control requires the availability of external force measurements and may provide high level of accuracy in non-contact tasks [7]. Its performance and stability characteristics largely depends on the quality of the underlying position controller and the environmental stiffness; in fact, stability is adversely affected in the presence of dynamic interactions with very stiff environments. As the majority of the available robotic equipment is commanded at the position or velocity level, admittance control is easily implementable and is preferred, when contacting a non-stiff environment.

Physical human-robot interaction involves the contact of the robot with the human, which can be viewed as a varying mechanical impedance of medium to low stiffness. As pHRI is central to this work, an admittance controller is utilized. The underlying position controller is a prescribed performance control law achieving preset transient steady state behaviour of position errors [8], thus providing high level of accuracy in non-contact tasks [7].

3   The Proposed Control Strategy

The proposed control scheme for both the robotic arm and hand is illustrated in Fig.1. The i-th model $(i = 1, ..., n)$ defines a pair of an arm and a hand target impedance model (TI) at the m-dimensional task space. Let us consider the task reference state x_r, and the respective admittance state d_x. The admittance model is generally described by the following second order differential equation:

$$H_d \ddot{d}(t) + B_d \dot{d}(t) + K_d d(t) = F \tag{1}$$

where $F \in \Re^m$ is the measured generalized contact force, $H_d, B_d, K_d \in \Re^{m \times m}$ are the target impedance parameters, reflecting the desired mass, damping and stiffness. In cases where no specific position reference is demanded, we set stiffness to zero $(K_d = 0)$, and the following mass-damper model is utilized:

$$M_d \ddot{d}(t) + D_d \dot{d}(t) = F \tag{2}$$

where $M_d, D_d \in \Re^{m \times m}$ are the target mass-damper impedance parameters. Such cases arise when the human desires to lead through the robot, by applying a force to its hand. Clearly, in case of model (1), $d_x = [d \ \dot{d}]^T$, $x_r = [p_r \ \dot{p}_r]^T$, with $d(t) \in \Re^m$ representing the displacement from the reference position trajectory $p_r \in \Re^m$ when in contact; in case of model (2), $d_x = \dot{d}$, with $\dot{d}$ representing deviations from a reference velocity $\dot{p}_r$. We assume that external forces and the arm-hand state are available for measurements. We further assume knowledge or availability of other sensorial inputs (e.g. visual) that are involved in a task e.g. human location. Given a task, planned or initiated at a higher decision level, the Target Impedance Switching controller (TIS) detects the task phase transitions, based on haptic cues as well as the state of the robotic system, switching the TI model accordingly. An example TI model values are shown in Table 1. For instance, when the end-effector enters the personal space of a human, TIS controller could switch from Model 2 to Model 1. The selected target impedance model triggered by the external force, is producing an output state d_x which together with the reference x_r forms a dynamically shaped desired state x_d to be followed by the position or velocity controlled robotic system via a kinematic level control law.

The admittance control strategy of the arm-hand system will allow a safe and natural human-robot interaction at each task phase. Stability and continuity regarding the overall switched admittance system are deferred to future extensions of this work. The controller produces a compliant motion of the fingers and the arm by reacting to the force applied at the fingertips and, consequently, indirectly at the arm. These forces can be the result of a human holding the

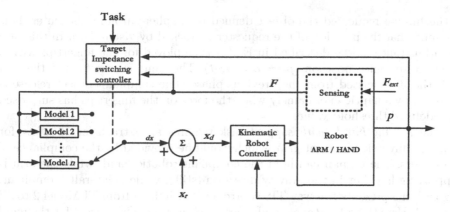

Fig. 1. Robotic arm-hand control scheme

Table 1. TI-Models in use

Models	Arm compliance	Hand compliance
Model 1	Medium	High
Model 2	Medium	Low
Model 3	non-uniform (high - low, lateral - vertical).	High
Model 4	Low	Low

robotic hand, or can be the result of a contact via an object. In this work, we consider three different tasks, typical of human-robot interaction. The *hand-in* task, the *hand-over* task and the *holding hands with support* task.

In the *hand-in task*, the human will pass an object to the hand by either applying a force on the compliant fingers, or by pressing the object on the hand's palm, applying an indirect force to the arm, as a haptic cue evoking the object grasp. Algorithm 1 presents the way TIS controller is switching over the compliant behavior of the arm-hand system, according to the task phase; F_{palm} denotes the force the arm is sensing along the normal to the palm direction, while p_{hand} is the position of the fingertips. The compliance in this scenario is critical for smoother and natural interaction. In this case the admittance model described in Eq.(2) is employed for the hand and a zero reference velocity ($\dot{p}_r = 0$). For the arm, the model described in Eq.(1) is employed, with $p_r = constant$.

In the *hand-over* task, the robot, having grasped an object, presents it to the human hand which acquires the object. Algorithm 2 manifests the way TIS controller is switching over the compliant behavior of the arm-hand system, according to the task phase. The basic idea is to maintain a stable and robust grasp during the transfer of the object (TI Model 2 and 4) and switch to a compliant behavior (TI Model 1) when the end effector enters the personal space of the human requested the object defined by a sphere around his center. It is assumed that the position of the requester is tracked by visual data. In this case the admittance model described in Eq.(1) is employed for the fingertips with a constant reference position ($p_r = constant$). The same is true for the the the arm with the p_r inherited from the previous phase. The compliant object release is initiated by a haptic cue, namely when the force on the fingertips has surpassed a predefined threshold value.

The *holding hands with support* task requires an extra layer of safety for avoiding any risk of a tight grip, which would be provided by the compliance of the fingers. The human can hold the compliant robotic hand, while the arm is supporting him/her in the gravity direction (stiff), while is laterally compliant. Algorithm 3 presents the way TIS controller is switching from TI Model 2 to TI Model 3. Here the admittance model described in Eq.(2) is utilized for the hand with $\dot{p}_r = constant$, since a similar response to the previous scenario is required with a constant reference velocity, due to the fact that we want the fingers to be moving (closing) with zero external force. That means the fingers will try to enfold the human hand, by preserving their compliance. For the arm, the same admittance model with the previous scenario is utilized with a non-homogeneous stiffness in task space (Model 3).

Algorithm 1. The pseudocode of the hand-in scenario

1: Initialize arm-hand system and locate human
2: Switch to TI Model 2
3: Preshape fingers and move the end effector to pose g
4: $p_{hand}(0) \leftarrow$ current p_{hand}
5: Switch to TI Model 1
6: **while** NOT ($\|p_{hand}(t) - p_{hand}(0)\|_\infty > Threshold_1$ OR $F_{palm} > Threshold_2$) **do**
7: **end while**   ▷ wait haptic cue
8: Switch to TI Model 2
9: Close grasp

---

4   Experimental Set-up and Results

The robotic arm KUKA LWR4+ and the robotic hand BarrettHand BH8-282 have been used for the experimental implementation of the arm-hand system. This is a 7 degrees of freedom (dof) robotic arm with rotational joints and anthropomorphic structure and a three finger hand with 4 actuated dofs; each finger has a proximal and distal finger link and has one dof making it an under-actuated two joint serial chain. The fourth dof refers to the spread of the two fingers symmetrically around the palm and the third fixed finger. The Barrett-Hand is equipped with fingertip torque sensors at the last underactuated joint of each of the three fingers which can measure safely up to a strain gauge value of 2kg at the tip (figure). On the other hand, the KUKA LWR4+ arm is equipped with joint torque sensors for measuring externally applied force and torques.

Algorithm 2. The pseudocode of the hand-over scenario

1: Initialize arm-hand system and locate object
2: Switch to TI Model 2
3: Preshape fingers and move the end effector to a point close to the object
4: Switch to TI Model 4
5: Approach the object with end effector
6: Close Fingers
7: Elevate the object
8: Measure and compensate object's mass
9: Switch to TI Model 2
10: **while** approaching human **do**
11:     **if** $||p - p_{human}||_2 < Threshold_1$ **then**
12:         Switch to TI Model 1   ▷ Pass object to human
13:         **if** fingertip force $> Threshold_2$ **then**   ▷ haptic cue
14:             Open fingers
15:         **end if**
16:     **end if**
17: **end while**

Algorithm 3. The pseudocode of the holding hands scenario

1: Initialize arm-hand system and locate human
2: Switch to TI Model 2
3: Move the arm to pose g and preshape fingers
4: Switch to TI Model 3
5: **while** trigger for exit not received **do**
6: **end while**   ▷ wait trigger for exit

The proprietary FRI library is being used for the communication with the controller KRC2 which is provided by the manufacturer for controlling KUKA LWR4+ and an open source driver was developed in C++, for controlling Bar-rettHand under Linux (BAD). Regarding the integration of BarrettHand with KUKA arm, a ROS-based implementation is being utilized by using ROS actions. In particular, a Master node implements the action clients and consists the mean of communication between the other two nodes, the BarrettHand node and the KUKA node, which implement the action servers . The BarrettHand action servers use the BAD driver functions, and the KUKA action servers use the FRI library. Each task can be performed by a ROS action and the corresponding client-server pair (Fig.2).

Regarding the BarrettHand, target impedance of Eq. (2) is in this implemen-tation simplified to direct proportional dependency $D_d \dot{d}(t) = F$, i.e. $M_d = 0$. Furthermore as the BarrettHand's fingers have 1-DOF, there is a one-to-one mapping between the joints and the fingertips. That means that we can per-form the control scheme at the joint space of each finger, rather than the task space and hence $\dot{d}(t)$ is indicating the finger joint reference velocity. Regarding the LWR4+, we consider the target impedance of Eq.(1). The motion of the LWR4+ robotic arm is performed by employing a 5-th order position trajectory

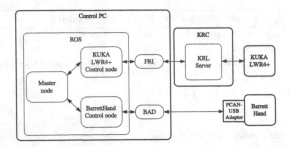

Fig. 2. The node communication structure of the experimental setup

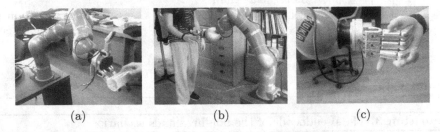

      (a)   (b)   (c)

Fig. 3. Experiment instances. (a) Hand-in task with finger force triggering, (b) Hand-over task, (c) Grip compliance in holding hands with support.

p_r from the initial point to the desired one, resulting in human-like bell shaped velocities in task space, under the following kinematic control law which is a joint reference velocity: $v_{qr} = J^\dagger(\dot{p}_d - k_p e_p)$, where $e_p = p - p_d$ and $J^\dagger$ is the Moore-Penrose right pseudoinverse of the Jacobian matrix of the arm, giving the minimum norm inverse kinematic solution.

In the arm the rotational impedance of the end-effector is set to high in all cases ($K_d = 2000 I_3 Nm/rad$, $B_d = 619.6 Nms/rad$ and $H_d = 48 Nms^2/rad$). The translational impedance of the end-effector is set to the following values for the low, medium and high compliance models of Table 1 respectively: arm low ($K_d = 2000 I_3 N/m$, $B_d = 252.98 I_3 Ns/m$, $H_d = 8 I_3 Ns^2/m$), arm medium ($K_d = 500 I_3 N/m$, $B_d = 126.5 I_3 Ns/m$, $H_d = 8 I_3 Ns^2/m$), arm high ($K_d = 200 I_3 N/m$, $B_d = 80 I_3 Ns/m$, $H_d = 8 I_3 Ns^2/m$) and arm non-uniform ($K_{dj} = 120 I_2 N/m$, $B_{dj} = 62 I_2 Ns/m$, $H_{dj} = 8 I_2 Ns^2/m$), where $j = \{x, y\}$ and I_n the identity matrices of dimension n and the arm low compliance values are employed in the gravity direction. Notice that the arm impedance parameters are selected to yields a critical response. The finger impedance parameter D_d is given by $0.204 Nms/rad$ for the high compliance models. The low compliance model for the fingers is not explicitly implemented owing to the unknown forces on the internal links of the BarrettHand in an envelope grasp; instead, we rely on the intrinsic characteristics of BarrettHand (non-backdrivability) for a stiff grasp. Regarding the hand-in haptic cue parameters of the corresponding Algorithm 1, the following values are used: $Threshold_1 = 0.123 rad$, $Threshold_2 = 20N$.

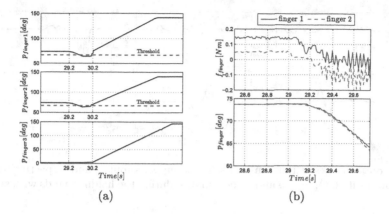

(a)   (b)

Fig. 4. (a) Proximal joint position responses (p_{finger}) of each finger, during the hand-in task by pushing the object on the fingertips (finger 1 and 2 in this experiment). The threshold utilized to change the admittance model is shown with the red dashed line. (b) Contact torques at the distal joints (f_{finger}, upper diagram) and respective detail from the proximal joint position response (p_{finger}, lower diagram).

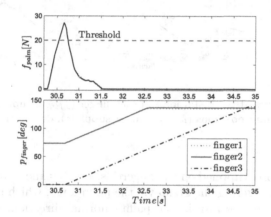

Fig. 5. Force normal to the palm's direction (f_{palm}, upper subplot) and finger proximal joint positions (p_{finger}, lower subplot), during the hand-in task by pushing the object on the hand's palm. The threshold utilized to change the admittance model is shown with the red dashed line.

Regarding the hand-over threshold parameters of the Algorithm 2, the following values are used: $Threshold_1 = 0.1m$, $Threshold_2 = 0.1Nm$.

A video is available showing the tasks at https://youtu.be/CKGSIjfBGNM . Fig.3 shows instances from each task. Fig. 4 shows the position force responses for the hand-in case with a fingertip haptic cue. Specifically, Fig. 4. a shows how the positions of fingers 1 and 2 are reshaped by the external force in the time span $[t_1, t_2] = [29.2s, 30.2s]$, resulting in the haptic cue of grasp triggering; Fig. 4.b shows details of the positions of fingers 1 and 2 in the above time span

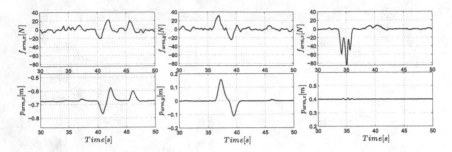

Fig. 6. Contact forces at the arm's wrist (f_{arm}, upper subplot) and respective position (p_{arm}, lower subplot), in x, y and z coordinates, during the holding hands with support task

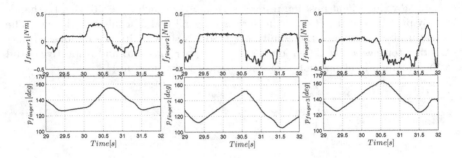

Fig. 7. Contact torques at the distal joint of each finger (f_{finger}, upper subplots) and respective proximal joint positions (p_{finger}, lower subplots), during the holding hands with support task

together with the forces applied on the fingertips. Fig. 5 demonstrates the haptic cue on the case of triggering the grasp via the palm force, which is shown in the force measurements of the arm in the palm's normal direction. Fig. 6, 7 show the compliant response of the end-effector and the fingers for the holding hands with support task. Notice the difference in compliance between the lateral and the vertical direction of the end-effector.

5   Conclusion

In this paper a task-based compliance switching strategy is presented and demonstrated by three typical to pHRI tasks; the hand-in, hand-over and holding hands with support tasks. The switching of the admittance parameters is related to each task phase transition triggered by the robotic system's state and/or a haptic cue. Experiments are conducted with a system consisted of a KUKA LWR4+ 7 dof robotic arm and a BarrettHand BH8-282, integrated by ROS, showing natural pHRI (https://youtu.be/CKGSIjfBGNM). Future extensions of this work will include stability analysis of the overall system.

References

1. Heinzmann, J., Zelinsky, A.: Quantitative Safety Guarantees for Physical Human-Robot Interaction. International Journal of Robotics Research **22**, 479–504 (2003)
2. Mortl, A., Lawitzky, M., Kucukyilmaz, A., Sezgin, M., Basdogan, C., Hirche, S.: The role of roles: Physical cooperation between humans and robots. International Journal of Robotics Research **31**, 1656–1674 (2012)
3. De Santis, A., Siciliano, B., De Luca, A., Bicchi, A.: An atlas of physical human-robot interaction. Mechanism and Machine Theory **43**, 253–270 (2008)
4. Howe, R.D., Popp, N., Akella, P., Kao, I., Cutkosky, M.R.: Grasping, manipulation, and control with tactile sensing. In: Proceedings of the IEEE International Conference on Robotics and Automation, pp. 1258–1263 (1990)
5. Hogan, N.: Impedance Control: An Approach to Manipulation: Part I - Theory. ASME Journal of Dynamic Systems, Measurement, and Control **107**, 1–7 (1985)
6. Hogan, N.: Impedance Control: An Approach to Manipulation: Part II - Implementation. ASME Journal of Dynamic Systems, Measurement, and Control **107**, 8–16 (1985)
7. Pelletier, M., Doyon, M.: On the implementation and performance of impedance control on position controlled robots. In: Proceedings of the IEEE International Conference on Robotics and Automation, vol. 2, pp. 1228–1233 (1994)
8. Atawnih, A., Papageorgiou, D., Doulgeri, Z.: Reaching for redundant arms with human-like motion and compliance properties. Robotics and Autonomous Systems **62**, 1731–1741 (2014)

Auction-Based Task Allocation for Multi-robot Teams in Dynamic Environments

Eric Schneider[1]([✉]), Elizabeth I. Sklar[1,2], Simon Parsons[1], and A. Tuna Özgelen[3]

[1] Department of Computer Science, University of Liverpool, Liverpool, UK
{eric.schneider,s.d.parsons}@liverpool.ac.uk
[2] Department of Informatics, King's College London, London, UK
elizabeth.sklar@kcl.ac.uk
[3] Department of Computer Science, Graduate Center,
City University of New York, New York, USA
aozgelen@gc.cuny.edu

Abstract. There has been much research on the use of auction-based methods to provide a distributed approach to task allocation for robot teams. Team members bid on tasks based on their locations, and the allocation is based on these bids. The focus of prior work has been on the optimality of the allocation, establishing that auction-based methods perform well in comparison with optimal allocation methods, with the advantage of scaling better. This paper compares several auction-based methods not on the *optimality* of the allocation, but on the *efficiency* of the execution of the allocated tasks giving a fuller picture of the practical use of auction-based methods. Our results show that the advantages of the best auction-based methods are much reduced when the robots are physically dispersed throughout the task space and when the tasks themselves are allocated over time.

1   Introduction

This paper is concerned with the *multi-robot routing* problem. This is a task allocation problem in which tasks require robots to reach particular *target* points, and each target must be visited by one robot only. Multi-robot routing is a standard task [9] that is part of problems such as search-and-rescue and humanitarian demining, and it is a common test domain for robot team coordination [2]. The problem is also computationally hard. The number of ways of allocating m target points to n robots is a Stirling number of the second kind:

$$S(m,n) = \frac{1}{n!} \sum_{j=0}^{n} (-1)^{n-j} \binom{n}{j} j^m$$

and the size of the problem quickly defeats attempts to use standard optimization techniques like integer programming [23].

The computational explosion in the multi-robot routing problem has led to researchers looking for more efficient solutions. Lagoudakis et al. [12] began

C. Dixon and K. Tuyls (Eds.): TAROS 2015, LNAI 9287, pp. 246–257, 2015.
DOI: 10.1007/978-3-319-22416-9_29

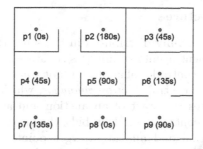

Fig. 1. The experimental environment — an office-like floor plan and set of target points. In dynamic allocation scenarios, points become available for allocation at the times indicated in the figure.

an interesting line of work in using *auction mechanisms* to allocate the target points in a multi-robot routing problem. This work showed that multi-round auctions, where each round consisted of bids for single tasks, had a performance that was experimentally determined to be close to that of an optimal allocation. This initial paper was followed by work on the way that rules for determining bids affected the performance of the auction mechanism, establishing bounds on the performance of different rules [13], and empirically comparing the performance of the auctions against an optimal allocation under different bidding rules [23]. Later work considered the *sequential single-item auction* [7,8] as an alternative to *combinatorial auctions* for multi-robot routing, again showing that the total path cost — the total distance travelled by all the robots in the team — is close to optimal.

Overall, this prior work shows the power of auction mechanisms in task allocation in this domain, and in this paper we build upon it. Our extension is twofold. First, we evaluate the effectiveness of auction mechanisms not only in terms of the total distance that the team expects to travel given the task allocation — what we might consider the theoretical cost of executing the tasks — but also in terms of a set of metrics that measure how the tasks are executed in practice. Second, we not only deal with static allocation — where the target points are all known at the beginning of the allocation — but also dynamic allocations, where target points are given to the team over time.

These two factors reveal some of the trade-offs between auction mechanisms in different scenarios, and show that there are situations in which the sequential single-item auction is no more efficient in practice than mechanisms which theory tells us should perform much worse.

2   Methodology

The work reported here is part of a larger effort [16,21] aimed at evaluating coordination techniques in terms of their potential for application in real robot systems. This influences a number of aspects of the experiments that we conducted. We discuss these aspects in this section.

2.1 System Architecture

Our multi-robot system is fully distributed in the sense that the controllers for the robots are independent agents running as separate processes. No controller has the ability to tell the controller of another robot what to do. The coordination of the team is through a central *auction manager*, which holds a list of target points and communicates the start of an auction and awards target points to *robot controllers*.[1] Each controller sends in bids, the auction manager determines the winner of the auction and allocates target points accordingly. Since the individual controllers only have access to their local information, their bids and the allocation don't explicitly attempt to optimize across all robots.

Our software architecture is agnostic about whether the team executes its tasks on real robot hardware or in simulation. That is, the robot controllers are capable of receiving information either from real robots operating in our lab, or from simulated robots operating on a model of the lab, and the same decisions are made and the same control signals output in both cases. For the work reported here, we ran the experiments using the Stage simulator [4]. Previous work [16] suggests that for our setup, there is close agreement between simulated results and results obtained on real robot hardware.

Our hardware platform is the Turtlebot 2, which has a differential drive base and a colour/depth-sensing Microsoft Kinect camera. The ROS [17] navigation stack provides communication, localisation and path planning capabilities. Our robot controller software works with ROS to implement bidding, task selection and collision avoidance behaviours using a finite state machine. Our simulated Turtlebot has the same properties as its physical counterpart (size, shape, velocity, acceleration, etc.). In the simulation experiments reported here, localisation is provided by Stage. An on-board sensor that approximates the Kinect camera is used for collision avoidance.

The environment in which our robots, both real and simulated, operate is an office-like environment with rooms opening off a central hallway. A sample layout is shown in Figure 1. It is a smaller environment, in terms of the number of rooms, than that studied by [8] (and in similar work by the same group of authors), a fact determined by the maximum size of the physical area available in our lab for the real robots and our desire to be able to run parallel experiments in both simulation and on the real robots (though here we only report work in simulation).

2.2 Metrics

To measure the performance of the robot team in practice — and hence the performance of the coordination mechanisms — we consider a number of metrics applicable to the performance of individual robots and the team as a whole.

---

[1] Though the bidding process and winner determination are managed centrally, there is no centralized control in the usual sense. The auction could also be distributed among the robots as in [3].

In any work with robots, an important consideration is power consumption. This is the fundamental scarce resource that a robot possesses. Robot batteries only last for a limited time, and so, all other things being equal, we prefer solutions to the multi-robot routing problem that minimize battery usage. As in [7,8,12,13,23], therefore, we measure the *distance travelled* by the robots in executing a set of tasks — both individually and as a group — since this is a suitable proxy for power consumption.

It is important to note that distance is not computed by looking at the shortest distances between the target points, but is (as closely as we can establish) the actual distance moved by the robots during task execution. We collect regular position updates from the simulator, compute the Euclidean distance between successive positions, and sum these up.

In many multi-robot team activities that relate to the routing problem, time to complete a set of tasks is also important. Time is important in exploration tasks — in search and rescue activities, in patrolling, and possibly in demining[2] — and so we measure *run time*, the time between the start of an experiment and the point at which the last robot on the team completes the tasks allocated to it. Also relevant is the *deliberation time*, the time that it takes for the tasks to be allocated amongst the robots. Deliberation time matters because it feeds into the overall time required to complete a set of tasks, but also because it allows us to establish how different allocation mechanisms compare in terms of the computational effort and communication resources required to run them. *Execution time*, is run time minus deliberation time, the time that the robots take to execute the set of tasks they are given.

We also measure *idle time*, the amount of time that robots sit idly during the execution of a set of tasks. We compute the idle time for a robot as the time that elapses between when that robot completes its last task and when all robots on the team have completed all of their assigned tasks. This gives us a way of quantifying how equally tasks are distributed among robots, and it also suggests the extent to which resources are being wasted by a particular allocation. Although a mismatch between the number of tasks and the number of robots means that idle time can be inevitable, idle time represents the use of precious power that is not being directed towards the completion of the tasks.

3   Experiments

3.1   Experimental Setup

As described earlier, the multi-robot routing problem that we are studying involves a team of robots tasked to visit a number of target points such that one robot visits each point. A *scenario* is defined by a specific set of parameters: the number of robots on the team (n), the starting locations for the robots,

---

[2] One can easily imagine demining happening against the clock — in humanitarian demining [6], for example, there may be the need to demine an area in order to allow refugees to move safely away from a dangerous situation.

the number of target points to visit (m), and the locations of the target points. Thus, a scenario can be described by a tuple

$$\langle n, \{(x_0, y_0), \ldots, (x_{n-1}, y_{n-1})\}, m, \{(x'_0, y'_0), \ldots, (x'_{m-1}, y'_{m-1})\}\rangle$$

The experiments reported here measured results in four different scenarios, all of which involved $n = 3$ robots. Two scenarios used static allocations and two scenarios used dynamic allocations. All scenarios used the environment in Figure 1 and had $m = 9$ target points. In static allocations scenarios, all target points were allocated at the beginning of each experiment in a single deliberation phase followed by a single execution phase. In dynamic allocation scenarios, target points were introduced over time according to a predefined schedule. Every introduction of target points triggered a new deliberation phase (possibly interrupting an execution phase). Labels in Figure 1 indicate when each point was introduced in seconds after the start of an experiment. There were two sets of starting locations, one which clustered the robots in the lower left room (as the environment appears in Figure 1), and one in which the robots were distributed in the lower left corner, the upper left corner and the upper right corner of the environment. Experiments were conducted with each of the four scenarios using each of the four task allocation mechanisms discussed below, and each was run 10 times. Thus 160 experimental trials were conducted in all:

$$2 \; start \; locations \times \{static, dynamic\} \times 4 \; allocation \; mechanisms \times 10 \; trials$$

Each experiment recorded the metrics described above: distance travelled; run time (including execution time); deliberation time; and idle time.

3.2 Mechanisms Tested

Our experiments involved four different mechanisms for task allocation: Round-robin (RR); Ordered single-item auction (OSI); Sequential single-item auction (SSI); and Parallel single-item auction (PSI). We describe these in turn.

In *round robin* allocation, we start with two ordered lists, one of target points and one of robots. The first target point is allocated to the first robot, the second target point to the second robot and so on. When one target point has been allocated to each robot, a second target point is allocated to the first robot. And so on. This is clearly not a particularly efficient way to approach task allocation, but it provides a baseline against which other mechanisms can be tested. We referred to this as the "greedy taxi" policy in our earlier work [16], because this policy emulates the behaviour of a taxi rank.

The approach we call the *ordered single-item auction* takes a simple step to improve on round robin allocation. In this case the target points are again placed in an ordered list, but this time each point in turn is offered to all the robots. Each robot makes a bid for the point, where the bid is the distance that the robot estimates (using an A* path planner) it will have to travel to reach the point from its current location. The point is allocated to the robot that makes the lowest bid, that robot updates its "bid-from" location to be the location of the target point it just acquired, and the next point is auctioned.

In the *sequential single-item auction* [8], all unallocated target points are presented to all the robots simultaneously. Each robot bids on the target point with the lowest cost (again computed as the A* distance to the point) and the point with the lowest bid is allocated to the robot that made the bid. The winning robot updates its "bid-from" location to that of the target point it just won and the process is repeated until all points have been allocated.

The *parallel single-item auction* (introduced as something of a strawman in [8]) starts like SSI with all robots bidding on all points from their starting locations. All the target points are allocated in one round, however, with each point going to whichever robot made the lowest bid on it.

3.3   Results

The results of the experiments can be seen in Table 1, Figure 2 and Figure 3. Table 1 gives the value of the metrics for each of the four task allocation mechanisms — RR, OSI, SSI and PSI — in each of the four scenarios — static and dynamic, clustered and distributed start. We give the average value across the ten runs, and the 95% confidence intervals. Figure 2 gives the average distances travelled by the team. This is the main metric we will consider here since it is the one that SSI is looking to optimise task allocations against (that is SSI seeks to minimise distance), it is the metric that we might expect it to perform best on. Figure 3 shows the remaining metrics.

The results for the clustered static allocation in Figure 2 shows SSI as we expect it to perform given the analysis in [8], generating allocations that result in shorter total distances for the team than any of the other mechanisms. The corresponding results from Figure 3 show SSI with lower run time, execution time, and idle time. Run time and execution time are obviously strongly related to the lower distance generated by SSI, while lower idle time suggests a more even allocation of tasks to robots than other mechanisms. PSI is notably poor in this regard in the static, clustered allocation, a point first noted in [8].

When we move to consider a static allocation with a distributed start, we note from Figure 2 that SSI doesn't provide an obviously better allocation in terms of distance than either OSI or PSI. (In fact, Table 1 reveals that SSI does fractionally worse than OSI.) The fact that the robots are distributed means that the target points fall into natural clusters that are obvious to the simpler allocation mechanisms without the exhaustive search through allocations that SSI provides. SSI however, continues to do better than the other mechanisms on other metrics in Figure 3, perhaps most noticeably on idle time.

Moving on to the dynamic allocations, in terms of distance (shown in Figure 2), we see SSI performing comparably with the other mechanisms (and maybe doing a bit worse than PSI). It seems to us that the dynamic allocation is spreading the robots out in much the same way as the distributed start does — because they start moving to the initial set of target points, the robots are physically spread out by the time that later points are auctioned. The same

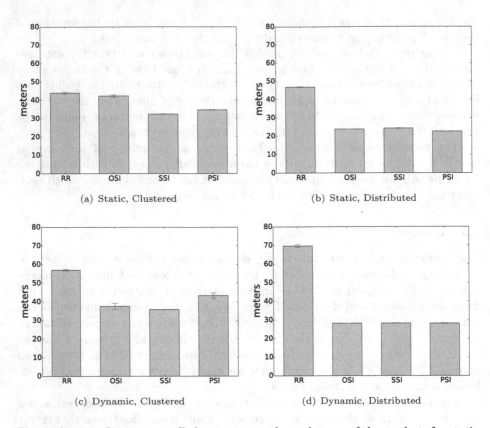

Fig. 2. Average distance travelled over ten runs by each team of three robots for static and dynamic scenarios, with clustered and distributed starts. The bars are ordered RR, OSI, SSI and PSI.

relationships holds for the distance results from the dynamic distributed start, though in this case all three auction mechanisms improve compared with RR, and across all the other metrics in Figure 3.

3.4   Discussion

The aim of these experiments is to understand the comparative performance of the allocation mechanisms across the different scenarios with the aim of establishing the suitability of the mechanisms for different kinds of multi-robot routing tasks.

Overall, for the scenarios we studied, our analysis supports the results in [7,8,12,13,23], showing the effectiveness of the sequential single-item auction in finding solutions to the multi-robot routing problem when the overall distance covered is the most important performance metric. For static allocation with clustered start points, SSI generated solutions which required the team to travel

Table 1. Team metrics for the static (top) and dynamic (bottom) allocations. Time is in seconds. Distance is in metres. The values given are means with 95% confidence intervals.

		Distance	Run time	Delib. time	Idle time
Clustered	RR	43.92 ± 1.12	262.62 ± 9.03	0.003 ± 0.0003	118.69 ± 20.95
	OSI	42.17 ± 1.23	238.17 ± 11.55	5.08 ± 0.05	70.87 ± 15.73
	SSI	32.30 ± 0.26	181.95 ± 7.08	7.87 ± 0.05	58.61 ± 10.42
	PSI	34.51 ± 0.24	405.74 ± 7.98	1.14 ± 0.03	389.80 ± 8.02
Distributed	RR	46.80 ± 0.39	233.77 ± 9.83	0.004 ± 0.0006	131.19 ± 16.49
	OSI	23.81 ± 0.15	138.93 ± 5.31	4.20 ± 0.09	97.62 ± 11.36
	SSI	24.23 ± 0.20	137.84 ± 5.41	6.43 ± 0.05	77.70 ± 10.78
	PSI	22.57 ± 0.15	169.12 ± 4.65	1.34 ± 0.02	207.94 ± 9.42

		Distance	Run time	Delib. time	Idle time
Clustered	RR	57.03 ± 0.93	234.65 ± 9.15	0.0153 ± 0.001	123.31 ± 18.69
	OSI	37.52 ± 3.64	211.79 ± 1.39	9.06 ± 0.34	125.56 ± 6.92
	SSI	35.78 ± 0.16	212.88 ± 1.61	9.38 ± 0.33	125.42 ± 3.08
	PSI	43.23 ± 3.26	220.62 ± 12.75	6.11 ± 0.18	127.4 ± 37.57
Distributed	RR	69.7 ± 1.69	233.54 ± 8.72	0.0143 ± 0.0007	101.04 ± 9.54
	OSI	28.2 ± 0.14	204.72 ± 2.43	9.17 ± 0.22	89.88 ± 4.42
	SSI	28.25 ± 0.12	204.0 ± 2.04	9.35 ± 0.24	88.45 ± 4.48
	PSI	28.14 ± 0.26	203.43 ± 0.52	6.05 ± 0.14	89.21 ± 2.08

the smallest overall combined distance on average. This means that the solutions generated by SSI were executed quickly in comparison to those generated by the other allocation mechanisms. In the remaining scenarios, the performance of SSI in terms of overall distance diminished in comparison to the other mechanisms.

SSI also performs well in terms of idle time. An individual robot accumulates idle time when it finishes visiting its allocated target points before other robots finish visiting theirs, so across the team it is a measure of wasted resource. For the static allocation scenarios, SSI has the lowest idle time of all the mechanisms. The cost for this performance can be seen in the deliberation times. SSI, which requires bids from all robots for all unallocated points on all rounds, involves more bidding than any of the other approaches, and this translates into the longest time spent in the allocation process (deliberation time). Only OSI in the dynamic allocations comes close to the deliberation time required by SSI.

In the scenarios we consider here, the deliberation times are never more than a small fraction of the total time to execute the set of tasks, but it is easy to imagine cases where deliberation time is more significant. The total number of bids to allocate m tasks to n robots is one bid from each robot for m tasks in the first round, one bid from each robot for $m - 1$ tasks in the second round, and so

Run Time:

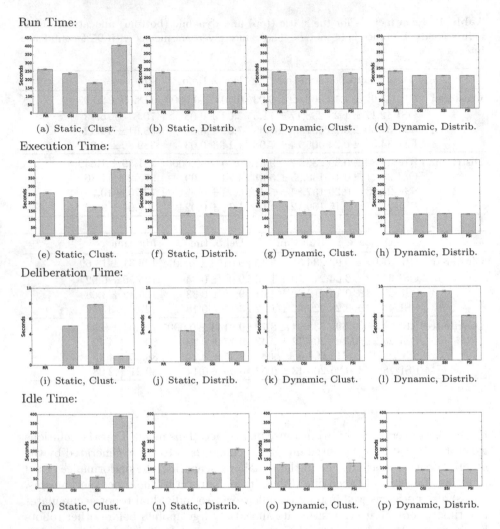

Execution Time:

Deliberation Time:

Idle Time:

Fig. 3. The remaining metrics for the teams. In order the rows give run time execution time, deliberation time, and idle time. The bars are ordered RR, OSI, SSI and PSI.

on,[3] and it is conceivable that this could become big enough to be problematic. For example, consider the case of one hundred robots — as in the Centibots project [10] — which have to allocate 500 target points. In such a case the SSI auction would require over 12 million bids, a 12,000-fold increase over what is required in our experiments, and enough to make deliberation time a significant contributor to the time for task completion. In addition, since each bid has to be transmitted wirelessly — either to a centralised auction manager, or to all

---

[3] Giving a total of $n\left(\frac{m(m+1)}{2}\right)$ bids.

other robots in a distributed auction — the number of messages can be a factor in robot deployments where communication bandwidth is limited [19].

4 Related Work

As mentioned above, there is much related work on auction mechanisms for task allocation in robotics. Some of this has already been mentioned, and here we take a look at the relevant work that we have not yet covered. The use of market mechanisms in distributed computing can be considered to start with Smith's contract net protocol [22], and this was followed by Wellman and Wurman's [24] *market-aware agents*. A primary strength of market-based approaches is their reliance only on local information or the self-interest of the agents to arrive at efficient solutions to large-scale, complex problems that are otherwise intractable [2].

The most common instantiations of market-based approaches in multirobot systems are auctions. Auctions are commonly used for distribution tasks, where resources or roles are treated as commodities and auctioned to agents. A significant body of work analyzes the effects of different auction mechanisms, [1,11,25], bidding strategies [18], dynamic task re-allocation or swapping [5] and levels of commitment to the contracts [14] on the overall solution quality.

In domains where there is a strong synergy between items, single-item auctions can result in sub-optimal allocations [1]. In the domain of interest here, multi-robot exploration, a strong synergy exists between target points that robots need to explore. Combinatorial auctions remedy this limitation by allowing agents to bid on bundles of items, and minimize the total travel distance because they take the synergies between target points into account [7]. Combinatorial auctions suffer, however, from the computational costs of bid generation and bundle valuation by the agents, and winner determination by the auction mechanism itself, all of which are NP-hard [8].

Of all the literature on auctions in multi-robot teams, [15] and [20] are the most closely related to our work. Both evaluate SSI in simulation and so could argue their work evaluates the practical cost of solutions generated using SSI (and [15] provides some results from real robots). However, the focus of both [15] and [20] is on finding optimal mechanisms for dynamic task reallocation during execution rather than, like our work, attempting to characterise auction mechanisms in terms of the tasks for which they are best fitted. In addition, neither consider the range of metrics that we do and so are unable, for example to comment on the load balancing that measuring idle time exposes.

5 Summary

This paper has studied the performance of a number of auction mechanisms on a version of the multi-robot routing problem. It has gone further than any work we are aware of in testing auction mechanisms across a range of performance metrics, in a way that evaluates how the mechanisms would work on a fleet of robots, and compares how mechanisms perform in dynamic versus static scenarios.

The main result is that while the sequential single-item (SSI) auction broadly outperforms other single item auctions in static scenarios where the robots start clustered together, its advantage is diminished when robots are distributed, and is also reduced when the target points are allocated dynamically. Given that we have only studied one specific set of target points in each of the static and dynamic scenarios, this result is tentative, and we are continuing to run experiments, gathering data on different configurations of target points, to ensure that our results are not an artefact of a specific configuration.

Finally, we are also running experiments with real robots on exactly the same scenarios studied here, but as yet do not have enough results to report.

Acknowledgments. This work was partially funded by National Science Foundation grants #IIS-1116843 and #IIS-1338884, and by a University of Liverpool Research Fellowship.

References

1. Berhault, M., Huang, H., Keskinocak, P., Koenig, S., Elmaghraby, W., Griffin, P.M., Kleywegt, A.: Robot exploration with combinatorial auctions. In: Proceedings of the IEEE/RSJ International Conference on Intelligent Robots and Systems (2003)
2. Dias, M.B., Zlot, R., Kalra, N., Stentz, A.: Market-based multirobot coordination: A survey and analysis. Proceedings of the IEEE **94**(7), 1257–1270 (2006)
3. Ezhilchelvan, P., Morgan, G.: A dependable distributed auction system: architecture and an implementation framework. In: Proceedings of the 5th International Symposium on Autonomous Decentralized Systems (2001)
4. Gerkey, B., Vaughan, R.T., Howard, A.: The player/stage project: tools for multi-robot and distributed sensor systems. In: Proceedings of the 11th International Conference on Advanced Robotics (2003)
5. Golfarelli, M., Maio, D., Rizzi, S.: A Task-Swap Negotiation Protocol Based on the Contract Net Paradigm. Technical Report 005–97, DEIS, CSITE - Università di Bologna (1997)
6. Habib, M.K.: Humanitarian Demining: Reality and the Challenge of Technology. International Journal of Advanced Robotic Systems **4**(2), 151–172 (2007)
7. Koenig, S., Keskinocak, P., Tovey, C.: Progress on agent coordination with cooperative auctions. In: Proceedings of the AAAI Conference on Artificial Intelligence (2010)
8. Koenig, S., Tovey, C., Lagoudakis, M., Kempe, D., Keskinocak, P., Kleywegt, A., Meyerson, A., Jain, S.: The power of sequential single-item auctions for agent coordination. In: Proceedings of National Conference on Artificial Intelligence (2006)
9. Koenig, S., Tovey, C.A., Zheng, X., Sungur, I.: Sequential bundle-bid single-sale auction algorithms for decentralized control. In: Proceedings of the International Joint Conference on Artificial Intelligence (2007)
10. Konolige, K., et al.: Centibots: very large scale distributed robotic teams. In: Ang Jr, M.H., Khatib, O. (eds.) Experimental Robotics IX. STAR, vol. 21, pp. 131–140. Springer, Heidelberg (2006)

11. Kraus, S.: Automated negotiation and decision making in multiagent environments. In: Luck, M., Mařík, V., Štěpánková, O., Trappl, R. (eds.) ACAI 2001 and EASSS 2001. LNCS (LNAI), vol. 2086, pp. 150–172. Springer, Heidelberg (2001)

12. Lagoudakis, M., Berhault, M., Koenig, S., Keskinocak, P., Kelywegt, A.: Simple auctions with performance guarantees for multi-robot task allocation. In: Proceedings of the International Conference on Intelligent Robotics and Systems (IROS) (2004)

13. Lagoudakis, M., Markakis, V., Kempe, D., Keskinocak, P., Koenig, S., Kleywegt, A., Tovey, C., Meyerson, A., Jain, S.: Auction-based multi-robot routing. In: Proceedings of Robotics: Science and Systems Conference (2005)

14. Mataric, M., Sukhatme, G., Ostergaard, E.: Multi-robot task allocation in uncertain environments. Autonomous Robots 14(2–3), 255–263 (2003)

15. Nanjanath, M., Gini, M.: Repeated auctions for robust task execution by a robot team. Robotics and Autonomous Systems 58(7), 900–909 (2010)

16. Özgelen, A.T., Schneider, E., Sklar, E.I., Costantino, M., Epstein, S.L., Parsons, S.: A first step toward testing multiagent coordination mechanisms on multi-robot teams. In: Proceedings of the Workshop on Autonomous Robots and Multirobot Systems (ARMS) at Autonomous Agents and MultiAgent Systems (AAMAS) (2013)

17. Quigley, M., Conley, K., Gerkey, B.P., Faust, J., Foote, T., Leibs, J., Wheeler, R., Ng, A.Y.: Ros: an open-source robot operating system. In: ICRA Workshop on Open Source Software (2009)

18. Sariel, S., Balch, T.: Efficient bids on task allocation for multi-robot exploration. In: Proceedings of the Nineteenth Florida Artificial Intelligence Research Society Conference (2006)

19. Savkin, A.V.: The problem of coordination and consensus achievement in groups of autonomous mobile robots with limited communication. Nonlinear Analysis: Theory, Methods & Applications 65(5), 1094–1102 (2006)

20. Schoenig, A., Pagnucco, M.: Evaluating sequential single-item auctions for dynamic task allocation. In: Li, J. (ed.) AI 2010. LNCS, vol. 6464, pp. 506–515. Springer, Heidelberg (2010)

21. Sklar, E., Ozgelen, A.T., Munoz, J.P., Gonzalez, J., Manashirov, M., Epstein, S.L., Parsons, S.: Designing the HRTeam framework: lessons learned from a rough-and-ready Human/Multi-Robot team. In: Dechesne, F., Hattori, H., ter Mors, A., Such, J.M., Weyns, D., Dignum, F. (eds.) AAMAS 2011 Workshops. LNCS, vol. 7068, pp. 232–251. Springer, Heidelberg (2012)

22. Smith, R.G.: The contract net protocol: high-level communication and control in a distributed problem solver. In: Bond, A.H., Gasser, L. (eds.) Distributed Artificial Intelligence. Morgan Kaufmann Publishers Inc. (1988)

23. Tovey, C., Lagoudakis, M.G., Jain, S., Koenig, S.: Generation of bidding rules for auction-based robot coordination. In: Proceedings of the 3rd International Multi-Robot Systems Workshop, March 2005

24. Wellman, M.P., Wurman, P.R.: Market-aware agents for a multiagent world. Robotics and Autonomous Systems 24, 115–125 (1998)

25. Zlot, R., Stentz, A., Dias, M.B., Thayer, S.: Multi-robot exploration controlled by a market economy. In: Proceedings of the IEEE Conference on Robotics and Automation (2002)

UAV Guidance: A Stereo-Based Technique for Interception of Stationary or Moving Targets

Reuben Strydom$^{(\boxtimes)}$, Saul Thurrowgood, Aymeric Denuelle, and Mandyam V. Srinivasan

The University of Queensland, Brisbane, Australia
reuben.strydom@uqconnect.edu.au

Abstract. We present a novel stereo-based method for the interception of a static or moving target, from an Unmanned Aerial Vehicle (UAV). This technique is directly applicable for outdoor applications such as search and rescue, monitoring and surveillance, and complex landing scenarios. Stereo vision is particularly useful for the interception of a moving target as it intrinsically measures the relative position and velocity between the UAV and the person or object. The target position is computed geometrically using its direction, as viewed by the vision system, and the UAV's stereo height. A Kalman filter computes a reliable estimate of relative position and velocity using the target centroid. The performance of this method is validated by conducting a number of closed-loop interceptions for both static and moving target cases. The mean interception error is found to be 0.01m with a standard deviation of 0.33m in tests with static targets and 0.14m with a standard deviation of 0.24m in tests with moving targets. Our method has been field-tested outdoors and provides results comparable to other vision-based techniques that have been tested under more controlled indoor conditions.

1 Introduction

Vision-based systems are well suited to interception applications where relatively little prior information is available about the position or size of the target. Examples include search and rescue, monitoring and surveillance, service robots, and complex landing scenarios. Search and rescue applications are often time sensitive and performed in dangerous situations – UAVs can work around the clock without compromising human safety. While there have been several vision-based approaches for the interception of static and moving objects, many rely on targets whose location, distance and orientation are specified by artificial markers. Many of these methods require additional sensors (e.g. pressure sensors and ultrasound range sensors) as well as information obtained from active external sources such as GPS or Vicon systems. Although interception is a well studied field, the present study has been motivated by the fact that relatively little attention has been devoted to the formulation of interception systems that rely solely on visual information; in particular the use of stereo. There are three main

© Springer International Publishing Switzerland 2015
C. Dixon and K. Tuyls (Eds.): TAROS 2015, LNAI 9287, pp. 258–269, 2015.
DOI: 10.1007/978-3-319-22416-9_30

stages to the interception of a target: (1) detection in image space, (2) computation of the relative distance and velocity of the target and (3) interception control strategy.

Although detection of the object is not the aim of this paper, it is a fundamental step in the process of intercepting a target. Infrared (IR) sensors have been used to detect a static target (e.g. [13] and [18]) in an indoor environment. However, such systems are also limiting as they are unsuitable for outdoor use. Motion contrast, measured from dense optic flow is another method utilised by [3] to detect a moving target. Although very little is assumed of the target, this method cannot detect static objects and is computationally expensive. Thus, this approach is unsuitable for search and rescue tasks. Colour segmentation has the ability to detect both static and moving targets with the only input required being a reference colour such as the vehicle colour of the party to be rescued. The method described here uses a seeded region growing algorithm combined with a refined segmentation step to ensure that the entire object is detected, even in varying lighting conditions. This method also allows for slight colour gradients across the object in an outdoor environment. Once the object has been detected, the next steps are to pursue and intercept it.

Currently, extensive research has been conducted on interception using ground (e.g. [7] and [11]) and aerial platforms (e.g. [2,4,9,14]) with the use of additional sensors such as pressure sensors, sonar and LiDAR. However, these sensors contribute an additional payload and point for failure – both of which can compromise mission reliability. Some methods as in [2,6] process data on a ground station, significantly limiting the range of the autonomous vehicle and thus decreasing the control responsiveness due to increased delays. There have been a number of vision only methods that utilise sparse optic flow for interception control (e.g. [5] and [17]). However, the disadvantage of using optic flow is that additional knowledge is required if the target is moving, such as its size or velocity. Artificial markers are commonly used to compute information on the target's distance, pose and relative velocity [10,19]. This could be unreliable in search and rescue missions, as artificial markers would be absent. The method described in [10], also relies on information about the motion of the ground vehicle, which is transmitted to the quadrotor – however, this requirement cannot be readily met in a search and rescue task because such information is unlikely to be transmitted by the party to be rescued.

In this paper we present a vision-based method, field tested in an outdoor environment, to address the task of guiding an aircraft to intercept both static or moving objects, which is essential for search and rescue tasks. The technique developed here uses stereo vision to determine the height of the aircraft without requiring altitude sensors – thus, limiting potential failure and allowing increased payload. Given that the size and shape of a target are often variable in search and rescue applications, colour information is used to detect the target in this paper. Our method does not use optic flow at all, thus saving on computational cost. Furthermore, unlike other interception techniques that use GPS or global optic flow, our method does not require additional information or assume a known target velocity for guiding the interception.

2 Flight Platform

The platform used in our study (see Fig. 1) is a custom built quadrotor, incorporating a MicroKopter flight controller. An Intel NUC is used for processing data. It features an Intel i5 core - 1.3GHz dual core processor, 8GB of RAM and a 120GB SSD. The platform also carries a MicroStrain Inertial Measurement Unit (IMU) and a custom-designed vision system.

The vision system consists of two fisheye cameras (Point Grey Firefly MV cameras, Sunex DSL216 lenses); each having a visual field of 190°. The cameras are mounted back-to-back and tilted toward each other by 10° to provide a stereo overlap of approximately 30° by 130°. The vision system is tilted downward by 45° to capture a larger portion of the ground. The cameras have a 25Hz frame rate and are software synchronized. Each image pair are stitched together using a camera calibration similar to that described in [8]. The resulting panoramic image has a 360° by 150° field of view, with 360 by 220 pixel resolution.

Fig. 1. Quadrotor used for tests   **Fig. 2.** Closeup of vision system

3 Target Detection Method

Our colour-based target detection method comprises (i) the reference colour (C_{REF}) computation, (ii) a segmentation phase, (iii) a validation phase and (iv) a refinement step. To segment the target, a seeded region growing method ([1]) was chosen over basic colour channel separation to increase the robustness to lighting changes and reflections caused by glare on the target.

(i) The RGB colour C_{REF} (1) – where the channel scaling and bias factors were determined through extensive testing – is used to detect the target and is dynamically adjusted for variations in image intensity (I):

$$C_{REF} = \begin{cases} R = 2.90I - 133 \\ G = 2.85I - 176 \\ B = 3.19I - 231 \end{cases} \tag{1}$$

here I is the mean intensity of the entire image (computed by using 'V' in the HSV colour space), and R, G, B are the reference intensities used in the segmentation phase (range: [0,255]).

(ii) Initial pixels (seeds) potentially contributing to the target are selected if their colour value is within ±15 intensity units from the reference colour defined in (1). In the case of insufficient seeds, the reference colour range is dynamically increased. Nearby detected seeds are then merged into candidates, and the smallest candidates (sparse pixels) are discarded. This *merge and discard* operation is repeated on the target candidates. Among the retained target candidates, depending on the knowledge of a previously detected centroid (previous frame), either the largest candidate or the candidate whose centroid is the closest to the target centroid computed from the previous frame, is selected. The current target object is then formed by recursively growing the seeds of the chosen candidate.

(iii) The grown target size and centroid are then examined against the detected characteristics in the previous frame. The current target is either updated or discarded, depending on the size and centroid consistency. To avoid large increases in processing time, the growing is limited to 10 iterations. Therefore, to increase the accuracy of the target position, the initial estimate (see Sect. 3) is refined by using a modified technique described in [12].

Fig. 3. Detection of target under various lighting conditions. (a) and (c) are the captured images. In (b) and (d) the green pixels represent the initial seed pixels, the blue pixels represent the grown pixels, and the red circle represents the centroid of the object, as computed at the end of the filling process.

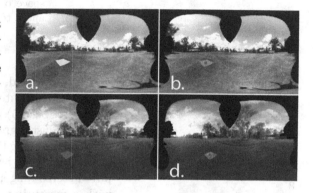

(iv) The refinement is further broken down into three stages: (1) The camera calibration is used to remap the target from Fig. 3a. to Fig. 4a. (2) The mean (μ_t) and standard deviation (σ_t) of the colour of the grown target (see Fig. 5) are used to separate the target from the background. The distance (D) in RGB space between the mean target colour and the selected pixel is computed using (2).

$$D = \begin{cases} D_R = R_t - R_p \\ D_G = G_t - G_p \\ D_B = B_t - B_p \end{cases} \tag{2}$$

Here R_t, G_t and B_t are the mean intensity values of the target. R_p, G_p and B_p are the intensity values for a particular pixel. A binary image is then created using an iterative process, where a pixel is classified as part of the target if $D < 5\sigma_t$ in Equation (2) (e.g. Fig. 4b). A refined target centroid is computed from this binary image. (3) The coordinates of the centroid are then fed into a Kalman filter, which provides an estimated target position in the next frame by combining the information on the target position from the current frame, with an extrapolated position based on the target's pixel velocity. This information is used to define the search region for the target in the following frame.

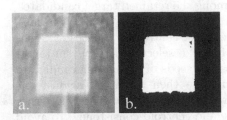

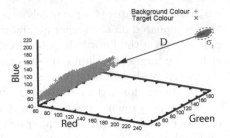

Fig. 4. Refinement of target segmentation. (a) is a localised unwarped target image of Fig. 3a. (b) is a binary image showing the segmented target.

Fig. 5. RGB colour space representation of each pixel. σ_t is the standard deviation around the mean target colour and D is the distance in RGB space between the mean target colour and a test pixel.

4 Stereo Calibration

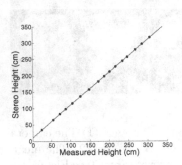

Fig. 6. Comparison between ground truth and stereo height measurements

A stereo method, as described in [15,16], is utilised to compute the relative distance to the target (see Sect. 5). It is essential to calibrate the stereo against a ground truth. This was accomplished by placing the flight platform on a tripod at various known heights and recording the computed stereo range; ground truth was measured with a measuring tape with a ±1cm accuracy. Two calibrations were conducted, each comprising more than 10 heights, ranging between [0.26, 3.09]m. The calibration involved a full system test, to ensure a realistic noise level comparable to that prevalent during flight testing.

The calibration results, shown in Fig. 6, demonstrate a linear relationship (with an $R^2 = 0.999$) between the ground truth and stereo-based height (h_s) above ground. Due to the precise fit of the linear regression, it is possible to infer the true height (h) with a high degree of confidence from the following relationship: $h = 1.06h_s - 0.09$ [m].

5 Relative Position and Velocity to Object

To intercept a target using the proposed method, it is necessary to first compute the relative position and velocity of the target. Consider the relative target position when the aircraft is at c_1. The unit vector $\hat{p}_1$ is the direction to the object. To determine the absolute target distance, the height above ground (h_1) is used. h_1 is computed using the vision-based stereo computation explained in [15,16]. Let p_h denote the projection of the vector $\hat{p}_1$ onto h_1 ($p_h = proj_{\hat{p}_1}(h_1)$). Here the ratio h_1/p_h is the scale factor required to compute the absolute position relative to the vision system in frame 1:

$$P_1 = \hat{p}_1(h_1/p_h) \tag{3}$$

The position vector P_1 is the vector connecting c_1 to o_1, and P_2 is the vector connecting c_2 to o_2. The difference between the position vectors P_1 and P_2 is used to compute the relative velocity (V_T).

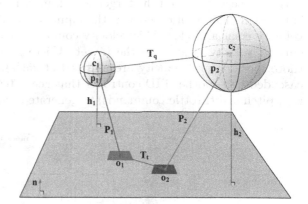

Fig. 7. Computing the motion of the target relative to the quadrotor. The vision system translates through a vector T_q from position c_1 to c_2 and the target motion is defined by the vector T_t. h_1 and h_2 are the heights above the ground. $\hat{p}_1$ and $\hat{p}_2$ are the unit vectors directed at target and P_1 and P_2 are the 3D vectors (m) to the target relative to the vision system.

6   Method Validation

To validate the accuracy of the full system, the target was moved on the ground over a distance of 6.5m at a constant velocity of 0.72 m/s, in a direction perpendicular to the front-back body axis. The quadrotor was stationary, and mounted on a tripod for the validation process. The difference between the measured velocity and the ground truth velocity of the target provided a measure of the errors associated with estimating the position and velocity of a moving target. 10 tests were conducted, in which the vision system was stationary and positioned at various heights above the ground ranging from [0.54, 3.09]m.

The target was moved by a winch that used a motor and pulley (1m/rev). The mean and standard deviation of the target velocity, as estimated by the vision system was computed over 100 frames of data for each of the 10 heights (number of samples > 1000 frames, to compute accuracy). Table. 1 provides the results for the validation, where a mean relative velocity error of only 0.01m s^{-1} (1.39%) was observed over the 10 trials between the actual target velocity (μ_{GT}) and the velocity (μ) as measured by the vision system.

Table 1. Moving target validation

μ_{GT} Velocity (ms^{-1})	μ Velocity (ms^{-1})	Mean Velocity Error (ms^{-1})	SD Velocity Error (ms^{-1})
0.72	0.73	0.01	0.04

7   Control of Interception

The interception control strategy is summarised in Fig. 8. Consider that the
aircraft is currently at a position P_i relative to the target, at time step 'i'. Note
that P_i is a 3D vector relative to the target. This vector defines the position error
between the aircraft and the target, which must be driven to zero if interception
is to occur. The position error is the input to a PID controller that computes a
velocity command (V_i). This velocity command is compared with the aircraft's
current velocity relative to the target (V_t) to generate a velocity error signal
(note: position and velocity are updated at 25Hz). The velocity error signal is
cascaded into another PID controller that generates appropriate changes to the
roll, pitch and throttle commands, to generate the desired aircraft motion.

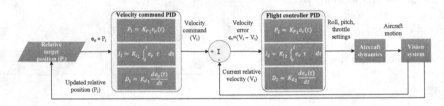

Fig. 8. Quadrotor control loop for target interception

Fig. 9 illustrates a quadrotor intercepting a moving target. The bearing angle
between the target and the quadrotor is γ. α and β are the elevation angles,
relative to the target, in the side-view and front-view planes. When α and β
equal zero, a perfect interception has occurred. Note that the goal is not landing
on the target, but following it at a prescribed distance and height.

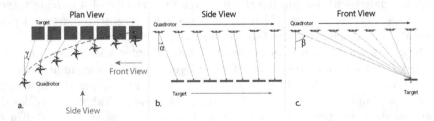

Fig. 9. Schematic illustration of a UAV intercepting a moving target. a. Plan view of
the interception trajectory, b. Side view and c. Front view.

8   Results: Closed-Loop Interception

Once the method demonstrated the ability to reliably determine the target's relative position and velocity, closed-loop flight testing was conducted in two kinds of experiments: (1) static target interception[1] and (2) moving target interception[2]. As the experiments were conducted outdoors, the UAV had to contend with wind speeds up to 5.0 knots – with an average wind speed of 3.2 knots over all flights.

The purpose of the static target interception was to test closed-loop interception with only one moving agent, where the target was initially 4m from the quadrotor. Once intercepted, the UAV hovers above the target, at a commanded height of 2.0m, until manual control is recovered. Fig. 10 illustrates a typical interception of a static target, where the prescribed goal position was offset by (-1,-1,2)m from the target center. The goal position for the aircraft was defined to be at a prescribed offset from the target position, in order to ensure that the target was always imaged with high resolution – the vision directly beneath the aircraft is slightly compromised because this region is imaged by the extreme periphery of the visual field of each camera.

The target was intercepted in an average time of 3.0 seconds, where the fastest and slowest intercepts were performed in 2.7 and 3.5 seconds respectively. Once intercepted, the static tests demonstrated a mean hover position of 0.01m with a standard deviation of 0.32m as an average of the x and y directions in Table 2. Here the object was deemed intercepted if the quadrotor was within 0.5m of the set point. The average angular error of the method was computed as:

$$Angular\ error = tan^{-1}\left(\frac{position\ error}{height}\right) \tag{4}$$

The moving target interception tests involved the same initial setup as the static target experiments. However, the target was moved at a constant velocity of 0.2, 0.4 or 0.6ms^{-1} in different experiments. The UAV's starting position was approximately (-2.5,-2.5,2)m relative to the target and the goal position was (-1,-1,2)m from the target center. To simplify the mechanical design for the target motion, it was moved in a straight line. However, the method proposed in this paper does not assume any velocity or direction and can intercept a target that has a dynamic trajectory.

Results for the moving target are shown in Table 3 and Fig. 10. The average interception time for the moving target was of 4.8 seconds, where the fastest and slowest interceptions were performed in 3.6 and 5.6 seconds respectively. Once intercepted, the static tests demonstrated a mean hover position error of only 0.14m with a standard deviation of 0.24m in the x and y directions.

For the moving target interception, panels (a)-(c) in Fig. 10 provides a direct comparison to the schematic illustration as shown in Fig. 9.

---

[1] See static target interception video: http://youtu.be/INi8OfsNIEA
[2] See moving target interception video: http://youtu.be/XP_snIBXws4

Table 2. Static target interception statistics

Trial	Mean Position Error(m)		SD Velocity Error (m)		Angular Error (Deg)	Interception Time (s)	Max wind speed (knots)
	X	Y	X	Y			
1	0.04	-0.05	0.26	0.31	7.33	3.48	2.00
2	0.01	0.03	0.27	0.35	8.85	2.68	5.00
3	-0.05	0.06	0.37	0.44	11.45	2.84	4.70
All	0.00	0.01	0.30	0.35	9.22	3.00	3.93

Table 3. Moving target interception statistics

Trial	Target Velocity (ms⁻¹)	Mean Position Error (m)		SD position error (m)		Angular Error (Deg)	Interception Time (s)	Max wind speed (knots)
		X	Y	X	Y			
1	0.2	-0.03	0.12	0.22	0.14	5.21	4.08	1.50
2	0.4	0.04	0.32	0.28	0.34	5.52	3.64	1.50
3	0.6	0.05	-0.03	0.18	0.20	6.11	5.60	2.00
4	0.6	0.34	0.11	0.21	0.22	8.82	3.64	3.90
All	Moving	0.15	0.13	0.22	0.25	6.81	4.76	2.47

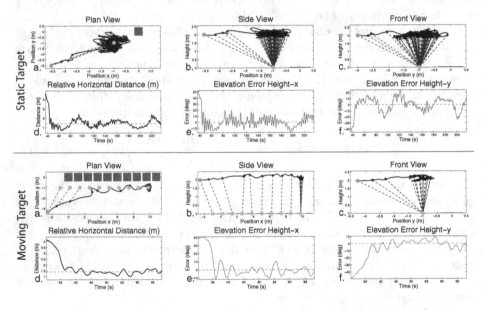

Fig. 10. Quadrotor intercepting a static target or a target moving at 0.6ms⁻¹. Figures (a)-(c) provide plan, side and front views of the interception with a (-1, -1, 2)m offset from the target. In (a)-(c) the black curves and crosses represent the quadrotor trajectory, and the cyan dots illustrate the prescribed goal locations. The red square in (a) and red line in (b),(c) represent the actual target position. The green dot is the starting point. (d) shows the relative horizontal distance between the target and the quadrotor. (e) and (f) depict the angular error in elevation for the height-x and height-y planes respectively. The dashed red line in (d) represents the commanded horizontal distance between the target and the prescribed goal location. The dashed red lines in (e) and (f) represent zero error in the elevation angles of the target in the side and front views.

9 Discussion

Although the detection algorithm, stereo ranging and pursuit strategies have previously been studied individually in separate contexts, to the best of our knowledge this is the first study that integrates the three modules to achieve purely vision-based interception. This strategy, whose working range is limited by the range of the stereo system, would be well suited for the last phase of a long-range interception system that might use other methods for sensing target ranges during the initial phase.

With the current modules, the static experiments demonstrated that the quadrotor could intercept the target with a mean error of 0.01m and a standard deviation of 0.32m. The moving target results demonstrate a following accuracy with a mean and standard deviation in positional error of 0.14m and 0.24m. As expected, the average interception times increased from 3.0 seconds for the static target to 4.8 seconds for the moving targets. The oscillations in the trajectory (see Fig. 10) were most likely due to wind disturbances, which influence the PID controllers. It is evident that the accuracy was higher for the moving target test, possibly because the wind conditions were not as severe as in the static target tests.

For comparison against other published interception methods (see Fig. 11) an angular error was computed using Eq. 4. This comparison was only possible for techniques that provided a position error and a test height, as discussed in Sect. 1. Other error metrics that were provided for various techniques, unfortunately could not be added to the comparison if an experimental height was not provided. We also note that this comparison is only a rough guide, as each method was tested under different conditions with varying hardware and software configurations. It is clear from Fig. 11, however, that our method, which was tested in outdoor conditions with wind and changing lighting conditions, performs as well as or better than many of the previously published techniques – most of which have been tested indoors under more controlled conditions.

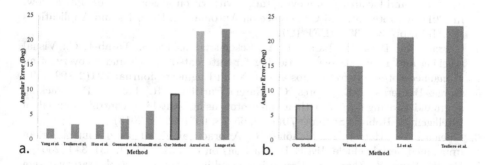

Fig. 11. Performance comparison for (a) static target and (b) moving target. The green bars represent the outdoor experiments and the blue bars illustrate indoor tests. The outlined bar shows the results of our method.

To demonstrate that the algorithm runs in real-time, the timing was determined for the system. The camera frame rate was 25Hz (40ms). On average, the total system time was under 20ms, which is well within the 40ms window between camera frames.

10   Conclusion

We present a new stereo-based method that intercepts a static or moving coloured target in outdoor conditions for applications such as search and rescue, monitoring and surveillance, and landing scenarios. Target detection under various lighting conditions, although not the main focus of this research, was achieved by using a seeded region growing detection algorithm that was only provided with information about the colour of the target. The results shown in Sect. 6 and 8 demonstrate that the method is capable of accurately detecting, tracking and intercepting an object without assuming its shape, texture or motion. The accuracy of target detection and tracking was validated by moving the target at a known velocity while keeping the UAV stationary – where the mean error was determined to be $0.01 ms^{-1}$ over 10 trials at heights varying between $[0.5, 3.09]$m. Finally, we present closed-loop flight results where the UAV intercepts and follows a static or moving target as well as or better than many other techniques that have been trialled in controlled indoor environments.

References

1. Adams, R., Bischof, L.: Seeded region growing. IEEE Transactions on Pattern Analysis and Machine Intelligence **16**(6), 641–647
2. Azrad, S., Kendoul, F., Nonami, K.: Visual servoing of quadrotor micro-air vehicle using color-based tracking algorithm. Journal of System Design and Dynamics **4**(2), 255–268, 1881–3046 (2010)
3. Choi, J.H., Lee, D., Bang, H.: Tracking an unknown moving target from uav: extracting and localizing an moving target with vision sensor based on optical flow. In: 2011 5th International Conference on Automation, Robotics and Applications (ICARA), pp. 384–389. IEEE (2011)
4. Garratt, M., Pota, H., Lambert, A., Eckersley-Masline, S., Farabet, C.: Visual tracking and lidar relative positioning for automated launch and recovery of an unmanned rotorcraft from ships at sea. Naval Engineers Journal **121**(2), 99–110
5. Gomez-Balderas, J.-E., Flores, G., García Carrillo, L.R., Lozano, R.: Tracking a ground moving target with a quadrotor using switching control. Journal of Intelligent & Robotic Systems **70**(1–4), 65–78, 0921–0296 (2013)
6. Guenard, N., Hamel, T., Mahony, R.: A practical visual servo control for an unmanned aerial vehicle. IEEE Transactions on Robotics **24**(2), 331–340 (2008)
7. He, W., Fang, Y., Zhang, X.: Prediction-based interception control strategy design with a specified approach angle constraint for wheeled service robots. IEEE (2013)
8. Kannala, J., Brandt, S.S.: A generic camera model and calibration method for conventional, wide-angle, and fish-eye lenses. IEEE Transactions on Pattern Analysis and Machine Intelligence **28**(8), 1335–1340 (2006)

9. Lange, S., Sunderhauf, N., Protzel, P.: A vision based onboard approach for landing and position control of an autonomous multirotor uav in gps-denied environments. In: International Conference on Advanced Robotics, ICAR 2009, pp. 1–6. IEEE (2009)
10. Li, W., Zhang, T., Kuhnlenz, K.: A vision-guided autonomous quadrotor in an air-ground multi-robot system. In: 2011 IEEE International Conference on Robotics and Automation (ICRA), pp. 2980–2985. IEEE (2011)
11. Low, E.M.P., Manchester, I.R., Savkin, A.V.: A biologically inspired method for vision-based docking of wheeled mobile robots. Robotics and Autonomous Systems 55(10), 769–784 (2007)
12. Moore, R.J.D: Vision systems for autonomous aircraft guidance. PhD thesis, The University of Queensland (2012)
13. Salazar-Cruz, S., Escareno, J., Lara, D., Lozano, R.: Embedded control system for a four-rotor uav. International Journal of Adaptive Control and Signal Processing 21(2–3), 189–204
14. Sanchez-Lopez, J.L., Pestana, J., Saripalli, S., Campoy, P.: An approach toward visual autonomous ship board landing of a vtol uav. Journal of Intelligent & Robotic Systems 74(1–2), 113–127, 0921–0296 (2014)
15. Strydom, R., Thurrowgood, S., Srinivasan, M.V.: Visual odometry: autonomous uav navigation using optic flow and stereo. In: Australasian Conference on Robotics and Automation (ACRA), pp. 1–10. Australian Robotics and Automation Association (2014)
16. Thurrowgood, S., Moore, R.J.D., Soccol, D., Knight, M., Srinivasan, M.V.: A biologically inspired, vision-based guidance system for automatic landing of a fixed-wing aircraft. Journal of Field Robotics 31(4), 699–727 (2014)
17. Watanabe, Y., Lesire, C., Piquereau, A., Fabiani, P., Sanfourche, M., Besnerais, G.L.: The onera ressac unmanned autonomous helicopter: visual air-to-ground target tracking in an urban environment. In: American Helicopter Society 66th Annual Forum (AHS 2010) (2010)
18. Wenzel, K.E., Rosset, P., Zell, A.: Low-cost visual tracking of a landing place and hovering flight control with a microcontroller. In: 2nd International Symposium on Selected Papers From the UAVs, Reno, Nevada, USA June 8–10, 2009, pp. 297–311. Springer (2010)
19. Yang, X., Mejias, L., Garratt, M.: Multi-sensor data fusion for UAV navigation during landing operations. Australian Robotics and Automation Association Inc., Monash University (2011)

A Novel Path Planning Approach for Robotic Navigation Using Consideration Within Crowds

Ross Walker(✉) and Tony J. Dodd

Department of Automatic Control and Systems Engineering,
The University of Sheffield, Sheffield, UK
{ross.walker,t.j.dodd}@sheffield.ac.uk

Abstract. This paper presents a novel approach towards mobile robotic path planning within a pedestrian environment, focused on being considerate towards crowd members. Through predicting pedestrian movement, with ellipses used to encompass the uncertainty, a modified Voronoi diagram is proposed to create a roadmap through the environment. Predictions of pedestrian trajectories are used to generate collision-free paths that minimise congestion and are considerate to the overall crowd flow. The results demonstrate that the robot's movement allows potential collisions to be recognised in advance and avoided before they can develop.

Keywords: Robotic navigation · Pedestrian Crowds · Path planning · Voronoi diagram · Consideration

1 Introduction

This paper proposes a novel approach to navigation through pedestrian environments. The unique use of 'consideration' creates a path planner that predicts pedestrian trajectories to assess the benefits of various paths. This results in the robot movement to be non-disruptive to the crowd by: avoiding potential collisions; not attributing to congestion when moving; and opting for movements that do not interfere with the pedestrians' trajectories.

Previous research involving path planning robots within crowds have used various systems. These range from simple evasion [1] to much complicated 'cooperation' [2]. However, with ever more complex systems the assumptions and limitations also increase. [1] uses potential fields for basic evasion, which produce indirect and uncoordinated movement, albeit in a real crowd. This is unsatisfactory for a navigation system as uncoordinated movement can interfere with other agents and make travel time too lengthy and inefficient. [2] uses 'joint collision avoidance' to simulate movement through a crowd by replacing a pedestrian in a data set with a model of their robot. This relies on increased assumptions on *a priori* knowledge of crowd members by including goal information, with movement predictions over a large time-horizon. However, as a crowd is partly stochastic [3] long prediction horizons become unreliable. As very short horizons do not require sophisticated models, linear extrapolation is sufficient [4].

© Springer International Publishing Switzerland 2015
C. Dixon and K. Tuyls (Eds.): TAROS 2015, LNAI 9287, pp. 270–282, 2015.
DOI: 10.1007/978-3-319-22416-9_31

The system we propose is based on knowledge that can be acquired from the environment and is very simple to implement, using only observed trajectory; velocity; and statistical estimation, to create very effective results. It can produce similar trajectories to [2] whilst also responding well to more unpredictable occurrences, coping well when moving 'across' the crowd by using the novelty of consideration when trajectories of robot and pedestrian directly interfere. A constant velocity model (CVM) is used to predict the crowd movement with an *uncertainty ellipse* to encompass abrupt stopping of pedestrians as well as lateral movement and velocity deviations. Using this method is superior to deterministic predictors, as if the prediction is wrong the system may not respond correctly, where as 'embracing' the uncertainty will help avoid collisions.

A further contribution of this paper is creating a 4-intersection Voronoi diagram (VD) of ellipses by modifying the VD of circles [5]; only a maximum of 2-intersections has been solved to date [6].

2   Related Work

2.1   Modelling Crowd Dynamics and Behaviours

Crowd modelling is key to successful path planning [7]. Designing an appropriate model that estimates where each pedestrian will be helps constrain uncertainties by assessing probabilities of movement [8]. An appropriate model relies on the choice between two forms of crowd motion: microscopic & macroscopic [9].

2.1.1 Macroscopic perspectives of crowd modelling focus on pedestrians being a continuum, typically using partial differential equations of fluid dynamics [10], using few parameters as individual agents are ignored. Distance between agents is considered negligible, referring to locally averaged quantities and ignoring local velocity fluctuations [11]. Models using continuum dynamics are poor choices for considerate movement as plotting a path *through* a crowd would be ineffective.

2.1.2 Microscopic perspectives of crowd modelling focus on pedestrians being individuals. Agent-based models are typically used to describe each pedestrian, simulating the actions and interactions of each person [12]. To design a model multiple parameters are available due the complex nature of the individual [7]. Assessing each individual separately can create a model to anticipate their next move, allowing the path planner to move *through* the crowd effectively. The novelty of the considerate path planner relies on this intimate knowledge.

To successfully plot a path through a crowd, microscopic evaluation must be used. Each crowd member must be assessed separately in order to exploit gaps between pedestrians. To move considerately the robot must be able to predict where each individual is going to be in order to best respond to the situation. Individual pedestrians are considered Markov in nature [13] and so no past state can be used to reinforce any future ones. This reinforces the use of a simple CVM and uncertainty ellipse to best estimate and capture the pedestrian movements.

2.2   Path Planning Techniques

To best exploit the gaps between pedestrians, and utilise the novelty of consideration, the Voronoi diagram (VD) provides an excellent framework [14]. The VD naturally creates a path of 'safest' distance from any object by creating vertex points (VP) of equal distance from any 3 neighbouring objects and minimum distance (MD) between them. Connecting these nodes (Fig. 1a) creates a roadmap that ensures no path ever gets closer to one object than the other (Fig. 1b), and through a modification the VD allows consideration to be utilised by connecting the MD of closest neighbours (Fig. 1c), instead of detouring via the VP.

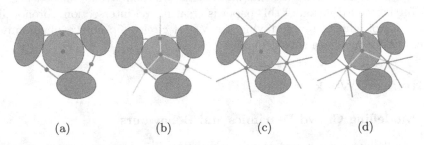

(a)   (b)   (c)   (d)

Fig. 1. Example of Euclidean vector roadmap. (a) Nodes (green) of VP (blue Apollonius circle) and MD (purple lines). (b) Approximated VD of E using vectors. (c) Shortcuts connecting MD-MD. (d) Roadmap of connected nodes.

The VD is advantageous over other methods [14] such as the probabilistic roadmap (PRM), which requires random sampling resulting in paths that are not diffuse enough between pedestrians to exploit all gaps. The visibility graph suffers by staying close to the objects, not creating enough space between the pedestrian and robot. Cell decomposition provides an inadequate framework similar to the PRM by subdividing the environment into adjacent squares, which may not have a resolution small enough to subdivide the small gaps in a crowd, resulting in gaps not being exploited.

2.3   Applied Systems

Museum tour guides: RHINO [15] and MINERVA [16], are robots specifically designed to manoeuvre within an environment populated by people. The path planning method uses occupancy grids with value iteration, creating a single shortest path from start to goal. The robots exhibit patience towards the crowd, using the μDWA [17] approach, however alternative paths are not seemingly chosen and it is up to the person to move away. The work presented in this paper differs by creating a path planner that seeks to always continue to the goal, checking multiple paths to negotiate a way around pedestrians more naturally.

Cooperative movement [2] replaces a pedestrian (P_{176} of data set [4]) with their robot model to show it engaging in 'joint collision avoidance' with other pedestrians. Bayesian inference calculates *a posterior* and despite referring to the crowd as stochastic, goal information is incorporated[1]. This is a major assumption as if goal prediction is incorrect then an effective path cannot execute. Also, the anecdotal results show the robot moving 3 time-steps (ts) behind the pedestrian, and when examining the video footage the pedestrian seems forceful when moving through the crowd; other pedestrians 'cooperate' by moving out of its way as it moves in a straight line. The crowd parts for the pedestrian and the robot follows, similar results could assume to be yielded using a pedestrian following robot [18].

3   Problem Formulation

To overcome the limitations of existing approaches microscopic pedestrian prediction is proposed to create a considerate path planner through a crowd of pedestrians. The path planner will be assessed using simulations, acquiring pedestrian data (P) from a video recording [4] (Fig. 2).

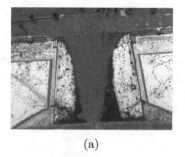 

(a)   (b)

Fig. 2. View from where the pedestrian data was recorded. (a) All pedestrian positions over footage. (b) Pedestrians moving within area they are recorded.

The challenges for this include how to model the agents in the environment. As P is observed from a plan perspective, the environment will be constricted to the Euclidean plane $(n \times m) \in \mathbb{R}^2$, with P and the robot (R) represented using a circle and ellipses (E), respectively (Fig. 3). We assume that the pedestrians cannot interact with the robot as P is pre-recorded.

To best plan a path, predicting the movement of P over a time-horizon of T-steps can be used to estimate where the crowd and robot could be in the future before executing any movement of R. This aids the main objective of moving with consideration, as the anticipation of the crowd's movement will allow: potential future collisions to be assessed and avoided; not move the robot into an over crowded area; and select a path that will move with the crowd. As stated in Section 2.1, microscopic prediction is required for a system concerned with the movement of individual agents, which combined with the Markov nature of pedestrians confirms that a CVM is suitable for a short prediction horizon.

---

[1] The pedestrians are heading towards a door.

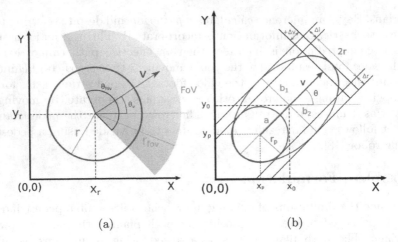

(a)   (b)

Fig. 3. Modelling the Agents. (a) The robot. (b) The dynamic pedestrians uncertainty ellipse with the zero velocity collapsed circle.

3.1   The Robot

The robot (Fig. 3a) is modelled to have a simple circular cross-section with centre (x_r, y_r); radii r_r; and holonomic kinematics that allows for instantaneous acceleration and onmi-directional movement. The robot selects only agents within a Field of View (FoV) with a viewing angle θ_{fov} and range r_{fov}.

3.2   The Crowd (Pedestrians)

For a pedestrian, P_p, (Fig. 3b) observed to have velocity, v_p, the movement over a time-horizon of T steps can be predicted using a CVM. The ellipse (Fig. 4) can account for where they are predicted to move, including the start position (x_p, y_p) to account for sudden stopping as well as velocity and lateral movement deviations ($\pm \Delta v_p$ & $+\Delta r_p / l_p$, respectively) based on statistical data [8]. A 2-DoF chi-square distribution confines these factors into a confidence interval of 90%, creating the ellipse. When a pedestrian is observed to have no velocity the ellipse collapses to a static circle with radii r_p and centre (x_p, y_p).

4   Moving Considerately

The main novelty we present is the concept of consideration when planning the robot's path. This includes: allowing P to move first and not have R directly cut in front; and giving P adequate room based on their spatial range [19]. This is advantageous as it reduces collision potential and allows P to move as if R was not there. This captures the objectives of not adding to congestion, allowing all agents to get to their destination as quickly and easily as possible.

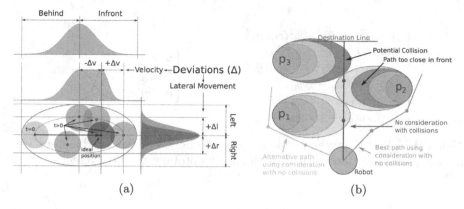

Fig. 4. Creation of the pedestrian uncertainty ellipse, with predicted movement using a CVM. (a) Construction of ellipse with possible pedestrian locations (circles). (b) Demo over 3 ts of direct path (red) vs. considerate paths (green/yellow) around pedestrians $p_{1,2,3}$.

To achieve this form of movement the algorithm must predict where P will be over T future time-steps. By assessing where P may be at time t, the best positions for where R could go at t can also be made. Trialling different locations until $t = T$ allows a number of paths to be assessed to find the most 'desirable' (Fig. 4b). Using the VD as a basis helps to form considerate connections as it creates an equidistant connectivity graph that helps prevent collisions by moving R away from P (Fig. 1b). Also 'shortcut' connections align more directly with E (Fig. 1c), giving the roadmap more versatility by aligning some vectors with the movement of the crowd, or allowing a connection to be made that moves behind individual pedestrians.

5   A Voronoi Diagram for Ellipses

Calculating the VD of arbitrarily placed fully intersecting ellipses cannot be resolved analytically; there are more unknown variables (9) than equations (3)[2]. Through a novel modification to the sweepline VD of circles [5] an approximation can be made using pseudo ellipses (PE) created from four tangentially aligned circles that produce the same semi-major a and semi-minor b axes as the original ellipse (Fig. 5a). A novel 'tangent' function is created, increasing the time complexity from $O(n^2 \log n)$ to $O([n^2 + 4n] \log n)$, which merges 2 parabolas of the sweepline together between the 2 tangential circles. The VP (x_{vp}, y_{vp}) of PE provide excellent initial approximations to VP of E, which can be found in only a few iterations with a quadratic rate of convergence

$$\mathbf{A}x^2 + \mathbf{B}xy + \mathbf{C}y^2 + \mathbf{D}x + \mathbf{E}y + \mathbf{F} = 0 \ . \tag{1}$$

---

[2] Minimum distance between just 2 ellipses requires a 12-dimensional polynomial

where

$$\begin{cases} A = a^2(\sin\theta)^2 + b^2(\cos\theta)^2 \\ B = 2(b^2 - a^2)\sin\theta\cos\theta \\ C = a^2(\cos\theta)^2 + b^2(\sin\theta)^2 \\ D = -2Ax_0 - By_0 \\ E = -Bx_0 - 2Cy_0 \\ F = Ax_0{}^2 + Bx_0y_0 + Cy_0{}^2 - a^2b^2 \end{cases} \quad \begin{cases} a > b \\ (x_0, y_0) \in \mathbb{R}^2 \\ 0 \le \theta \le 2\pi \end{cases}$$

Using the VP of E as starting points, the MD between associated ellipses are found using the bisection method, with an asymptotic rate of convergence.

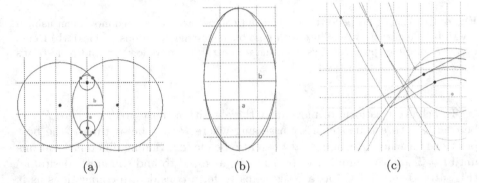

(a)   (b)   (c)

Fig. 5. (a) Construction of PE (magenta) from circles, aligned at red points. (b) Comparison of PE and E. (c) Application of 'tangent' function. From sweepline locations (yellow). Top: reached the tangent point, p_t, (green). Middle: Tangentially aligned parabola points (red and blue). Bottom: Merged 'super parabola'.

The 'Tangent' Function. This is a novel adaptation of [5], which creates a reliable approximation of fully intersecting and arbitrarily placed ellipses and requires only one extra step to the original (Fig. 5c). At the point where the circles tangentially align, the parabolas of both circles that associate with the sweepline at that point are also tangential, p_t. As the sweepline continues past the 'tangent' event, p_t continues to move with the 2 parabolas at their tangential point. Keeping the 2 parts of the parabola that associate with the circles creates a 'super parabola', describing the convex polygon of PE (Fig. 5a).

6   Algorithm

The algorithm proposed for this paper uses pre-recorded footage of a crowd, to process the simulations, with each pedestrian's position calculated using a tracking algorithm [4] and stored in a .txt file. Each frame of the video (2.5Hz) have manually annotated pedestrian numbers to uniquely identify them.

6.1   Pedestrians in the Environment

The P selection extracted from the .txt file at each frame will only be those within the FoV range, replicating a real-life system, from a slightly raised perspective in order to remove occlusion. As data for P can only be obtained through observation, each agent must appear within 2 consecutive frames of the video. The observations at those 2 frames allow velocity v_p to be calculated using a CVM for each P_p.

$$v_p = \frac{\sqrt{(x_{(t-1)} - x_t)^2 + (y_{(t-1)} - y_t)^2}}{0.4}. \tag{2}$$

To predict P over the time-horizon the chosen maximum potential deviations are added to the ideal position as defined by v_p, with deviations being cumulative and increasing at each ts (Fig. 6).

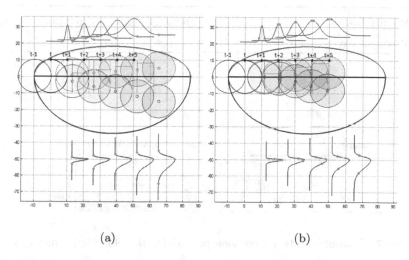

(a)   (b)

Fig. 6. Matlab execution of predicting the uncertainty ellipse over 5 time-steps. Velocity, left/right lateral movement deviations are: $\pm 25\%$, $+5\%$, $+10\%$, respectively. (a) Maximum deviations at each ts. (b) Random deviations at each ts.

6.2   Planning a Path

The connections that form the roadmap have been specifically designed to create a network of Euclidean vectors, with the VP and MD forming nodes that are minimum and equal distances from associated ellipses (Fig. 1). These include 2 forms: 1) an approximate VD created by connecting VPs to their associated MDs. 2) 'shortcuts' created by connecting MD-MD of the same associated VP.

A search algorithm sends out a wave to find all available paths by moving the crest along connections of the roadmap beginning from R, moving at the maximum speed of R for 1 ts. At the crest of the wave, will be all possible locations of where R could have moved to. E of P are then evolved using the CVM to $t+1$ and the VD of E is reproduced. From a selection of the 'best' positions at the end of the wave a new wave commences, repeating until $t = T$.

Applying Consideration. This is achieved by adding resistance Ω to the roadmap vectors, reducing the wave's speed, based on the relationships between R and E involving: distance, d, between them and their normalised relative trajectory, ϕ_n (θ_r w.r.t. θ_p). Two functions calculate Ω (Table 1) for: proximity p; and a Membership Function (MF) to calculate collision potential ζ.

Table 1. Resistances Ω applied to the search algorithm wave

Resistance Ω (Robot w.r.t. Pedestrian)	Intersecting	Non-Intersecting
Moving Towards	$p.\phi_n.(\zeta_1 + \zeta_2)$	$p.\phi_n.\zeta_1$
Moving Away	$p.\phi_n.\zeta_2$	0

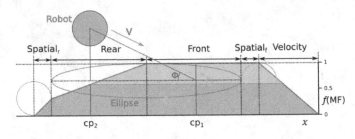

Fig. 7. Example of MF for collision potential with defuzzified values $\zeta_{1,2}$

Proximity. Ω is modified based on the proximity between R and E. If R becomes close to E an inverse square is applied ($p = \frac{1}{d^2}$). If R intersects E a direct square is applied ($p = d^2$), dramatically increasing resistance due to a probable collision.

Membership Function. Ω is modified based on the relative trajectory and position of R w.r.t. E (Fig. 7). Collision points of their current trajectories cp_1 and the perpendicular intersection to E from (x_r, y_r) are calculated. Defuzzified values ζ_1 and ζ_2 are calculated for cp_1 and cp_2, respectively. The MF divides into specific zones regarding factors contributing to collision potential (Table 2).

Table 2. MF (Fig. 7) zones used to calculate $\zeta_{1,2}$, based on collision potential

Zone	Description	$f'(\text{MF})$	$f(\text{MF})$
Spatial$_r$	Radius r_r of robot behind pedestrian	1	$\int_0^{r_r} x\,\mathrm{d}x$
Rear	Collision potential increases	$\frac{b_1+b_2}{a}$	$\int_0^a \frac{b_1+b_2}{a} x\,\mathrm{d}x + r_r$
Front	Most likely area to collide with pedestrian	0	$r_r + (b_1 + b_2)$
Spatial$_f$	Radius r_r of robot behind pedestrian	0	$r_r + (b_1 + b_2)$
Velocity	Consider velocity of pedestrian over next ts	$-\frac{b_1+b_2+a}{vp}$	$\int_{v_p}^0 \frac{b_1+b_2+a}{vp} x\,\mathrm{d}x$

7   Results

The results shown highlight the movements of R and P over selected frames (f) of the video footage, all sequences begin at $f = 1$. For all figures R and P are represented as follows. R is shown with: purple dots for t prediction locations over T; an orange line for destination goal; and a pink path taken. P is represented as ellipses, created from CVM extending over one ts, with the green and red dot representing the front and rear, respectively.

7.1   Comparison to 'Cooperative' Navigation

To test our model against [2][3] R begins where P_{176} does, and is given the same speed as P_{176} at each ts. When $T = 1$ the model follows a similar direct path to P_{176} and [2]. Increasing T demonstrates how R responds to P considerately. As R approaches the first cluster of P (Figs. 8a and 8b) it chooses to move straight forward through a smaller space between the P in the cluster rather than deviate left or right and directly cut in front of others. As P is within a set area (Fig. 2), pedestrians just outside are not stored in P, even if inside the FoV. As R approaches the edge (Fig. 8c) 'incoming' pedestrians are not registered in P until they enter the area (Fig. 8d), resulting in pedestrians appearing directly in front of R without any prior detection. By increasing T, allowing different paths to be assessed, a path is found that moves R slightly to the right (Figs. 8e and 8f) exploiting a larger gap and avoid forcing pedestrians off their original trajectories, or create a head-on collision. Moving with the flow of the crowd demonstrates that R can navigate in a manner similar to humans in this scenario.

7.2   Demonstrating Consideration when Cutting Across a Crowd

A scenario is chosen within the pedestrian data set that incorporates alternating trajectories of neighbouring pedestrians (Fig. 9a), with the robot's path set to take it directly across P. As R cuts directly through the crowd, it demonstrates

---

[3] See Fig. 5c and Fig. 5d in [2].

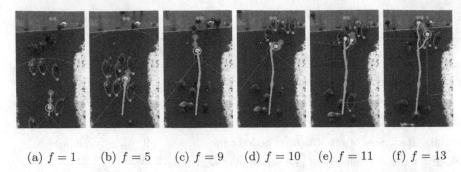

(a) $f = 1$ (b) $f = 5$ (c) $f = 9$ (d) $f = 10$ (e) $f = 11$ (f) $f = 13$

Fig. 8. Using consideration to move with crowd flow. Robot parameters used: $v_r = v_p$, $\theta_r = 270°$, $r_{fov} = 5m$, $T = 3$. P_{176} is shown with a blue path taken.(a,b) R moves through gaps in crowd, avoiding collisions with oncoming P. (d-f) Robot exploits gap to move around P.

a desire to move with the direction of crowd flow (Fig. 9b). However this can be seen to change in the next move (Fig. 9c) when later predicted movements ($t > 3$) change the direction of the path, as moving behind P_2 is preferable for advancing to the goal. From later crowd movements this idea is reinforced as all ts choose a path behind (Fig. 9d), with this repeated as R moves in between the alternating trajectories (Figs. 9e and 9f). When moving in front of P_5 (Fig. 9g) R has not disrupted P_5 as R passes by before entering collision zones (eg. Fig. 7 and Table 1). The robot successfully navigates to its goal (Fig. 9h), with only minimal deviations from a direct path, causing no disruption to the crowd.

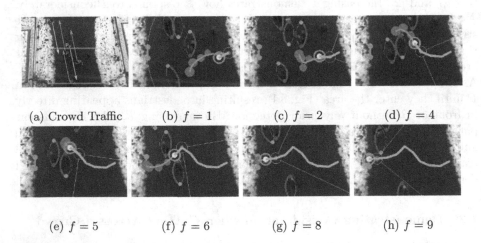

(a) Crowd Traffic   (b) $f = 1$ (c) $f = 2$ (d) $f = 4$

(e) $f = 5$ (f) $f = 6$ (g) $f = 8$ (h) $f = 9$

Fig. 9. Using consideration to move across the crowd flow. Robot parameters: $v_r = 10ms^{-1}$, $\theta_r = 270°$, $r_{fov} = 5m$, $T = 5$. (a) P used with directional arrows (blue: 'up', yellow: 'down'). Rectangle highlights area used in following figures.

8 Evaluation

The system presented has demonstrated a novel application of 'consideration' for use in Human-Robot navigation systems. The results presented demonstrate a path planner capable of moving swiftly through a crowd, predicting the evolution of each crowd member using very simple methods and acquiring very effective results. All executed paths were collision-free, causing no disruption to other pedestrians. When moving with the crowd, replacing a pedestrian, the robot moved in very similar ways (Fig. 8) with a small prediction horizon. Increasing T resulted in more considerate paths being planned and executed, not disrupting any pedestrians by making use of space and reacting to potential collisions before they can occur. When cutting through alternating trajectories (Fig. 9), the system made excellent use of crowd movement predictions. Finding ways to move directly to the goal and not cause any disruption to passing pedestrians.

References

1. Zeng, S., Weng, J.: Obstacle avoidance through incremental learning with attention selection. In: Proceedings of the 2004 IEEE International Conference on Robotics and Automation, ICRA 2004, vol. 1, pp. 115–121 (2004)
2. Trautman, P., Krause, A.: Unfreezing the robot: navigation in dense, interacting crowds. In: IEEE/RSJ International Conference on Intelligent Robots and Systems (IROS), pp. 797–803 (2010)
3. Helbing, D.: Models for Pedestrian Behavior (1998). eprint arXiv:cond-mat/9805089
4. Pellegrini, S., et al.: You'll never walk alone: modeling social behavior for multi-target tracking. In: 2009 IEEE 12th International Conference on Computer Vision, pp. 261–268 (2009)
5. Jin, L., et al.: A sweepline algorithm for euclidean voronoi diagram of circles. Computer-Aided Design 38(3), 260–272 (2006)
6. Emiris, I.Z., et al.: Exact voronoi diagram of smooth convex pseudo-circles: General predicates, and implementation for ellipses. Computer Aided Geometric Design 30(8), 760–777 (2013)
7. Lhner, R.: On the modeling of pedestrian motion. Applied Mathematical Modelling 34(2), 366–382 (2010)
8. Huber, M., et al.: Adjustments of speed and path when avoiding collisions with another pedestrian. PLoS ONE 9(2), e89589 (2014). http://dx.doi.org/10.1371%2Fjournal.pone.0089589
9. Pelechano, N., et al.: Virtual Crowds: Methods, Simulation, and Control. Morgan & Claypool Publishers (2008)
10. Hughes, R.L.: The flow of human crowds. Annual Review of Fluid Mechanics 35(1), 169–182 (2003)
11. Bellomo, N., Dogbé, C.: On the modelling crowd dynamics from scaling to hyperbolic macroscopic models. Mathematical Models and Methods in Applied Sciences 18(supp01), 1317–1345 (2008)
12. Helbing, D., Molnár, P.: Social force model for pedestrian dynamics. Phys. Rev. E 51, 4282–4286 (1995)

13. Henry, P., et al.: Learning to navigate through crowded environments. In: IEEE International Conference on Robotics and Automation, pp. 981–986 (2010)
14. Choset, H., et al.: Principles of Robot Motion: Theory, Algorithms, and Implementations. MIT Press, Cambridge (2005)
15. Burgard, W., et al.: The interactive museum tour-guide robot. In: Proc. of the Fifteenth National Conference on Artificial Intelligence (AAAI 1998) (1998)
16. Thrun, S., et al.: Minerva: a second-generation museum tour-guide robot. In: Proceedings of the 1999 IEEE International Conference on Robotics and Automation, vol. 3, pp. 1999–2005 (1999)
17. Fox, D., et al.: A hybrid collision avoidance method for mobile robots. In: Proc. of the IEEE International Conference on Robotics and Automation 1998 (1998)
18. Yoshimi, T., et al.: Development of a person following robot with vision based target detection. In: 2006 IEEE/RSJ International Conference on Intelligent Robots and Systems, pp. 5286–5291, October 2006
19. Walters, M., et al.: The influence of subjects' personality traits on personal spatial zones in a human-robot interaction experiment. In: IEEE International Workshop on Robot and Human Interactive Communication, pp. 347–352 (2005)

Adaptive Sampling-Based Motion Planning for Mobile Robots with Differential Constraints

Andrew Wells and Erion Plaku[✉]

Department of Electrical Engineering and Computer Science,
Catholic University of America, Washington, DC, USA
plaku@cua.edu

Abstract. This paper presents a sampling-based motion planner geared towards mobile robots with differential constraints. The planner conducts the search for a trajectory to the goal region by using sampling to expand a tree of collision-free and dynamically-feasible motions. To guide the tree expansion, a workspace decomposition is used to partition the motion tree into groups. Priority is given to tree expansions from groups that are close to the goal according to the shortest-path distances in the workspace decomposition. To counterbalance the greediness of the shortest-path heuristic, when the planner fails to expand the tree from one region to the next, the costs of the corresponding edges in the workspace decomposition are increased. Such cost increases enable the planner to quickly discover alternative routes to the goal when progress along the current route becomes difficult or impossible. Comparisons to related work show significant speedups.

Keywords: Sampling-based motion planning · Differential constraints · Workspace decomposition · Discrete search

1 Introduction

In motion planning, the objective is to compute a collision-free and dynamically-feasible trajectory which enables the robot to reach a desired goal region while avoiding collisions with obstacles. Motion planning arises in numerous robotics applications ranging from autonomous vehicles, manufacturing, surveillance to robotic-assisted surgery, air-traffic control, and computer animations [1,5,13].

In order to facilitate the execution of the planned motions, the motion planner needs to take into account the underlying robot dynamics. Robot dynamics express physical constraints on the feasible motions, such as minimum turning radius, directional stability, bounded curvature, velocity, and acceleration. Robot dynamics are often modeled as differential equations of the form $\dot{s} = f(s, u)$ to indicate the changes that occur in the state s as a result of applying the input control u, e.g., setting the acceleration and steering angle. Such differential constraints make motion planning challenging as the planner needs to find controls that result in a motion trajectory to the goal that is not only collision-free but also dynamically-feasible so that it can be followed by the robot.

© Springer International Publishing Switzerland 2015
C. Dixon and K. Tuyls (Eds.): TAROS 2015, LNAI 9287, pp. 283–295, 2015.
DOI: 10.1007/978-3-319-22416-9_32

To take into account the constraints imposed by the robot dynamics and obstacles, this paper draws from sampling-based motion planning which has shown promise in solving challenging problems [5,13]. In a sampling-based formulation, motion planning is defined as a search problem. To conduct the search, a motion tree is rooted at the initial state and is incrementally expanded by adding collision-free and dynamically-feasible trajectories as branches. The proposed approach uses a workspace decomposition to obtain a simplified planning layer which serves to guide the sampling-based search. The workspace decomposition is used to induce a partition of the motion tree into groups and to determine routes along which to expand each group. In particular, each group is defined by a region r in the decomposition and the motion-tree vertices that are in r. The search is driven by procedures to (i) select a group, (ii) expand the motion tree from the selected group, and (iii) update heuristic costs to reflect the progress made. To promote expansions toward the goal, priority is given to groups that are close to the goal according to the shortest-path distances in the workspace decomposition. After selecting a group, attempts are made to expand the motion tree along the shortest path in the workspace decomposition to the goal region. To counterbalance the greediness of the shortest-path heuristic, when the group expansion fails, the costs of the corresponding edges in the workspace decomposition are increased. Such cost increases enable the proposed approach to quickly discover alternative routes to the goal when progress along the current route becomes difficult or impossible. Experimental validation is provided using high-dimensional models with nonlinear dynamics where the robot has to operate in complex environments and wiggle its way through narrow passages in order to reach the goal. Comparisons to related work show significant speedups.

2   Related Work

The discussion of related work focuses on sampling-based motion planning since it is the basis for the proposed approach. Over the years, numerous approaches have been proposed on how to conduct a sampling-based search. RRT [15], which is one of the most widely used planners, and its variants [11,14] expand the motion tree from the nearest vertex to a random sample to bias the search toward the largest uncovered Voronoi regions. Other approaches rely on probability distributions [10], subdivisions [6], principal-component analysis [7], stochastic transitions [8], and density estimations [9].

Even though considerable progress has been made, it remains challenging to effectively incorporate robot dynamics into planning. As the problem dimensionality increases and the robot dynamics become more complex, the search in many of these approaches becomes less and less efficient [5,13].

To improve the computational efficiency, Syclop [17,19] coupled discrete search over a workspace decomposition with sampling-based motion planning. Subsequent work [16,18] offered further improvements by partitioning the motion-tree into groups and using shortest-path distances as heuristic costs to effectively guide the motion-tree expansion. Comparisons to RRT and other planners showed significant speedups. The approach proposed in this paper builds upon this line of work which couples discrete search with sampling-based motion planning. The proposed approach addresses the shortcomings observed in Syclop [19] and its variants [16,18] when dealing with complex problems. In particular, these planners, due to the greediness of the shortest-path heuristic, spend considerable time before realizing that the current guide should be abandoned due to constraints imposed by obstacles and robot dynamics. Fig. 1 shows an illustration where the shortest-path heuristic leads Syclop and its variants along an infeasible route. Other examples are shown in Fig. 3. The proposed approach addresses such shortcomings by counterbalancing the greediness of the shortest-path heuristic with edge penalties. When the tree expansion fails to make progress, the costs of the corresponding edges in the workspace decomposition are increased. Such cost increases enable the proposed approach to quickly discover alternative routes to the goal. As illustrated in Fig. 1, after failing to move the robot through the impassible openings at the top and on the right, the increases in the edge costs quickly enable the proposed approach to discover a feasible route to the goal by passing through the opening on the left side. Comparisons to prior work show significant speedups.

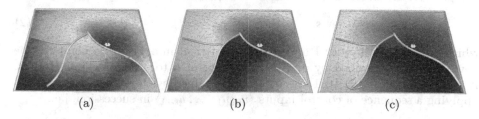

(a)   (b)   (c)

Fig. 1. Illustration of how the proposed approach changes the heuristic costs over time to account for expansion failures: (a) initial heuristic costs; (b) updated heuristic costs after the snake-like robot fails to pass through the opening at the top; and (c) updated heuristic costs after robot fails to pass through the opening on the right side. The color of each region r represents the shortest-path distance in the workspace decomposition from r to the goal (red: near, blue: far). Figures are best viewed in color and on screen.

3   Problem Formulation

A robot model is defined in terms of its state space S, control space U, and dynamics f. A state $s \in S$ defines the position, orientation, velocity, steering angle, and other components related to motion. A control $u \in U$ defines the external inputs that are used to control the robot. Dynamics are specified as a set of differential equations $\dot{s} = f(s, u)$ which indicate the changes that occur to the state s when applying the control u. An example is provided below.

Example: The snake-like robot used in the experiments is modeled as a car pulling several trailers. This provides a high-dimensional model with second-order differential equations, defined as (adapted from [13, pp.731])

$$\dot{x} = v\cos(\theta_0)\cos(\psi) \qquad \dot{y} = v\sin(\theta_0)\cos(\psi) \qquad \dot{\theta}_0 = v\sin(\psi)/L$$

$$\dot{v} = u_a \qquad \dot{\psi} = u_\omega \qquad \dot{\theta}_i = \frac{v}{d}(\sin(\theta_{i-1}) - \sin(\theta_0))\prod_{j=1}^{i-1}\cos(\theta_{j-1} - \theta_j) \qquad (1)$$

The state $s = (x, y, \theta_0, v, \psi, \theta_1, \ldots, \theta_N)$ defines the position (x, y), orientation θ_0, velocity v, and steering angle ψ of the car, and the orientation θ_i of each of the N trailers. The control $u = (u_a, u_\omega)$ defines the acceleration and the rotational velocity of the steering angle. The body and hitch lengths (L and d) are set to small values so that the robot resembles a snake, as shown in Fig. 1.

Let W denote the workspace in which the robot operates. Let $O_1, \ldots, O_m \subset W$ denote the obstacles and let $G \subset W$ denote the goal region. A state $s \in S$ is considered valid if there are no self-collisions or collisions with obstacles when the robot is placed according to the position and orientation specified by s. Such function, COLLISION : $S \rightarrow \{\texttt{true}, \texttt{false}\}$, can be implemented efficiently by using available collision-detection packages, such as PQP [12]. The goal is reached when the position defined by s, denoted by POS(s), is in G. From a motion-planning perspective, the effect of the dynamics is captured by a function

$$s_{new} \leftarrow \text{MOTION}(f, s, u, dt), \qquad (2)$$

which uses a fourth-order Runge-Kutta numerical integration to compute the new state s_{new} obtained by applying the control u to s for one time step dt. A motion trajectory $\zeta : \{0, 1, \ldots, \ell\} \rightarrow S$ is obtained by starting at a state s and applying a sequence of control inputs $\langle u_0, u_1, \ldots, u_{\ell-1}\rangle$ in succession, i.e.,

$$\zeta(0) = s \text{ and } \forall i \in \{0, \ldots, \ell-1\} : \zeta(i+1) = \text{MOTION}(f, \zeta(i), u_i, dt) \qquad (3)$$

The motion-planning problem can now be stated as follows: Given a workspace W, obstacles $O_1, \ldots, O_m \subset W$, goal region $G \subset W$, robot model $\langle S, U, f\rangle$, and an initial state $s_{init} \in S$, compute a sequence of control inputs $\langle u_0, u_1, \ldots, u_{\ell-1}\rangle$ such that dynamically-feasible trajectory $\zeta : \{0, 1, \ldots, \ell\} \rightarrow S$ obtained by applying the control inputs in succession starting at s_{init} is collision-free and reaches the goal, i.e., $\forall i \in \{0, 1, \ldots, \ell\} : \text{COLLISION}(\zeta(i)) = \texttt{false}$ and POS$(\zeta(\ell)) \in G$.

4   Method

Pseudocode for the approach is shown in Alg. 1. To facilitate presentation, the workspace decomposition is described first in Section 4.1. The overall guided search is described in Section 4.2.

4.1   Workspace Decomposition

To obtain a simplified planning layer, the unoccupied area of the workspace, i.e., $W \setminus (O_1 \cup \ldots \cup O_m \cup G)$, is decomposed into nonoverlapping triangular regions

---

Algorithm 1. Pseudocode for the proposed approach

Input: W: workspace; $O_1, \ldots, O_m \subset W$: obstacles; $G \subset W$: goal region
$\langle S, U, f \rangle$: robot model; $s_{init} \in S$: initial state; t_{max}: upper bound on runtime
Output: collision-free and dynamically-feasible trajectory to goal or **null** if no
solution is found within t_{max} runtime

1: $D = (R, E, C) \leftarrow$ WORKSPACEDECOMPOSITION$(W, O_1, \ldots, O_m, G)$
2: $\langle h(r_1), \ldots, h(r_n) \rangle \leftarrow$ HEURISTICCOSTS(D, G)
3: $\langle \mathcal{T}, \Gamma \rangle \leftarrow$ INITMOTIONTREE(s_{init})
4: **while** TIME $< t_{max}$ **do**
5:   **for** κ iterations **do**
6:     $\Gamma_r \leftarrow$ SELECTGROUP$(\Gamma, \langle h(r_1), \ldots, h(r_n) \rangle)$
7:     $\langle status, v_{last} \rangle \leftarrow$ EXPANDMOTIONTREE$(\mathcal{T}, \Gamma, \Gamma_r)$
8:     **if** $status =$ **goalReached then return** TRAJ$(\mathcal{T}, v_{last})$
9:     **if** $status =$ **collisionEncountered then**
10:       $r_{next} \leftarrow$ FIRST(SHORTESTPATH$(D, region(v_{last}), G)$)
11:       **for** $r_{neigh} \in$ NEIGHS(D, r_{next}) **do** INCREASEEDGECOST(D, r_{next}, r_{neigh})
12:   $\langle h(r_1), \ldots, h(r_n) \rangle \leftarrow$ UPDATEHEURISTICCOSTS(D, G)
13: **return null**

$t_1, \ldots, t_n$ [20]. Figs. 1–3 show some examples. The workspace decomposition is represented as an undirected, weighted, graph $D = (R, E, C)$, where R, E, C denote the regions, edges, and edge costs, respectively. The set of regions contains the triangles and the goal region, i.e., $R = \{t_1, \ldots, t_n, G\}$, since these regions can be used to generate collision-free motions. The set of edges captures the physical adjacency of the regions in the workspace decomposition, i.e.,

$$E = \{(r, r') : r, r' \in R \text{ share an edge}\}. \tag{4}$$

Edge costs are used to provide the approach with short routes to the goal region. Initially, the cost of an edge is set to the Euclidean distance between the centroids of the corresponding regions, i.e., $C(r, r') = ||centroid(r) - centroid(r')||_2$. As described later, the approach increases $C(r, r')$ when it fails to make progress expanding the motion tree from r to r'. These cost increases enable the approach to discover alternative routes to the goal. The approach also relies on a function LOCATEREGION : $W \rightarrow R \cup \{\perp\}$ which maps a point $p \in W$ to the region $r \in R$ that contains p. The symbol $\perp$ denotes the case when p falls inside an obstacle. LOCATEREGION is implemented efficiently to run in polylogarithmic time [3].

4.2 Guided Motion-Tree Search

The search for a collision-free and dynamically-feasible trajectory from the initial state s_{init} to the goal G is conducted by expanding a motion tree $\mathcal{T} = (V_{\mathcal{T}}, E_{\mathcal{T}})$. Each vertex $v \in V_{\mathcal{T}}$ is associated with a collision-free state, denoted by $state(v)$. Each edge $(v, v') \in E_{\mathcal{T}}$ is associated with some control input $u \in U$ such that $state(v') =$ MOTION$(f, state(v), u, dt)$. A solution is found when a vertex v that has reached G is added to $\mathcal{T}$, i.e., POS$(state(v)) \in G$. In that case, the solution

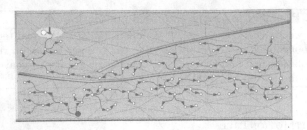

Fig. 2. Illustration of a motion tree. Initial state is shown as a blue circle; the other vertices are shown as white circles. Goal region is shown in yellow.

corresponds to the trajectory obtained by concatenating the motions associated with the edges in $\mathcal{T}$ from the root to v. An illustration is shown in Fig. 2.

To effectively guide the search, the motion-tree vertices are grouped together based on the corresponding regions in the workspace decomposition. In fact, for a region $r \in R$, let Γ_r denote all the vertices that map to r, i.e.,

$$\Gamma_r = \{v : v \in V_{\mathcal{T}} \wedge \text{POS}(state(v)) \in r\}. \tag{5}$$

This mapping induces a partition of the motion tree into groups, i.e.,

$$\Gamma = \{\Gamma_r : r \in R \wedge |\Gamma_r| > 0\}. \tag{6}$$

Since the decomposition $D = (R, E, C)$ provides a simplified planning layer, the shortest-path distance from r to G is used as a heuristic cost to estimate the feasibility of reaching G by expanding $\mathcal{T}$ from Γ_r. The search is then driven by (i) selecting a group Γ_r from Γ based on the heuristic costs, (ii) expanding $\mathcal{T}$ from Γ_r, and (iii) updating the heuristic costs to reflect the progress made. These steps are repeated until a solution is found or an upper bound on the runtime is reached. The rest of the section describes these procedures in more detail.

Group Selection. The group-selection strategy combines the heuristic costs with selection penalties in order to promote expansions from groups that are close to the goal or groups that have not been frequently explored in the past. Let $h(r)$ denote the shortest-path distance in $D = (R, E, C)$ from $r \in R$ to G. Let $nsel(\Gamma_r)$ denote the number of times Γ_r has been selected for expansion. The heuristic cost and the number of selections are combined to define a weight

$$w(\Gamma_r) = (\epsilon + 1 - h(r)/h_{max})^\alpha \beta^{nsel(\Gamma_r)}, \tag{7}$$

where $h_{max} = \max_{r' \in R} h(r')$, $\epsilon > 0$, $\alpha \geq 1$, and $0 < \beta < 1$. Among all the groups in Γ, the one with the maximum weight is selected for expansion, i.e.,

$$\text{SELECTGROUP}(\Gamma) = \arg \max_{\Gamma_r \in \Gamma} w(\Gamma_r). \tag{8}$$

Note that α serves to tune the strength of the heuristic by promoting selections of those groups that are close to the goal according to shortest-path distances in

---

Algorithm 2. EXPANDMOTIONTREE($\mathcal{T}, \Gamma, \Gamma_r$)

Input: $\mathcal{T}$: motion tree; Γ: partition of $\mathcal{T}$ into groups;
Γ_r: group from which to expand $\mathcal{T}$
Output: Function attempts to add a collision-free and dynamically-feasible trajectory from a vertex in Γ_r. It returns $\langle$status, $v_{last}\rangle$ indicating the status of the
expansion and the last vertex added to $\mathcal{T}$

1: $v \leftarrow$ select vertex at random from Γ_r
2: $r' \leftarrow$ select region at random from shortest path in $D = (R, E, C)$ from r to G
3: $p \leftarrow$ generate point at random inside r'
4: **for** several steps **do**
5:    $u \leftarrow$ CONTROLLER($state(v), p$); $s_{new} \leftarrow$ MOTION($f, state(v), u, dt$)
6:    **if** COLLISION(s_{new}) = **true then return** $\langle$collision, $v\rangle$
7:    $v_{new} \leftarrow$ NEWVERTEX(); $state(v_{new}) \leftarrow s_{new}$;
        $region(v_{new}) \leftarrow$ LOCATEREGION(POS(s_{new})); $V_{\mathcal{T}} \leftarrow V_{\mathcal{T}} \cup \{v_{new}\}$;
8:    $(v, v_{new}) \leftarrow$ NEWEDGE(); $control(v, v_{new}) \leftarrow u$; $E_{\mathcal{T}} \leftarrow E_{\mathcal{T}} \cup \{(v, v_{new})\}$
9:    $\Gamma_{r_{new}} \leftarrow$ FIND($\Gamma, region(v_{new})$)
10:   **if** $\Gamma_{r_{new}}$ = **null then** { $\Gamma_{r_{new}} \leftarrow$ NEWGROUP(r_{new}); INSERT($\Gamma, \Gamma_{r_{new}}$) }
11:   INSERT($\Gamma_{r_{new}}, v_{new}$)
12:   **if** $region(v_{new}) = G$ **then return** $\langle$goalReached, $v_{new}\rangle$
13:   **if** NEAR(POS(s_{new}), p) = **true then return** $\langle$targetReached, $v_{new}\rangle$
14:   $v \leftarrow v_{new}$
15: **return** $\langle$targetNotReached, $v\rangle$

$D = (R, E, C)$. To balance the greediness of the heuristic, β provides a penalty
factor which reduces the weight each time Γ_r is selected for expansion. This
guarantees that Γ_r will not always be selected, since, after a number of weight
reductions, some other group will have larger weight than Γ_r. This enables the
approach to avoid becoming stuck when expansions from Γ_r are infeasible due
to constraints imposed by the obstacles and robot dynamics. Finally, a small
nonzero value is used for ϵ to ensure that each Γ_r has a positive weight, which
guarantees that every Γ_r will be eventually selected for expansion. This enables
the approach to be methodical during the search.

Group Expansion. After selecting Γ_r, the objective is to expand $\mathcal{T}$ from a
vertex $v \in \Gamma_r$ along the shortest path in $D = (R, E, C)$ from r to G. Pseudocode
is shown in Alg. 2. The vertex v is selected uniformly at random from the vertices
in Γ_r (Alg. 2:1). Random selections are commonly used in sampling-based motion
planning as a way to promote expansions along different directions [5, chap. 7].
After selecting v, a target region r' is selected uniformly at random from the
first few regions along the shortest path in D from r to G, and a target point
p is generated uniformly at random inside r' (Alg. 2:2–3). This ensures that the
target will not be too far away from r, which increases the likelihood of successful
expansions. The objective is then to expand $\mathcal{T}$ from v toward p in order to get
closer to G (Alg. 2:4–11). Specifically, a PID controller [2] is used to determine an
input control u that would steer the robot toward p (Alg. 2:5). A new state, s_{new},
is obtained by integrating the motion equations f when the control u is applied

to $state(v)$ for one time step (Alg. 2:5). If s_{new} is in collision, the expansion from v terminates. Otherwise, a new vertex v_{new} and a new edge (v, v_{new}) are added to $\mathcal{T}$ (Alg. 2:6–8). At this time, the motion-tree partition is also updated. If $region(v_{new})$ had not been reached before, then a new group $\Gamma_{region(v_{new})}$ is created and added to Γ (Alg. 2:9–10). Otherwise, the group $\Gamma_{region(v_{new})}$ is retrieved from Γ. In each case, the vertex v_{new} is added to $\Gamma_{region(v_{new})}$ (Alg. 2:11). These updates give the approach the flexibility to expand the motion tree from new groups.

If s_{new} reached the goal region G, the motion-tree expansion terminates successfully (Alg. 2:12). The expansion from v also terminates if the target point p is reached (Alg. 2:113). Otherwise, the expansion continues from v_{new} (Alg. 2:14). In this way, EXPANDMOTIONTREE$(\mathcal{T}, \Gamma, \Gamma_r)$ expands the motion tree $\mathcal{T}$ from a vertex $v \in \Gamma_r$ along the shortest path in D from r to G.

Updating the Heuristic Costs. To counterbalance the greediness of the shortest-path heuristic, when a collision is encountered during the motion-tree expansion, the costs of the corresponding edges in the workspace decomposition $D = (R, E, C)$ are increased. More specifically, let v_{last} denote the last vertex that was added to $\mathcal{T}$ during the expansion from Γ_r (Alg. 1:6). Let r_{next} denote the first region of the shortest path in $D = (R, E, C)$ from $region(v_{last})$ to G (Alg. 1:8). The costs of the edges in D that have r_{next} as a vertex are then increased in order to reduce the likelihood of future expansions passing through r_{next} since the previous expansion resulted in a collision (Alg. 1:9). Specifically, the cost of each edge $(r_{next}, r_{neigh}) \in E$ is increased as

$$C(r_{next}, r_{neigh}) \leftarrow C(r_{next}, r_{neigh}) + \gamma / C(r_{next}, r_{neigh}), \qquad (9)$$

where $\gamma > 0$. The intuition is that our estimation of the difficulty of passing through a region should increase when motion-tree expansions fail. The increase is large when $C(r_{next}, r_{neigh})$ is small, and it is slowly reduced over time as the edge cost becomes larger. The parameter γ can be tuned to provide appropriate cost increases based on the workspace size. As the experiments indicate, the approach works well for a wide range of parameter values and workspace sizes.

The heuristic costs $h(r_1), \ldots, h(r_n)$ are also updated in order to reflect the increases in the edge costs (Alg. 1:10). These updates enable the approach to discover alternative routes to reach G, as shown in Fig. 1. Note that $h(r_1), \ldots, h(r_n)$ can be computed by a single call to Dijkstra's shortest-path algorithm using G as the source. When dealing with large scenes, in order to save computation time, the heuristic costs can be updated after several attempts have been made to expand the motion tree instead of doing the updates after each attempt.

5   Experiments and Results

Experiments are conducted with a high-dimensional snake-like robot model whose dynamics are expressed by second-order differential equations, as described in Section 3. As shown in Figs. 1 and 3, the robot is required to

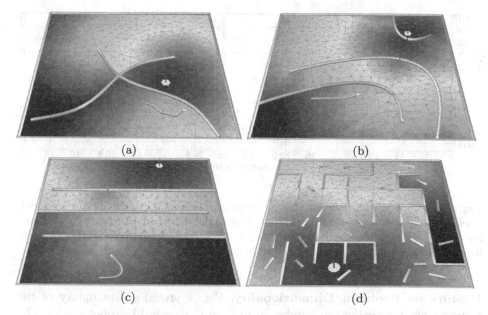

(a)   (b)

(c)   (d)

Fig. 3. The other four scenes used in the experiments (first scene is shown in Fig. 1). The color of each region r represents the distance of the shortest path in the workspace decomposition from r to the goal (red: near, blue: far). Videos showing solutions obtained by the approach are available for download [21].

move in complex scenes characterized by narrow passages, some of which are so small that it is impossible for the robot to pass through them. These scenes demonstrate the ability of the approach to quickly discover alternative routes to the goal. Experiments are also conducted with a scene which does not have impassible narrow passages (Fig. 3(d)) to show that the approach works well in different scenarios.

The approach is compared to RRT [14], which is one of the most popular sampling-based motion planners. As recommended, the RRT implementation uses the connect version, goal bias, and efficient data structures for nearest neighbors [4]. The approach is also compared to prior work on coupling motion planning with discrete search [18], using a highly efficient implementation.

Experiments are also conducted to measure the scalability of the approach when varying the dimensionality of the robot state (by changing the number of links) or the size of the scene. Results are also presented to demonstrate that the approach works well for a wide range of parameter values.

Due to the randomized nature of sampling-based motion planning, each method is run 60 times for each scenario. A time limit of 100s is set for each run. Running time measures everything from reading the input until finding a solution. Results show the mean runtime and standard deviation after discarding the five best and worst runs in order to reduce the influence of outliers. Experiments were run on an Intel Core i7 machine using Fedora 20 and GNU g++-4.9.2.

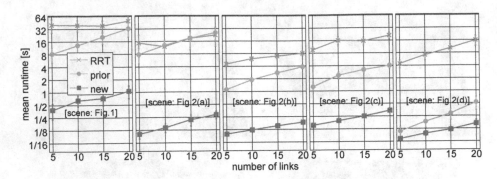

Fig. 4. Runtime results when varying the number of links of the snake-like robot model. Each bar indicates one standard deviation. Due to the significant differences in runtime, logscale is used for the y-axis with the label showing the actual value rather than its logarithm. New refers to the proposed approach and prior refers to prior work [18].

Results on Problem Dimensionality. Fig. 4 provides a summary of the results when increasing the number of links in the snake-like robot model. The results indicate that the proposed approach significantly outperforms RRT and the prior work [18]. RRT is known to have difficulties when dealing with narrow passages. In fact, since RRT expands the motion tree from the nearest neighbor to a randomly-sampled state, RRT often becomes stuck attempting to expand from vertices that are close to obstacles. The motion planner from prior work [18], although significantly faster than RRT, still has difficulty solving the problems with multiple narrow passages, some of which are impassible. The prior planner spends considerable time before realizing that it should consider alternative routes to the goal. In contrast, the proposed approach increases the edge costs of the workspace decomposition when it fails to make progress from one region to the next. These cost increases enable the approach to update the shortest-path heuristics and to quickly discover alternative routes to the goal.

Results on Scene Size. Table 1 shows the results when varying the scene size. During the scaling, the dimensions of the narrow passages are kept the same as in the original scene. In this way, the difficulty of the problem due to narrow passages remains the same in the scaled versions as in the original scene. When considering larger scenes, an increase in runtime is expected since longer trajectories have to be planned. Results in Table 1 show that RRT quickly times out in the large scenes. There is also a significant increase in the runtime of the prior planner [18] since it now spends more time before realizing that it should consider alternative routes. The running time of the proposed approach also increases, but even in the large scenes it remains computationally efficient.

Parameter Selection. Table 2 summarizes the impact of the parameter selection on the performance of the proposed approach. The results indicate that the approach works well for a wide selection of parameter values. When starting

Table 1. Mean runtime (std in parentheses) when scaling the scene. During scaling, the sizes of the narrow passages are kept the same as in the original scene. Entries marked with X indicate failure. Results are shown for the scene in Fig. 3(c) where the snake-like robot model has 15 links.

scene scaling factor	0.75	1.0	2.0	3.0	4.0
new	0.18s (0.03)	0.25s (0.05)	0.93s (0.12)	3.38s (0.30)	7.60s (0.69)
prior	1.52s (0.93)	3.20s (1.06)	32.68s (17.02)	97.34 (5.10)	X
RRT	7.20s (2.93)	15.56s (7.08)	95.14 (9.30)	X	X

Table 2. Results when varying the parameters of the approach. Results along each row are obtained by changing the corresponding parameter as indicated and keeping the other parameters to their default values (shown in bold). Recall that α tunes the strength of the heuristic (Eqn. 7), β provides the group-selection penalty (Eqn. 7), γ impacts the increase in the edge costs (Eqn. 9), and κ indicates the frequency of updating the heuristic costs (Alg. 1:5). Results are shown for the scene in Fig. 3(b) where the snake-like robot model has 15 links.

param	param values					mean runtime[s] (std)				
α	2	4	**8**	12	16	0.18 (0.02)	0.15 (0.02)	0.15 (0.02)	0.16 (0.02)	0.15 (0.02)
β	0.5	0.65	0.75	**0.85**	0.95	0.18 (0.02)	0.17 (0.02)	0.16 (0.02)	0.16 (0.02)	0.15 (0.02)
γ	15	25	**50**	75	100	0.17 (0.02)	0.15 (0.02)	0.15 (0.02)	0.15 (0.02)	0.15 (0.02)
κ	1	5	10	**20**	30	0.99 (0.17)	0.28 (0.05)	0.19 (0.03)	0.15 (0.02)	0.14 (0.02)

with a new problem, our recommendation is to use the default values shown in Table 2 as they have worked well for a variety of problems.

6 Discussion

This paper presented an efficient approach for planning dynamically-feasible trajectories that enable a mobile robot to reach a goal region while avoiding collisions with obstacles. The approach used a workspace decomposition to partition a motion tree into equivalent groups and guide the expansion of the motion tree along shortest-path routes. A key component of the approach was its adjustment of the edge costs in order to quickly discover alternative routes to the goal when expansions along the current route failed to make progress. Comparisons to related work showed significant speedups. Directions for future work include further improvements to the interplay between sampling-based motion planning and discrete search, investigation of machine-learning techniques to find optimal parameter values, and extensions of the approach to multiple robots.

Acknowledgement. This work was supported by NSF IIS-1449505 and NSF ACI-1440587.

References

1. Alterovitz, R., Goldberg, K.: Motion Planning in Medicine: Optimization and Simulation Algorithms for Image-Guided Procedures. Springer Tracts in Advanced Robotics (2008)
2. Åström, K.J., Hägglund, T.: PID controllers: theory, design, and tuning. The Instrumentation, Systems, and Automation Society (1995)
3. de Berg, M., Cheong, O., van Kreveld, M., Overmars, M.H.: Computational Geometry: Algorithms and Applications, 3rd edn. Springer-Verlag (2008)
4. Brin, S.: Near neighbor search in large metric spaces. In: International Conference on Very Large Data Bases, pp. 574–584 (1995)
5. Choset, H., Lynch, K.M., Hutchinson, S., Kantor, G., Burgard, W., Kavraki, L.E., Thrun, S.: Principles of Robot Motion: Theory, Algorithms, and Implementations. MIT Press (2005)
6. Şucan, I.A., Kavraki, L.E.: A sampling-based tree planner for systems with complex dynamics. IEEE Transactions on Robotics 28(1), 116–131 (2012)
7. Dalibard, S., Laumond, J.-P.: Control of probabilistic diffusion in motion planning. In: Chirikjian, G.S., Choset, H., Morales, M., Murphey, T. (eds.) Algorithmic Foundation of Robotics VIII. STAR, vol. 57, pp. 467–481. Springer, Heidelberg (2009)
8. Devaurs, D., Simeon, T., Cortés, J.: Enhancing the transition-based RRT to deal with complex cost spaces. In: IEEE International Conference on Robotics and Automation, pp. 4120–4125 (2013)
9. Gipson, B., Moll, M., Kavraki, L.E.: Resolution independent density estimation for motion planning in high-dimensional spaces. In: IEEE International Conference on Robotics and Automation, pp. 2437–2443 (2013)
10. Hsu, D., Kindel, R., Latombe, J.C., Rock, S.: Randomized kinodynamic motion planning with moving obstacles. International Journal of Robotics Research 21(3), 233–255 (2002)
11. Karaman, S., Frazzoli, E.: Sampling-based algorithms for optimal motion planning. International Journal of Robotics Research 30(7), 846–894 (2011)
12. Larsen, E., Gottschalk, S., Lin, M., Manocha, D.: Fast proximity queries with swept sphere volumes. In: IEEE International Conference on Robotics and Automation, pp. 3719–3726 (2000)
13. LaValle, S.M.: Planning Algorithms. Cambridge University Press (2006)
14. LaValle, S.M.: Motion planning: The essentials. IEEE Robotics & Automation Magazine 18(1), 79–89 (2011)
15. LaValle, S.M., Kuffner, J.J.: Randomized kinodynamic planning. International Journal of Robotics Research 20(5), 378–400 (2001)
16. Le, D., Plaku, E.: Guiding sampling-based tree search for motion planning with dynamics via probabilistic roadmap abstractions. In: IEEE/RSJ International Conference on Intelligent Robots and Systems, pp. 212–217 (2014)
17. Plaku, E., Kavraki, L.E., Vardi, M.Y.: Impact of workspace decompositions on discrete search leading continuous exploration (DSLX) motion planning. In: IEEE International Conference on Robotics and Automation, pp. 3751–3756 (2008)
18. Plaku, E.: Robot motion planning with dynamics as hybrid search. In: AAAI Conference on Artificial Intelligence, pp. 1415–1421 (2013)
19. Plaku, E., Kavraki, L.E., Vardi, M.Y.: Motion planning with dynamics by a synergistic combination of layers of planning. IEEE Transactions on Robotics 26(3), 469–482 (2010)

20. Shewchuk, J.R.: Delaunay refinement algorithms for triangular mesh generation. Computational Geometry: Theory and Applications **22**(1–3), 21–74 (2002). http://www.cs.cmu.edu/~quake/triangle.html

21. Wells, A., Plaku, E.: Supplementary material, http://faculty.cua.edu/plaku/TAROS15Videos.zip

Collaborating Low Cost Micro Aerial Vehicles: A Demonstration

Richard Williams$^{(\boxtimes)}$, Boris Konev, and Frans Coenen

Department of Computer Science, University of Liverpool, Liverpool, UK
R.M.Williams1@liv.ac.uk

Abstract. In this paper we demonstrate our Distributed Collaborative Tracking and Mapping (DCTAM) system for collaborative localisation and mapping with teams of Micro-Aerial Vehicle's MAVs. DCTAM uses a distributed architecture which allows us to run both image capture and frame-to-frame tracking on-board the MAV while offloading the more computationally demanding tasks of map creation/refinement to an off-board computer. The low computational cost of the localisation components of our system allow us to run additional software on-board such as an Extended Kalman Filter (EKF) for full state estimation and a PID-based Position Controller. This allows us to demonstrate complete cooperative autonomous operation.

1 Introduction and Motivation

Autonomous aerial vehicles are becoming pervasive in many diverse application domains from search and rescue to aerial transportation. The small size and robust nature of Micro Aerial Vehicles (MAVs) mean they have numerous applications, particularly in indoor environments (for exploration or remote inspection tasks) where it becomes difficult to rely on external positions systems such as GPS and Motion Capture systems or carry heavy sensor payloads (e.g. Laser Rangefinders). Monocular vision-based localisation systems offer advantages as they provide a very light-weight, low power sensor solution. Much recent work has addressed the issue of localisation using monocular vision[1–6]. However fewer works address the problem from a multi-robot perspective. In this paper we demonstrate a distributed framework for collaborate multi-robot localisation and mapping for teams of low-cost, light-weight ($\approx$ 500 grams) MAV platforms. We demonstrate a complete state estimation and position control solution that allows teams of MAVs to perform autonomous, collaborative localisation and mapping tasks.

1.1 Related Work

The multi-robot SLAM problem has been previously explored for ground-based robots with range sensors (such as laser range-finders, stereo vision)[7–9]. There is comparatively less work on the use of monocular vision as the only extrospective sensor, or involving agents capable of omni-directional (6DOF) motion

© Springer International Publishing Switzerland 2015
C. Dixon and K. Tuyls (Eds.): TAROS 2015, LNAI 9287, pp. 296–302, 2015.
DOI: 10.1007/978-3-319-22416-9_33

such as flying robots or hand-held devices (e.g. mobile phones). Foster et al.[10] introduce a centralised system in which each agent tracks their local position using a Visual Odometry (VO) algorithm and sends image features from selected keyframes to a centralised mapping server. Foster et al. report real-time performance with up to 3 MAVs. Riazuelo et al. [11] is the closest to the proposed approach in terms of components and architecture. Specifically they also build on Parallel Tracking and Mapping (PTAM) and use a distributed architecture, however they focus on multiple map merging using a RGB-D camera-based solution. In our work we focus more on cooperative navigation where robots start from the same location (a common assumption in most practical deployments). We assume all robots localise themselves within the global map before proceeding which allows them to perform cooperative tasks like exploration and robot-to-robot collision avoidance immediately without waiting for a map merge/rendezvous to occur. Additionally our work focuses on using RGB cameras only (we only use grey-scale images) which are more lightweight and consume less bandwidth than the RGB-D cameras. This allows our system to operate on low power ARM and ATOM processor-based MAV clients. Our previous work [12] featured a highly centralised approach with both tracking and mapping for each MAV running on a single ground-station computer and only image and sensor capture running on-board the MAVs. The sensitivity our previous approach to wireless interference and its limitations in terms of scalability (a maximum of 4 MAVs) motivated the development of the distributed approach we demonstrate in this paper.

2   System Overview

Our goal is to enable cooperative multi-robot navigation tasks using light MAVs with very low on-board computing resources. Our DCTAM system is based on the PTAM system developed by Klien and Murray[13]. While Klien and Murray separate the tasks of real-time motion estimation and map creation/refinement into separate threads running on the same computer for tracking a hand-held camera we split these components into a distributed system where the tracking component operates on-board several MAVs in parallel and the map creation/refinement component runs on a more powerful ground-station computer.

An overview of the system is shown in Figure 1. Each MAV has a low-level flight controller responsible for attitude estimation, stabilization and motor control. The flight controller provides sensor data via a Serial link to the companion computer (Odroid U3/C1). The companion computer runs the main state estimation and control components of the system. The DCTAM Tracker is responsible for real-time camera pose estimation and selecting the keyframes to be used for global map construction. Pose estimates are fused with orientation and acceleration data from the flight controller to generate a complete state estimate. This state estimate is used by the High-level controller for real-time velocity and position control. The position controller sends raw control (pitch, roll, yaw and throttle) commands to the flight controller via the Serial link. The ground-station computer communicates with each MAV via a WiFi link and runs the DCTAM

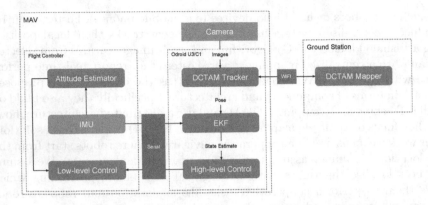

Fig. 1. System overview showing the components of the system(blue), sensors (green) and communication links (grey)

Fig. 2. AR. Drone/PX4-based MAV Platform

Mapper component which is responsible for map creation and optimization.Our framework has been implemented in C++ and integrated into the Robot Operating System (ROS) [14].

3   Hardware Platforms

We have developed two separate hardware platforms to demonstrate our system. While comprised of different hardware each has the required components to run our system i.e. flight controller, on-board computer with WiFi link and a single camera. The first demonstration platform is based on the popular AR.Drone[1] frame. We replaced the AR.Drone electronics with a PX4-based flight-control system consisting of a PX4 Flight Management Unit (PX4FMU) and AR.Drone

---

[1] http://ardrone2.parrot.com/

adapter board (PX4IOAR) which interfaces with the AR.Drone motors. The on-board companion computer is an Odroid U3, an ARM-based single board computer with a 1.7 Ghz Quad-core processor with 2 GB of RAM. The MAV has a single MatrixVision mvBlueFOX-MLC200w 752x480 pixel monochrome camera fitted with a 100° wide-angle lens. The total cost of this platform is £750 and the average flight time is 10 minutes. Our second platform is based around an open source 3D-printed quadcopter frame (The T4 Mini²). It uses low cost flight controller (Ardupilot APM2.6) and a low cost on-board computer Odroid C1. The C1 is an ARM-based single-board computer with a 1.5 Ghz Quad-core processor and 1 GB of RAM. The MAV also features the same MatrixVision monochrome camera. The total cost of this platform is £400 and the average flight time is 12 minutes.

4 Experimental Evaluation

We conducted experiments to verify the performance of our DCTAM system on the demonstration platforms. Figure 4 (left) shows the tracking times plotted against map size for the ODROID U3. The very short spikes in runtime coinciding with the arrival of a map update (resulting in an increase in map size). The average tracking time for the U3 was 0.01 milliseconds(ms) and 0.04 ms for the C1. Both platforms show good performance and the trackers demonstrate near constant-time performance on map sizes from small (5mx5m, ≈ 2000 map-points) indoor environments to large (20mx20m, ≈ 20000 map-points) outdoor environments. This is as a result of running the costly bundle adjustment procedure on the ground-station.

We show in Figure 4 the bandwidth requirements of a single MAV exploring an indoor environment; the final map for this experiment was 52 keyframes and

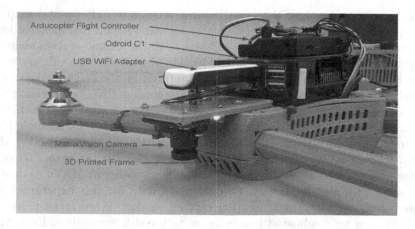

Fig. 3. 3D Printed MAV Platform

---

² https://www.thingiverse.com/thing:408363

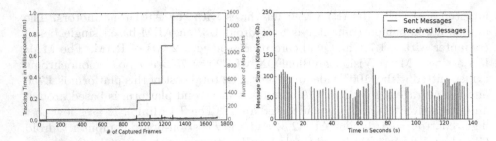

Fig. 4. Results showing tracking performance plotted against map size for the ODROID U3 (right). Bandwidth requirements for a single drone operating as part of a team. Recieved messages consist of keyframes and map-points generated by other MAV's (Left).

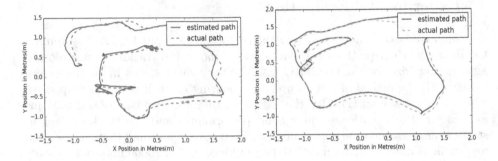

Fig. 5. Results of a physical experiment with a two MAVs navigating a 2mx2m area under manual control. The RMS Error for both MAVs is 0.13 metres(reft) and 0.15 metres (right).

7240 map-points. We show that even with a very large(7240 points) map the required bandwidth remains low (42 Kb/s) for our system. Significantly lower than streaming colour video directly (28 MB/s) to the ground-station as in [12] and even lower than the 1 MB/s required by [11] who use the same library as our system but who send the full colour image captured by the camera. We instead send only a compressed grey-scale image (we use lossless PNG compression with a low compression rate to limit computation time) and are able to achieve a requirement of only 9 Kb/s for a single MAV and 42 Kb/s for a single MAV operating as part of a team. As stated previously, the additional bandwidth is required when sending keyframes to the other trackers in the team. To verify tracking performance we performed an experiment with two MAVs navigating in the same area simultaneously. Ground truth data for this experiment was captured using an OptiTrack[3] motion capture system. Position set-points for both MAVs we were adjusted manually using joystick controllers. Each MAV was flown on a (roughly) square path around a 2 metre by 2 metre area. Figure 5

---

[3] http://www.optitrack.com/

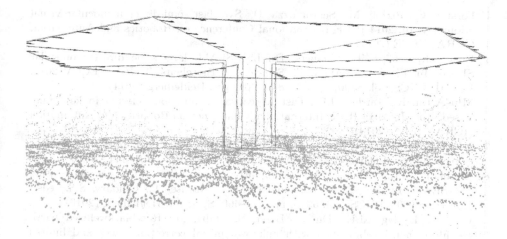

Fig. 6. The output of a large-scale simulated exploration experiment. Each MAV's estimated path is shown in red, the actual path in green and each 3D mappoint is shown in orange.

shows the results of this experiment. For clarity we have separated the position plots for both MAVs and do not include vertical position. The RMS Error for both MAVs was 13cm (left) and 15cm(right) (including Z position error).

5 Demonstration

In this demonstration we show multiple MAVs performing a cooperative localisation task. Each MAV estimates its global position (using the DCTAM Tracker and EKF) with respect to a shared global map produced by the DCTAM Mapper. Each MAV is given a series of position goals, to explore their environment and expand the existing map. Examples of the expected output of the system are show in Fig. 5 and Fig 6 Additional details including videos and source code can be found at the following web-page: http://cgi.csc.liv.ac.uk/~rmw/DCTAM.html

References

1. Weiss, S., Scaramuzza, D., Siegwart, R.: Monocular-slam-based navigation for autonomous micro helicopters in gps-denied environments. J. Field Robot. **28**(6), 854–874 (2011)
2. Engel, J., Sturm, J., Cremers, D.: Camera-based navigation of a low-cost quadrocopter. In: 2012 IEEE/RSJ International Conference on Intelligent Robots and Systems (IROS), pp. 2815–2821. IEEE (2012)
3. Wendel, A., Maurer, M., Graber, G., Pock, T., Bischof, H.: Dense reconstruction on-the-fly. In: 2012 IEEE Conference on Computer Vision and Pattern Recognition (CVPR). IEEE

4. Forster, C., Pizzoli, M., Scaramuzza, D.: Svo: fast semi-direct monocular visual odometry. In: 2014 IEEE International Conference on Robotics and Automation (ICRA). IEEE (2014)
5. Engel, J., Schöps, T., Cremers, D.: LSD-SLAM: large-scale direct monocular SLAM. In: Fleet, D., Pajdla, T., Schiele, B., Tuytelaars, T. (eds.) ECCV 2014, Part II. LNCS, vol. 8690, pp. 834–849. Springer, Heidelberg (2014)
6. Mur-Artal, R., Tardós, J.D.: Fast relocalisation and loop closing in keyframe-based slam. In: 2014 IEEE International Conference on Robotics and Automation (ICRA). IEEE (2014)
7. Fox, D., Burgard, W., Kruppa, H., Thrun, S.: A probabilistic approach to collaborative multi-robot localization. Autonomous robots 8(3), 325–344 (2000)
8. Özkucur, N.E., Akın, H.L.: Cooperative multi-robot map merging using fast-SLAM. In: Baltes, J., Lagoudakis, M.G., Naruse, T., Ghidary, S.S. (eds.) RoboCup 2009. LNCS, vol. 5949, pp. 449–460. Springer, Heidelberg (2010)
9. Carlone, L., Ng, M.K., Du, J., Bona, B., Indri, M.: Rao-blackwellized particle filters multi robot slam with unknown initial correspondences and limited communication. In: 2010 IEEE International Conference on Robotics and Automation (ICRA), pp. 243–249. IEEE (2010)
10. Forster, C., Lynen, S., Kneip, L., Scaramuzza, D.: Collaborative monocular slam with multiple micro aerial vehicles. In: 2013 IEEE/RSJ International Conference on Intelligent Robots and Systems (IROS), pp. 3962–3970. IEEE (2013)
11. Riazuelo, L., Civera, J., Montiel, J.: C 2 tam: A cloud framework for cooperative tracking and mapping. Robotics and Autonomous Systems 62(4), 401–413 (2014)
12. Williams, R., Konev, B., Coenen, F.: Multi-agent environment exploration with AR.Drones. In: Mistry, M., Leonardis, A., Witkowski, M., Melhuish, C. (eds.) TAROS 2014. LNCS, vol. 8717, pp. 60–71. Springer, Heidelberg (2014)
13. Klein, G., Murray, D.: Parallel tracking and mapping for small ar workspaces. In: ISMAR 2007. 6th IEEE and ACM International Symposium on Mixed and Augmented Reality, 2007, pp. 225–234. IEEE (2007)
14. Quigley, M., Conley, K., Gerkey, B., Faust, J., Foote, T., Leibs, J., Wheeler, R., Ng, A.Y.: Ros: an open-source robot operating system. In: ICRA Workshop on Open Source Software, vol. 3 (2009)

Vision-Based Trajectories Planning for Four Wheels Independently Steered Mobile Robots with Maximum Allowable Velocities

Zahra Ziaei$^{(\boxtimes)}$, Reza Oftadeh, and Jouni Mattila

Department of Intelligent Hydraulics and Automation (IHA),
Tampere University of Technology, Tampere, Finland
{zahra.ziaei,reza.oftadeh}@tut.fi

Abstract. In this paper, we extend our previous work to introduce a novel vision-based trajectories planning method for four-wheel-steered mobile robots. Relying only on the overhead camera and by utilizing artificial potential fields and visual servoing concepts, we simultaneously, generate the synchronized trajectories for all wheels in the world coordinates with sufficient number of trajectories midpoints. The synchronized trajectories are used to provide the robot's kinematic variables and robot-instantaneous-center of rotation to reduce the complexity of the robot kinematic model. Therefore, we plan maximum allowable velocities for all wheels so that at least one of the actuators is always working at maximum velocity. Experiment results are presented to illustrate the efficiency of the proposed method for four-wheel-steered mobile robot called iMoro.

Keywords: Vision-based · Trajectories planning · Steering velocity planning · Driving velocity planning

1 Introduction

Industrial and scientific applications of mobile robots are continuously growing particularly under accessibility considerations such as inspection and manipulation in environments that are inaccessible to human [1]. An extensive example of such hazardous and out-of-reach environments in large-scale scientific infrastructures is CERN[1] with a 27-km LHC[2] tunnel, that requires high mobility and maneuverability. While the use of mobile robots in such environments minimizes the danger of radiation exposure to human, the robots themselves are vulnerable to the radiation. Therefore efficient time-optimal motion and trajectory planning algorithms are required to find a safe trajectory with maximum bounded velocities from a given start configuration to a goal configuration. The main motivation of this work is to take advantage of the proposed method for inspection and manipulation tasks for four-wheel-steered (4WS) mobile robots with high maneuverability in confined and out-of-reach work spaces such as CERN tunnel.

---

[1] The European Organization for Nuclear Research.
[2] Large Hadron Collider.

© Springer International Publishing Switzerland 2015
C. Dixon and K. Tuyls (Eds.): TAROS 2015, LNAI 9287, pp. 303–309, 2015.
DOI: 10.1007/978-3-319-22416-9_34

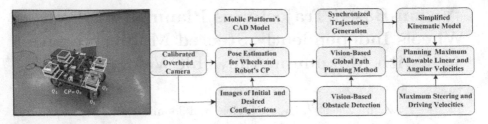

Fig. 1. Four wheels and robot's center point (CP) mapped in the ground are denoted by $\{Q_0, Q_1, ..., Q_4\}$ in the image space (right). System architecture of proposed vision-based trajectory planning for mobile robot with 4WS (left).

2   Vision-Based Trajectories Planning

In this paper, we present the vision-based method for generating the synchronized trajectories for all wheels of mobile robots with 4WS, to plan maximum allowable velocities so that at least one of the actuators is always working at maximum velocities. A mobile robot with 4WS is a type of the non-holonomic mobile robots with omni-directional steering wheels that are able to move along any direction simultaneously and attaining desired orientation. Such robots are more flexible than ordinary mobile robots which make them suitable for confined workspace such as manipulation task in camera field of view.

Figure 1 provides an overview of desired system architecture and the image space that determine the levels of proposed trajectories generation method. Unlike the conventional trajectory planning that consider the mobile robot as only one point, we consider 4WS mobile robot with its center-point (CP) plus the wheels CP with the same rotation and different positions, (see Fig. 1). CAD-based pose estimation method is used to recognize and estimate the position and orientation of the desired robot's points in the image space [2]. Approach for image-based obstacle segmentation method which explained in [2] are used to extract color features of the obstacle from the background.

In this paper, we defines the relationship between the motion of a mobile robot with 4WS, in the image space and in the world coordinates by utilizing the Artificial potential field (APF) integrated with visual servoing concepts, while the overhead camera is fixed. Inertial potential force is obtained by $\mathbf{F}(\boldsymbol{\Upsilon}) = \mathbf{F}_{at}(\boldsymbol{\Upsilon}) + \mathbf{F}_{rp}(\boldsymbol{\Upsilon})$, where attractive force $\mathbf{F}_{at}(\boldsymbol{\Upsilon})$ and repulsive force $\mathbf{F}_{rp}(\boldsymbol{\Upsilon})$ induced by proposed potential fields. The desired discrete-time trajectories for $\{Q_0, Q_1, ..., Q_4\}$ along the direction of 6×1 pose vector $\boldsymbol{\Upsilon}$ are obtained by the transition equation which generate trajectory midpoints for all desired points, simultaneously. A desired path along the direction of the artificial potential force is calculated via following equation [2]:

$$\boldsymbol{\Upsilon}_{k+1} = \boldsymbol{\Upsilon}_k + \epsilon_k \frac{\mathbf{F}(\boldsymbol{\Upsilon}_k)}{\parallel \mathbf{F}(\boldsymbol{\Upsilon}_k) \parallel}, \tag{1}$$

where k is an index that increases during the generation of a trajectory, and ϵ_k is a positive scaling factor. We allow the algorithm to generate sufficient number of midpoints by increasing ϵ_k. Then, we can easily transform each k-th generated trajectory to the camera center-frame or mobile robot inertial frame in the world coordinates. To turn the generated path into a trajectory, we must append a velocity component for each midpoint. The time values are chosen by spacing proportionally to the distance between two positions of the mobile robot's desired points. Thus, $\Delta t_k = t_k - t_{k-1}$ determine the constant time between two consecutive mobile robot positions in the ground.

3   Kinematic Model

Unsynchronized wheels of mobile robots result in slippage and misalignment in translation and rotation movement [3]. To overcome these problems, the points on the k-th generated trajectories for all wheels have to form a unique instantaneous center of rotation (ICR) (see Fig. 2). Therefore, by knowing the synchronized trajectories midpoints belonging to at least two wheels, we easily obtain ICR point. As illustrated in Fig. 2, generated synchronized trajectories determine the main kinematic variables of mobile robots. The direction of the robot linear velocity vector is determined by the unit vector $\hat{\mathbf{v}}$. Scalar value v is the magnitude of the linear velocity in CP. The angle θ_B is the rotation angle of robot-CP in X direction. θ_B and ψ_v show the angles $\hat{\mathbf{x}}_0$ and $\hat{\mathbf{v}}$ respectively. Variables $\omega_B = \dot{\theta}_B$ and $\omega_v = \dot{\psi}_v$ are the angular velocities of the robot-CP and of $\hat{\mathbf{v}}$ and $\hat{\mathbf{x}}_0$ respectively. The following kinematic constraint describes the relation between robot-CP velocity and the wheels velocities:

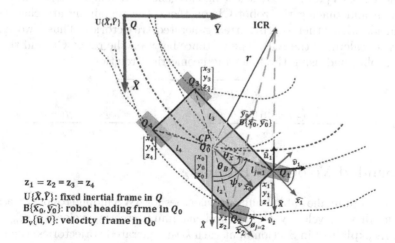

Fig. 2. Generated synchronized trajectories for desired points $\{Q_0, Q_1, ..., Q_4\}$, provide the robot kinematic variables to simplify its kinematic model and estimate ICR using at least two synchronized generated trajectories for instance (Q_1, Q_2).

$$^{B}\hat{\mathbf{v}} = \mathbf{R}(\psi_v - \theta_B)[1,0,0]^{T}, \quad v_j^{B}\hat{\mathbf{v}}_{\mathbf{j}} = v^{B}\hat{\mathbf{v}} + \omega_B(\hat{\mathbf{z}} \times^{B} \hat{\ell}_{\mathbf{j}}), \tag{2}$$

where $\mathbf{R}(\psi_v - \theta_B)$ is the rotation matrix with angle $(\psi_v - \theta_B)$ around the z-axis, which is frame $\mathbf{B}_v$ in $\mathbf{B}$. $\{^{B}\hat{\ell}_{\mathbf{j}}, j = 1..4\}$ in frame $\mathbf{B}$ show the constant vectors $\overrightarrow{Q_0 Q_j}$. Angles $\{\phi_j, j = 1..4\}$ are the steering angles of j-th wheel, then $^{B}\hat{\mathbf{v}}_{\mathbf{j}}$ can be written as $[cos(\phi_j)\ sin(\phi_j)\ 0]$. $v_j^{B}\hat{\mathbf{v}}_{\mathbf{j}}$ is the velocity vector of $\{Q_j, j = 1..4\}$.

To plan the mobile robot velocities, it is necessary to obtain the mobile robot's ICR with zero velocity while undergoing planar movement in k-th generated trajectories. The relation between ω_B and v_B in the frame $\mathbf{B}$ is:

$$\omega_B = \frac{v_B}{r}, \tag{3}$$

where r is the distance between ICR and CP. v_B and ω_B are the liner and angular velocities of CP respectively. Therefore substituting r in (2) gives the relation between the velocity of the robot-CP and the velocity of its four wheels:

$$v_j = v_B \| \hat{\mathbf{v}} + \frac{1}{r}(\hat{\mathbf{z}} \times \overrightarrow{\ell_{\mathbf{j}}}) \|, \{j = 1...4\}. \tag{4}$$

By utilizing the definition of the curvature $\kappa_j = \frac{\dot{\phi}_j}{v_j}$, we can obtain the relation between the driving velocity of each wheel (v_j) and its steering velocity $(\dot{\phi}_j)$:

$$\omega_j = \dot{\phi}_j = (\kappa_j.v_B) \| \hat{\mathbf{v}} + \frac{1}{r}(\hat{\mathbf{z}} \times \overrightarrow{\ell_{\mathbf{j}}}) \|, \{j = 1...4\}, \tag{5}$$

As illustrated in Fig. 2, the tangent in each point of generated trajectory for each wheel gives $\hat{\mathbf{v}}_{\mathbf{j}}$ and $\hat{\mathbf{u}}_{\mathbf{j}}$. By selecting two generated synchronized trajectories out of four and placing the robot-CP on them, the remaining attached points will coincide with other synchronized generated trajectories. Thus, two v_js are sufficient to calculate the ICR. The distance between the robot-CP and the ICR point r is obtained using the following geometrical relation:

$$r = \sqrt{(x_0 - x_r)^2 + (y_0 - y_r)^2},$$

$$x_r = \frac{v_1.x_2 - v_2(x_1 + v_1(y_1 - y_2))}{v_1 - v_2}, \quad y_r = \frac{x_1 - x_2 + v_1.y_1 - v_2.y_2}{v_1 - v_2}. \tag{6}$$

4 Bounded Velocities Planning

To plan maximum allowable bounded velocities for 4WS mobile robot, we assume maximum driving velocity (V_{max}^{D}) and maximum steering velocity $(\dot{\phi}_{max}^{S})$ are known. As explained in Section 3, at each k-th generated trajectories, the curvatures (κ_j), and the relation between v_B and v_j can be calculated. While a mobile robot has four independently steered wheels, there are eight velocity candidates to fulfill four driving and steering constraints. Thus, at each k-th generated trajectories, only one of the driving or steering velocities of each wheel may have maximum value. Therefore, bounded velocities for each wheel are obtained using following steps [4]:

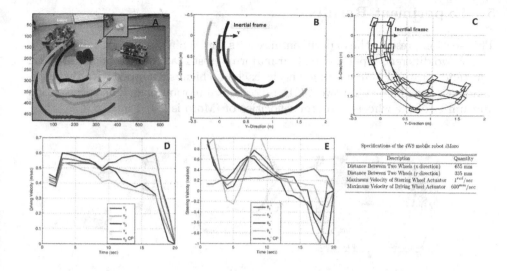

Fig. 3. Experiment 1, while *i*Moro must avoid two types of obstacles including recognized static obstacles in the image space (two bricks) and the camera field of view. The generated trajectories between the initial and the desired poses in the image space are illustrated in A, and corresponding trajectories and footprints in the world coordinates are illustrated in B and C respectively. Maximum allowable driving and steering velocities for CP and wheels are illustrated in D and E respectively.

– Evaluate maximum driving velocity of the wheel j ($V_{j,max}$), that keeps its driving and steering velocities bounded:

$$V_{j,max} = min(V_{max}^D, \frac{\dot{\phi}_{max}^S}{|\kappa_j|}) \qquad (7)$$

– Evaluate $v_{B,j}$ as a candidate for the base velocity v_B, that is divided based on the maximum driving velocities of the j wheel:

$$v_{B,j} = \frac{V_{j,max}}{\|(\hat{\mathbf{v}} + \frac{1}{r}(\hat{\mathbf{z}} \times \overrightarrow{\ell_{\mathbf{j}}}))\|} \qquad (8)$$

– Evaluate the velocity of CP v_B:

$$v_B = min(v_{B,j}), j \in 1, 2, 3, 4. \qquad (9)$$

If κ_j becomes small enough then $V_{j,max}^S = \dot{\phi}_{j,max}^S$ will exceed V_{max}^D. Thus, driving velocity should be bounded by V_{max}^D. If the curvature increases, $V_{j,max}^S$ decreases and steering velocity becomes critical. In this case, driving velocity should not exceed $V_{j,max}^S$. Therefore, steering velocity stays below $\dot{\phi}_{max}^S$.

5   Experiment Results

This section presents the experimental results for 4WS mobile robot called *i*Moro under two different experiments. Our major interest is a mobile robot comprising of a rigid base and four steering wheels, each of which is equipped with two independent servo drives with two DOF in horizontal motion plan. The specification and limits of the wheels actuator velocities of *i*Moro is listed in Fig. 3.

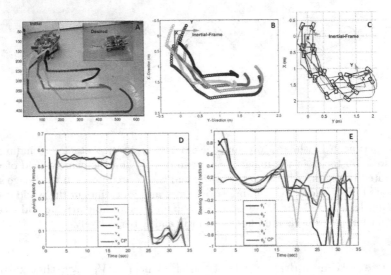

Fig. 4. Experiment 2, while *i*Moro avoids only image boundaries. Generated trajectories in the image space are illustrated in A and corresponding trajectories and footprints in the world coordinates are illustrated in B and C respectively. Planned maximum allowable driving and steering velocities are shown in D and E respectively.

6   Conclusion and Future Work

In this paper, a vision-based trajectory planner (based on an overhead camera) is presented for wheel-steered mobile robots in a confined workspace and in the presence of obstacles. The generated synchronized trajectories simplify the complex kinematic model of mobile robot and plan maximum allowable steering and driving velocities so that at least one of the actuators velocities is at its maximum bound. In the future, this method will be extended to present a vision-based path coordination method for multiple mobile robots in shared and confined workspace to avoid mutual collision.

References

1. Trevelyan, J.P., Kang, S.-C., Hamel, W.R.: Robotics in hazardous applications. In: Springer Handbook of Robotics, pp. 1101–1126. Springer (2008)
2. Ziaei, Z., Oftadeh, R., Mattila, J.: Global path planning with obstacle avoidance for omnidirectional mobile robot using overhead camera. In: 2014 IEEE International Conference on Mechatronics and Automation (ICMA), pp. 697–704. IEEE (2014)
3. Schwesinger, U., Pradalier, C., Siegwart, R.: A novel approach for steering wheel synchronization with velocity/acceleration limits and mechanical constraints. In: 2012 IEEE/RSJ International Conference on Intelligent Robots and Systems (IROS), pp. 5360–5366. IEEE (2012)
4. Oftadeh, R., Aref, M.M., Ghabcheloo, R., Mattila, J.: Bounded-velocity motion control of four wheel steered mobile robots. In: 2013 IEEE/ASME International Conference on Advanced Intelligent Mechatronics (AIM), pp. 255–260. IEEE (2013)

Author Index

Printed in the United States
By Bookmasters